*The Floracrats*

# NEW PERSPECTIVES IN
# SOUTHEAST ASIAN STUDIES

# The Floracrats

*State-Sponsored Science and the Failure of the Enlightenment in Indonesia*

## Andrew Goss

THE UNIVERSITY OF WISCONSIN PRESS

Publication of this volume has been made possible, in part,
through support from the **Association for Asian Studies, Inc.** and the
**Center for Southeast Asian Studies at the University of Wisconsin–Madison.**

The University of Wisconsin Press
1930 Monroe Street, 3rd Floor
Madison, Wisconsin 53711-2059
uwpress.wisc.edu

3 Henrietta Street
London WCE 8LU, England
eurospanbookstore.com

Printed in the United States of America

Library of Congress Cataloging-in-Publication Data
Goss, Andrew.
The floracrats: state-sponsored science and the failure of the Enlightenment
in Indonesia / Andrew Goss.
p.   cm. — (New perspectives in Southeast Asian studies)
Includes bibliographical references and index.
ISBN 978-0-299-24864-2 (pbk.: alk. paper)
ISBN 978-0-299-24863-5 (e-book)
1. Botany, Economic—Political aspects—Indonesia.
2. Science and state—Indonesia—History.
3. Civil service—Indonesia—History.   4. Botanists—Indonesia—History.
5. Netherlands—Colonies—Indonesia—Administration.
6. Enlightenment—Indonesia—Influence.
7. Federal aid to research—Indonesia—History.
I. Title.   II. Series: New perspectives in Southeast Asian studies.
SB108.I5G67 2011
509.598—dc22
2010012967

# Contents

# Illustrations

# Preface and Acknowledgments

This book began in the University of Michigan Library in 1997 when I stumbled across vast numbers of biology journals from Dutch Indonesia, alongside scientific books and pamphlets on Indonesian nature, going back to the 1840s. The scientific journals and books from the Netherlands East Indies were much like those published in research centers and universities in Boston, London, and Berlin. In my excitement I thought I had made a stunning discovery about colonial Indonesia; not only did the Netherlands East Indies have numerous scientific institutions interpreting and explaining tropical nature, but by 1900 these institutions were doing biological research on a world-class level. As I stood in the library in Ann Arbor, I believed I was holding the evidence of an overlooked success story in Indonesian history. Over the years of pursuing this lead, I have come to see that the opposite is true. In researching and writing this book, I have established that the Enlightenment ideal of dispassionate, free, and useful inquiry into the workings of nature has been a failure in Indonesia. Nonetheless, my original hypothesis has been fruitful in suggesting a new line of reasoning about the history of science, knowledge, and power in Indonesia. In this book I argue that it is the failure of the Enlightenment that is the key to understanding the history of science and the politics of knowledge in modern Indonesia.

This book could not have been completed without a lot of help. My heartfelt thanks go to my teachers at the University of Michigan. I am especially indebted to Rudolf Mrázek, Ann Laura Stoler, Geoff Eley, Gabrielle Hecht, Nancy Florida, and Victor Lieberman. I am grateful for their advice, guidance, and support. I also wish to thank my botany teachers at the University of Michigan, who very kindly guided a historian through plant systematics: William Anderson, Daniel DeJoode, David Taylor, Herb Wagner, and Michael Wynne.

Research in Indonesia and the Netherlands was made possible with a generous Fulbright-Hays Dissertation Research Grant. At the beginning of my research, a travel fellowship from the United States–Indonesia Society allowed me to investigate the viability of a history of biology project. I also wish to thank the University of Michigan History Department and Horace H. Rackham School of Graduate Studies for support while in Ann Arbor. Thanks go to the College of Liberal Arts at the University of New Orleans for a June Scholars award in 2005. And thanks to an American Historical Association Bernadotte E. Schmitt Grant, I was able to return to the Dutch archives in 2005.

My research in Indonesia during 2001 was made possible through the sponsorship of the Bogor Botanical Gardens and the Indonesian Institute of Sciences. Thanks go to the staff at the National Archives in Jakarta for helping me locate sources for the history of Indonesian science. Thanks also go to the Indonesian biologists who agreed to sit down with me for an interview: Sampurno Kadarsan, Mien Rifai, Didin Sastrapradja, Setiyati Sastrapradja, Otto Soemarwoto, Soetomo Soerohaldoko, S. Somadikarto, Blasius Sudarsono, and Aprilani Sugiarto.

Friends in Jakarta made a productive and enjoyable life thousands of miles from home. I am grateful to Jeffrey Hadler, Paul Haenen, Piet Hendrardjo, Frank Jotzo, David Linnan, Sarah Maxim, Greta N. Morris, Cornelia Paliama, Richard Pineda, Kumi Sawada, Stacey Sowards, Karl Stoltz, Samantha Tate, Sally White, and Julianna Wilson. Many thanks go to Bambang Hidayat, whose hospitality, personal library, and good cheer have rejuvenated me countless times.

Thanks go to the staff at the National Archive in The Hague, the Boerhaave Museum in Leiden, and the Naturalist Museum in Leiden for helping me access archival records. Special thanks go to Rob Visser at the Institute for History and Foundations of Science at Utrecht University, who shared with me his extensive knowledge about the history of biology and made available to me archival records in the institute's collection. And thanks go to Cees Lut and Ludmila Frankova at the library of the National Herbarium of the Netherlands in Leiden, who opened the archives of the herbarium to me.

Over the past few years, I have had the good fortune to discuss the history of Indonesian science with many friends and colleagues. My sincerest thanks go to Peter Boomgaard, David Chandler, Michael Charney, Fernando Coronil, Jennifer Gaynor, Piotr Gorecki, Frances Gouda, David Henley, Hans de Jonge, Prakash Kumar, Anna Lawrence, Sean Lopez, Ian McNeely, Suzanne Moon, Atsuko Naono, Gary Paoli, Daniel Puskar, Alice Ritscherle, Eric Stein, Jeffery Wilson, and Sarah Womack, who commented on earlier versions of the manuscript. And special thanks go to Michael Wintroub for last-minute advice and guidance.

The University of Wisconsin Press has been a joy to work with. My thanks go to the editors of the New Perspectives in Southeast Asian Studies series for their support and guidance and to Gwen Walker, Sheila McMahon, and the editorial and production staff at the University of Wisconsin Press for bringing this book to publication. In particular I wish to thank Janet Oopdyke for her meticulous copyediting. And many thanks go to the two reviewers, Robert Cribb and Eric Tagliacozzo, who saved me from many embarrassing errors and omissions and whose comments, questions, and suggestions made my argumentation and writing stronger in all respects.

Any errors, including those of interpretation, are my responsibility alone.

This book was finished on the campus of the University of New Orleans. For the past six years, its Department of History has been my professional home, and I cannot imagine a better group of colleagues with whom to work on teaching and researching the past. I thank the faculty, staff, and students at the University of New Orleans have who encouraged and supported me.

My family has nurtured this project from the beginning. My two children, Sage and Galen, continue to be enthusiastic that Dad is writing a book—I am deeply appreciative of their loving interest. My wife, Pax Bobrow, has traveled around the world with me for the past fifteen years, and at every stage she has supported me and my study of Indonesian history. With her gentle, patient, and critical prodding, this book has blossomed. Thank you.

Indonesian spelling conventions have changed considerably in the last century. In the text I have chosen to render Indonesian terms and place-names in their current standard spelling. But for other Indonesian proper names, including personal names, institutional names, and publications, I have maintained the contemporary spelling, thus Tjipto Mangoenkoesomo (not Cipto Mangunkusumo), Boedi Oetomo (not Budi Utomo), and *Kemadjoean Ra'jat* (not *Kemajuan Rakyat*). For Indonesia's first president, I have chosen to use the spelling Sukarno, not Soekarno.

# *Abbreviations*

| | |
|---|---|
| ANRI | Arsip Nasional Republik Indonesia (National Archive of the Republic of Indonesia, Jakarta, Indonesia) |
| BIOTROP | Southeast Asian Regional Centre for Tropical Biology |
| Bt. | Gouvernement Besluit (Government Executive Order) |
| CNWS | Centrum voor Niet-Westerse Studies (Research School for Asian, African, and Amerindian Studies) |
| CSIRO | Commonwealth Scientific and Industrial Research Organization |
| DEPERNAS | Dewan Perantjang Nasional (National Development Council) |
| Exh. | Exhibitum (Agenda Item) |
| f. | Dutch guilder |
| HS | Handschrift (Manuscript) |
| ICWO | Indische Comité voor Wetenschappelijke Onderzoekingen (Indies Committee for Scientific Research) |
| ITB | Institut Teknologi Bandung (Bandung Technical University) |
| KIT | Koninklijk Insituut voor de Tropen (Royal Tropical Institute) |
| KITLV | Koninklijk Instituut voor Taal-, Land- en Volkenkunde (Royal Netherlands Institute of Southeast Asian and Caribbean Studies) |
| KNV | Koninklijke Natuurkundige Vereeniging (Royal Natural Association) |
| KOSTRAD | Komando Cadangan Strategis Angkatan Darat (Army Strategic Reserve) |
| LBN | Lembaga Biologi Nasional (National Biological Institute) |
| LIPI | Lembaga Ilmu Pengetahuan Indonesia (Indonesian Institute of Sciences) |
| LNH | Departement van Landbouw, Nijverheid, en Handel (Department of Agriculture, Industry, and Trade) |
| LPPA | Lembaga Pusat Penjelidikan Alam (Central Institute for the Research of Nature) |
| Manipol | Manifesto Politik (Political Manifesto of the 1945 constitution) |
| Mgs. | Missive Gouvernements Secretaris (Government Clerk's Missive) |
| MIPI | Madjelis Ilmu Pengetahuan Indonesia (Indonesian Council for Science) |
| MPRS | Madjelis Permusjawaratan Rakyat Sementara (Provisional People's Consultative Assembly) |

NGO         nongovernmental organization
NHN         Nationaal Herbarium Nederland (National Herbarium of the Netherlands)
OSR         Organization for Scientific Research
PKI         Partai Komunis Indonesia (Indonesian Communist Party)
PNI         Partai Nasional Indonesia (Indonesian Nationalist Party)
PUSPITEK    Pusat Penelitian, Ilmu Pengetahuan, dan Teknologi (Center for Science, Technology, and Research)
REPELITA    Rencana Pembangunan Lima Tahun (Five-Year Development Plan)
RI          Republik Indonesia
STOVIA      School tot Opleiding van Inlandsche Artsen (School for the Training of Native Physicians)
UNESCO      United Nations Education, Scientific, and Cultural Organization
USDEK       Undang-undang 1945; Sosialisme à la Indonesia; Demokrasi Terpimpin; Ekonomi Terpimpin; and Kepribadian Indonesia (1945 Constitution; Indonesian socialism; Guided Democracy; Guided Economy; and Indonesian Identity)
V.          Verbaal (Minute)
VOC         Vereenigde Oost Indische Compagnie (Dutch East India Company)

*The Floracrats*

# Introduction

The tropical rain forests of Indonesia have lured naturalists for over two hundred years. Ever since Carl Linnaeus sent disciples to the islands that lie between the Pacific and Indian oceans, naturalists have sought fame and glory by building collections from its tropical flora and fauna.[1] Most famously, Alfred Wallace explored Indonesia between 1854 and 1862. During a bout of fever in 1858, Wallace formulated a theory of speciation through natural selection, which when written up he sent to Charles Darwin for comment.[2] In the widely read narrative of his trip, *The Malay Archipelago*, Wallace chronicled the natural world of Indonesia and made a long-lasting contribution to the ecological understanding of the region by demonstrating the division between the Asian and Pacific flora and fauna, now known as the Wallace line.[3] In the decades thereafter Dutch naturalists traveled to their colony for adventure, for natural treasures, and to advance their careers.[4] These naturalists, as well as Japanese, European, American, and Australian scientists more recently, have seen and used Indonesia as a storehouse of nature's great bounty and have repeatedly returned for opportunities to research evolution unfolding in its tropical rain forests. The most recent example of this has been the discovery of an approximately thirty-thousand-year-old skeleton on Flores, classified as *Homo floresiensis*, that has challenged the consensus theory of hominid evolution.[5] Their tropical adventures in Indonesia shaped these biologists' careers and, in instances like that of Wallace, even the history of biology.

This long history of international scientific interest in the natural bounty of Indonesia raises a question: what have Indonesian naturalists and natural scientists accomplished in the world of science? Answering questions about the natural history of the archipelago has been of consistent interest to the inhabitants of colonial and postcolonial Indonesia. Starting in Dutch colonial outposts in the seventeenth century with the merchant G. E. Rumphius of the Vereenigde Oost Indische Compagnie (Dutch East India Company, VOC), colonial naturalists have explored, collected, and written about their tropical environs.[6] Rumphius, and his successor François Valentijn, took indigenous knowledge, gathered from local expertise, and translated it for European audiences. The historiography of Indonesian natural science, works written by Dutch colonials, Indonesian citizens, and Western historians, has stressed how these naturalists' early achievements were superseded by professional institutions, which beginning in the nineteenth century created modern scientific disciplines not based on local natural knowledge. These histories have examined the creation of a class of professional scientists who worked in research institutions based on European models.[7] At the end of a review of the *longue durée* of Indonesian science, summarizing its varied achievements, Lewis Pyenson concludes optimistically that "the best of science in Indonesia is the equal of research anywhere."[8] Still, despite the growth of professional science and research in the last century, there have been few notable international achievements for Indonesian science. Looking from the outside in, Indonesian natural science has been largely unremarkable. Among the Asian scientists who have, no Indonesian scientist has ever won a Nobel Prize, and only one colonial scientist, the Dutch chemist Christiaan Eijkman, won, in 1929 for research about the causes of beriberi in chickens done in the colony during the 1890s. American universities do not recruit faculty in Indonesia. Only a select few Indonesian graduate students study or have studied in natural science departments outside of Indonesia, and those few return home to careers as scientific administrators. Most American Indonesia watchers would be hard pressed to name even a single contemporary Indonesian natural scientist. And even though Indonesia remains an important research site for tropical biologists from around the world, that research happens on a parallel track with Indonesian biology and the two worlds are distances apart. Although Indonesian biologists have cooperated sporadically with international scientists (notably in the mid-1950s and 1970s), American, Japanese, and European biologists today interact minimally with their Indonesian counterparts.[9] Perhaps because of this, Indonesian science, even that extending back to the colonial period, has become a footnote in the history of science. In the eighteen-volume *Dictionary of Scientific Biography*, only two scientists who made their careers in Indonesia (Melchior Treub from 1880 to 1909 and Jacob Clay from 1920 to 1929) receive a full-length entry.[10] Since the 1950s, Indonesian science and scientists have drawn scant attention from scholars, scientists or otherwise, outside of Indonesia.

Neither the Dutch and Indonesian narratives of scientific achievement nor the absence of Indonesian science within the larger history of science answer the harder question of the role science played in Indonesian society. Recent historical scholarship about political power in Indonesia provides a fruitful line of research. Over the last twenty years, scholars have proven convincingly that in the Netherlands East Indies, scientists, ranging from anthropologists and doctors to cartographers and geophysicists, were tentacles of colonial power, creating knowledge about the people and places of the archipelago, as well as suggesting ways in which to control them.[11] Colonial officials turned to scientists when confronted with new problems created by recent territorial acquisitions, economic demands, or social unrest. Naturalists, from the colonial state's point of view, were useful as expert managers of tropical nature in all its complexity. The colonial government hired foresters, trained in German scientific forestry techniques, to extend state control over Java's forests by devising legal and bureaucratic mechanisms, backed by science, to block local access to the forests. They were some of the first professional scientists in colonial Indonesia.[12] By the 1890s every export crop, including quinine, sugar, coffee, tea, and indigo, had at least one experiment station dedicated to that crop's cultivation staffed by European-trained scientists.[13] And in the early twentieth century the colonial state hired agricultural experts for the development of peasant society through the spread of scientific rice cultivation techniques.[14]

In this book I examine how naturalists interpreted the physical world of Indonesia with Enlightenment techniques of description and taxonomy, and institutionalized natural history in the Indies by creating a professional discipline of biology. This history parallels the transition from Enlightenment natural history to biology in Europe. In the nineteenth century, natural history scholarship was formalized under the influence of Linnaean taxonomy and George Cuvier's paleontology. Naturalists continued to focus on collecting, describing, and classifying nature, though with increasing interest in explaining the origins, history, and changes of nature. During the late nineteenth century, laboratory-based research, which stressed experimental approaches, began addressing the anatomy, development, and morphology of living beings. With the expansion of new research methods in genetics, organic chemistry, and crystallography in the early twentieth century, scientists created disciplines of biology separate from natural history. The evolutionary synthesis of the 1930s and 1940s reconciled the different strains of natural history and the biological sciences, which lessened the distance between field and laboratory biologists and brought genetics, Darwinism, and natural history together. In Europe and North America this synthesis created a unified field of biology.[15]

Botany as Enlightenment knowledge was linked closely to the growth of the European empires. Eighteenth-century voyages and expeditions, and the natural collections they brought back, taught Europeans how to envision tropical nature.[16] At the same time, government patronage of natural scientists and botanical gardens created close ties between European botanists and government officials; both groups were interested in applying this knowledge to invigorate society, for example by introducing new crops.[17] This close relationship was further enhanced with the enlargement of the European empires in the nineteenth century, which turned some botanists into important imperial agents, especially in the British Empire.[18] At the same time, the European botanical gardens' herbarium collections of dried plants, which grew rapidly as a result of collecting done in the French and British empires, became the institutional basis for standardizing and universalizing nineteenth-century botanical knowledge.[19] Thus Enlightenment traditions were associated with giving Europeans the ability to classify, to understand, and to govern imperial nature. This arrangement continued until the collapse of the European empires in the twentieth century. Botany was allied with imperial governments, and research about tropical nature invigorated European thinking about evolution, ecology, and the environment.[20]

Although the development of Indonesian natural history and biology interacts with Europe's, I argue that its history was shaped largely by local concerns. To be sure, for Indonesian colonial and postcolonial scientists, the European scientific traditions of the Enlightenment have been influential models. For example, classifying the plants and animals of Indonesia has remained central to the research programs of Indonesian biologists. And like their European forebears, they sought government patronage and built herbarium collections filled with plants collected in the empire, and later the nation. In fact, even as Indonesian biologists created biological research institutions in the twentieth century, these natural history traditions have remained central. Answering theoretical questions about function, morphology, or evolution has never been as important as creating accurate and complete descriptions of the tropical species in Indonesia.

Indonesian professional naturalists and biologists have been guided by the desire to improve society by enhancing access to, and control of, nature. For this reason, the local colonial and postcolonial context structured how the biological discipline developed. In the colonial period this happened outside of Dutch metropolitan control. The Dutch botanical institutes never rivaled the power of their British and French counterparts, and after 1850, scientific institutes in the Indies were largely independent of European scientific oversight.[21] Moreover, the transition from natural history to biology in Indonesia took place outside of universities; no Indonesian higher educational institution offered biology as a field of study until the late 1940s.[22] And while popular interest in natural history supported colonial walking groups, environmental protection clubs, and natural history instruction in high schools, as well as indigenous herbal medicine, these pursuits did not significantly influence professional science after 1900.[23] This distinguishes Indonesian popular science from what, for example, happened in Germany, where populist biological initiatives in the late nineteenth century created the discipline of ecology in the early twentieth century.[24] And while in the 1920s an astronomy observatory in west Java was endowed by the Dutch planter K. A. R. Bosscha, this private patronage of science was a singular exception.[25] For most of the last two hundred years, Indonesian elites have judged the value of natural history and biology based on how well its knowledge could be applied in the agricultural economy. Moreover, the professionalization of Indonesian natural history happened inside state institutions, and this context determined the intellectual pursuits of Indonesian biology. The definition of *amateur* was anyone outside of the state scientific system. This led to the complete marginalization of privately funded research as well as native expertise inside Indonesian professional science.

By contrast, in Europe in the late eighteenth century, botany had become a way for amateur gentlemen and ladies, and working-class men and women also, to participate in generating knowledge about nature.[26] The simplicity of Linnaean nomenclature and taxonomy made botany a popular Enlightenment tradition in England.[27] In the nineteenth century this Enlightenment tradition spread to the European colonies in Asia carried by educated European enthusiasts who took to classifying and ordering the tropical world of Asia. In the Netherlands East Indies this was more than a passing interest. As the Dutch expanded their territorial control of the archipelago, European inhabitants developed a sustained interest in reading about a colleague's exploration of a region of the islands newly opened to Dutch control.[28] These explorers thereby created Enlightenment learned societies in the early nineteenth century. Although the amateur learned societies were superseded in the late nineteenth century by scientific institutes, popular interest in natural history continued. For example, after 1911 the Netherlands-Indies Natural History Association catered to continued amateur interest in natural history. Still, even this popular movement was confined to a small slice of Europeans; at its peak in 1929 their journal *De Tropische Natuur* (Tropical Nature) reached about a thousand subscribers.[29]

Many naturalists in colonial Indonesia had greater ambitions for natural history than observing, describing, and cataloging nature. Like their European Enlightenment forebears, they sought to create knowledge for public utility.[30] During the nineteenth century, Dutch university professors positioned natural science as a discipline of useful knowledge. They sought to create an enlightened society by educating an elite intellectual class that would be the motor of progress for the Dutch nation and its people.[31] Many of the colonial naturalists in the Netherlands East Indies after 1840 were influenced by what they had been taught in the Netherlands was the functional and social role of scientists. They came to the colony seeing themselves as apostles of enlightenment, as I call them, believing their knowledge was capable of transforming the colony. They cast themselves as trailblazers who would lead the colony toward modernity. This required more than writing about their observations; it meant making their expertise locally relevant by fashioning their knowledge as not just authoritative but meaningful, useful, and necessary to the colony and its peoples. While it is largely educated elites who have been the enthusiastic apostles of enlightenment, they have seen themselves as serving social and political purposes beyond the generation of knowledge itself. Since the eighteenth century these apostles of enlightenment have tried to root their expertise in the Indonesian soil by generating knowledge of local relevance and then spreading it broadly.

In addition to individual efforts, apostles of enlightenment tried to expand the reach of their knowledge by institutionalizing scholarly efforts. Since the end of the eighteenth century, apostles of enlightenment have seen professional science as the best way to bring to fruition their ideals of creating useful knowledge. Recent research by Huib J. Zuidervaart and Rob H. Van Gent chronicles the first attempts to institutionalize Enlightenment learning in late eighteenth-century Batavia (now Jakarta). Spurred on by a European expedition to observe the transit of Venus across the sun in 1761, the Reverend J. M. Mohr built an observatory with instruments imported from Europe and presented himself as the colony's astronomer and meteorologist. He sent his observations of the subsequent 1769 transit of Venus, an event of great interest to eighteenth-century Enlightenment Europe, for publication in Europe and gave tours to visiting explorers, including Joseph Banks, James Cook, and Louis-Antoine de Bougainville. Despite his personal efforts, however, he never institutionalized Enlightenment scholarship in Batavia and was unable to win official recognition as an astronomer from the Dutch East India Company. It was only after Mohr's death in 1778, when a sympathetic governor general held sway, that J. C. M. Radermacher established an Enlightenment learned society, the Batavian Society of Arts and Sciences. In its first decade it attempted, with little success, to revive Mohr's astronomy, but Radermacher did manage to establish a journal, as well as a museum and cabinet of natural curiosities overseen by a botanist. But with key backers of the Batavian Society dying or departing and a change to a colonial administration with little appetite for the Enlightenment, the Batavian Society faded away after 1789. Long-term institutionalization of the Enlightenment would require more than individual interest or patronage.[32]

After the 1840s, apostles of enlightenment fared better than Mohr and his colleagues in that they were able to found permanent scientific institutions. In particular they were successful with formal organizations devoted to research in natural history. Botany especially garnered broad colonial support. In the nineteenth century, among the European inhabitants Enlightenment knowledge about tropical flora made it easier to envision building a permanent European settlement in the colony. After the 1870s the importance of export agriculture created scientific institutions devoted to economic botany. And after independence, Indonesian scientists promised to reveal the national nature while biology also promised to be the easiest way to connect Indonesia with the international world of science. Still, these institutions were like the early Batavian Society in that Enlightenment knowledge and science remained elite affairs even after Indonesian independence. Many naturalists and scientists from Indonesia, both before and after independence, would have agreed with Mohr's argument that they should devote themselves to the "useful sciences . . . with the intention of raising its status and promoting these sciences as most suited to fighting ignorance and superstition, so much prevalent in this country."

And, while not all would have concurred with Mohr's Christian argument that they do this "with the purpose of filling the hearts of the people with reasonable and lofty thoughts about the works of their Creator, worthy of making them susceptible to religion," the broader missionary argument that spreading Enlightenment knowledge would make the original inhabitants more susceptible to Western ideas has remained a persistent claim from the apostles of enlightenment.[33] Notwithstanding their stated ambition of creating enlightened inhabitants, though, Mohr and his successors have had little success fomenting a popular enlightenment.

In chapter 1 I analyze the process through which the earliest permanent institutions of professional science emerged in the 1840s. New Dutch settlers and colonial officials brought interest in Enlightenment knowledge and intellectual energy, and revived the Batavian Society in the late 1830s. A community of intellectuals interested in natural history and other Enlightenment disciplines built the first substantial scholarly community with regular meetings, multiple journals, and sustained research areas. They, too, had larger ambitions, which they pursued by organizing a popular fair in Batavia in 1853, which would show publicly the fruits of Enlightenment knowledge. Still, despite their attempts to popularize the Enlightenment, these Batavian apostles of enlightenment failed to establish themselves as leaders of civil society in the hierarchy of the colony. The 1853 exhibition was the event of the season, but it did not lead to the long-term institutionalization of Enlightenment knowledge and had no long-term impact. At the same time naturalists, working independently of metropolitan control, initiated a new model of professional science, working at the Botanical Gardens connected to the governor general's palace in Buitenzorg (now Bogor).

Individual, and often amateur, natural history projects continued throughout the nineteenth century, but the extent of the audience for this scholarship was limited to a small number of educated Europeans. It was ignored by the colonial government and its officials, and thus by professional scientists. Separate from this, however, some apostles of enlightenment found employment with the colonial state, which had particular needs for useful knowledge, especially about export crops. In chapter 2 I examine the first generation of what I call floracrats, naturalists working as colonial officials on state projects, in this case the acclimatization of cinchona, the quinine tree, in the highlands of Java. I argue that the technical problems associated with acclimatizing cinchona were overcome not just through applied science but by the creation of a professional identity that aligned with the goals of the Dutch Liberal Party and the colonial state. The professional capacity of these floracrats was as state experts responsible for aiding the Dutch cinchona planters. The institutionalization of scientific expertise pioneered by these cinchona floracrats became the norm for the professional biological sciences in the Dutch colony.

Professional science reached its apex in the decades around 1900, in particular through the expansion of the Botanical Gardens in Buitenzorg. In chapter

3 I examine how the biologist Melchior Treub created authority for science among colonial officials. After becoming director of the Buitenzorg Botanical Gardens in 1880, he opened a visitors' laboratory, which hosted regular occupants from Europe and North America. He thus raised the profile of the Buitenzorg Botanical Gardens by convincing other colonial inhabitants, in particular colonial officials, that the Indies now housed world-class institutions of enlightened thought. In chapter 4 I follow Treub in the second part of his career when he created a scientific empire out of the Botanical Gardens. Like earlier apostles of enlightenment, he did this by arguing that his knowledge was useful to the colony, particularly to agriculture. He was persuasive, and because he had already secured local authority for his scientific knowledge, he became director of the newly created Department of Agriculture in 1905. There he controlled dozens of scientific institutes, employing hundreds of scientists full time. After Treub, careers in colonial science, especially in various biological disciplines, became routine and even well paid. I further examine how the ethical policy goals of expanding native education and developing native agriculture allowed the power of the state to grow, with far greater authority handed over to professional scientists such as those who worked inside Treub's scientific empire.

After the turn of the century, a new group of apostles of enlightenment sprung up among members of the educated native elite who had been through the medical and agricultural training programs of the colonial state. In chapter 5 I examine how the intellectual programs of these apostles of enlightenment sought to engage with nationalist goals. Beginning with the organization Boedi Oetomo in 1908, these apostles of enlightenment believed that spreading knowledge of Western scientific traditions and practices among the native inhabitants would lead the colony and its people toward greater independence from the Dutch. Initially they expected that professional science, under the colonial state, would advance their ideals. But once they found that the colonial state had use for native professional scientists only to the degree that they advanced colonial interests, they looked for ways to expand the Enlightenment on their own. Yet their endeavors, in schools and publications, did not create the kind of political movement needed to dislodge Dutch colonialism. Ultimately it would be the Japanese invasion, and then the subsequent national revolution, that would establish an independent Indonesia. The apostles of enlightenment would not be important figures in the fight over political power with the Dutch.

The Great Depression of the 1930s and the war and revolution of the 1940s disrupted professional science, drastically shrinking the number of scientists working in Indonesia. As a result of the Depression and the war, the colonial state rethought the place of science. In chapter 6 I visit the Botanical Gardens during the 1930s depression era. At that time, the rising star of the colonial bureaucracy, H. J. van Mook, decided to reinvent state-sponsored science in the colony. His vision, first implemented at the Botanical Gardens, was to have scientists in charge of all research decisions, including which research received funding. Their findings would then guide economic policy, thus turning them into technocrats. Although this goal was unfulfilled because of the war and revolution, it left a powerful precedent for the future Republic of Indonesia. In chapter 7 I examine the origins of the professional culture of independent Indonesia, in particular the first generation of Indonesian biologists. When Indonesian science was born in 1950, Indonesian scientists revived the colonial scientific institutions but now with Indonesian national concerns front and center. At the Botanical Gardens in Bogor, they turned to professional development, which especially included training the next generation of biologists. They did so without government intervention until 1959 and the onset of the revolutionary authoritarianism of Sukarno's Guided Democracy. In the 1960s the state took a much more active role in shaping science, particularly by aligning it with nationalist ideology. Some Indonesian professional scientists benefited from the political shift toward authoritarian rule. The biologists in Bogor saw the state as a consistent patron and aligned their scientific goals with the nationalist ideology of the New Order government of President Suharto.

The history of Indonesian professional scientists is part of a larger story of Indonesian elites' attempts to shape a modern future for their country through cultural innovation.[34] In this study of how apostles of enlightenment, particularly those interested in natural history, became professional scientists, I make two larger claims about the intellectual history of modern Indonesia. First, scientists established authority and legitimacy through the state. Since the 1850s the colonial and postcolonial Indonesian state has controlled the scientific agenda by effectively absorbing the individuals, institutions, and ideals of science. Scientific funding originated almost exclusively from the colonial and postcolonial governments, and scientists consciously tried to fit their research and work into state programs. The state has been the sole, consistent consumer of professional scientists' research. And the state has required that either science must have no politics or if it does it must be under state control. Most apostles of enlightenment have taken state opportunities when they appear. As a result these scientists have created certain kinds of knowledge, in particular science that could be shown to enhance the economy or in some way contribute to the government's prestige. A career in Indonesian botany has meant working in Leviathan's garden, that is, in a government institution employed as a floracrat. In its garden the state has been able to call upon its floracrats for specific needs such as acclimatizing the quinine tree, combating a coffee-culture pest, or helping to achieve national self-sufficiency in rice. But this also means that the state has controlled biology as a discipline. Officialdom has directed that floracrats work toward small results rather than radical transformations. From the vantage point of the dreams, ideals, and sheer, creative potential of the professional scientists, state employment has led to the failure of Indonesian science.

Second, scientists have found it very difficult to connect to, let alone lead, Indonesian society. Apostles of enlightenment have not stimulated popular

Enlightenment knowledge despite their desire to spread the benefits of science broadly to the population. Central to the idealism of many scientists is the hope of bringing the benefits of Enlightenment thinking to Indonesian society. Still, scientists, who idealistically see themselves as representing a bridge to civil society, have few venues in which to connect to society. Instead of fomenting Enlightenment knowledge among the inhabitants of Indonesia, they have worked toward creating professional institutions of science. Moreover, professional scientists have made very little progress in obtaining scientific authority from civil society or gaining patronage from wealthy individuals. While the apostles of enlightenment continue to be idealistic about what they can accomplish, they have not succeeded in bridging the distance between themselves and civil society. They work in elite state institutions of professional science with few lines of communication open to the Indonesian population. Colonial scientists channeled their expertise and results via state institutions on the assumption that this was the best way to reach society. Nationalist scientists, for whom the creation of Indonesian science is a patriotic duty, have been able to create certain forms of national science, but they have become intertwined with the state's nationalist ideology. In this manner I argue that the Enlightenment has failed in Indonesia because its apostles have not been able to gain authority from the general populace of Indonesia.

Whenever professional scientists have begun to demonstrate possibilities for leading civil society, the state has co-opted this potential by employing them as full-time bureaucrats, often with defined tasks and research agendas. Yes, science is a tool of the state, but more broadly the state has absorbed science, scientists, and scientific scholarship. Since 1840, Indonesian colonial and postcolonial science has not lacked for capable, ambitious, and innovative scientists. And many have adeptly navigated between the global world of science and the colonial and national cultural contexts in which they work. But on the whole they have failed to establish themselves as leaders of civil society who can bridge the distance between citizens on the one hand and the state on the other, no matter the era. Science has remained an elite affair practiced at times with flair and interest but unable to jump the fence that surrounds Leviathan's garden.

# 1

# Apostles of Enlightenment

In the spring of 1848, as Dutch Batavia began to read about the February 1848 Paris uprisings in the European newspapers and the Dutch king's concession to government reforms, some were confident enough to test the strength of civil society in the Indies. In May of 1848 Baron W. R. van Hoëvell organized a meeting in which five to six hundred inhabitants met to discuss better access to government jobs for locally born colonizers through expanded educational offerings in Batavia. A committee was quickly created to prepare a petition for the king. Those at the meeting expected more, and the mood was chaotic, with widely different anticipations among the participants. The organizers adjourned the meeting quickly. This display of politics, which was not repeated, was hardly revolutionary. Van Hoëvell and others had received permission from Governor General J. J. Rochussen to hold it. Nonetheless, this meeting was an effort to politically organize colonial civil society, the first such attempt since the return to Dutch control of the colony after the British interregnum ended forty years earlier. After the meeting Rochussen worried that the European community had found a way to lead other elements of colonial society, and he immediately banned all future political meetings.[1] And with that the 1848 Batavian revolution came to an end.

The political premise of the May meeting—that the Dutch inhabitants of Batavia could directly challenge government authority—was the result of a heightened awareness among European Batavians about civil participation in the colony's leadership. In this chapter I argue that this change was an outgrowth of an explosion of interest in Enlightenment culture in the previous ten years. Since the late 1830s certain Dutch thinkers had argued that through investigations of its geography, people, and environment the Dutch could shape a suitable political order in the colony. Fleshing out knowledge of the colony

13

using Enlightenment methods was for them the first step in crafting a political culture appropriate for a European society living in the tropics. These apostles of enlightenment promoted the contagious idea that generating scholarly knowledge of the colonial environment was itself productive of civil society. It was this context that had Van Hoëvell, one of the leading scholars of the 1840s, heading the May 1848 meeting.

The meeting served as a wake-up call for the colonial administration. After the meeting was over, Rochussen began to worry that it had allowed popular sentiments to run toward treason even though he concurred with some of the demands about local education. Although the meeting had passed without incident, it showed Van Hoëvell's ability to channel the demands of the poorer, and more dangerous, Eurasian community. Van Hoëvell had in the process emerged as a rival leader outside of the normal colonial hierarchy. The state took swift action to prevent civil society from growing stronger, not only by prohibiting further meetings but by expelling Van Hoëvell and the other leaders of the meeting. Van Hoëvell returned to the Netherlands later in 1848, where he quickly began an illustrious career as a parliamentarian. But Rochussen and the colonial state saw the threat as broader than the activities of individuals such as Van Hoëvell. They also took care to prevent any future popular expression of civil society by co-opting the apostles of enlightenment and placing them in the administrative structure of the colonial state. Thus, by the early 1850s scholarship had become divided between amateur hobbyists in Batavia and the official scholarship of the Netherlands East Indies state. Dreams of building a civil society separate from the state had disappeared. Crafting a more productive colonial culture through science and scholarship was now a state project.

## Enlightenment Scholarship

Before the clever taming of passionate enlightened civilians, however, back in the 1840s, a community of European apostles of enlightenment had begun enthusiastically talking, lecturing, and writing about the colony. These men, coalescing around the meeting rooms of the Batavian Society of Arts and Sciences, were largely recent arrivals in the colony, most employed as colonial officials. None had resumes of scholarly or even professional achievement in Europe. Yet in the colony they reinvented themselves as scholars. Eschewing the label of amateur, they had great ambitions for themselves and the colony they lived in, believing they had an opportunity to shape colonial culture. These apostles of enlightenment brought the message not just of superior natural knowledge but of civil and popular participation in the colony's future. To this end they advocated using the tools of European natural history to quantify, translate, and explain the peculiarities of the colony's society, culture, and nature. This cultural awakening would, they hoped, set the stage for the shaping of a European, enlightened society where their children could build a permanent home. They did not take direction from the colonial administration; instead they believed that their scholarship itself would usher in a suitable political order.

The mouthpiece of the apostles of enlightenment was the *Tijdschrift voor Nederlandsch-Indië* (Journal for the Netherlands Indies), founded in 1838. It was modeled on the Dutch journal *De Gids* (The Guide) and aimed to foster a general exchange of ideas about the colony. Right from the start it stressed the importance of scholarship. The Batavian protestant minister Van Hoëvell, one of the driving forces behind the journal, in a programmatic 1839 article about the history of arts and sciences in the colony, explained such local scholarship as an effort to "spread enlightenment and civilization." This, he argued, would establish "something good and permanent" for future generations of colonizers.[2] Van Hoëvell encouraged a wide range of scholarship, from the philological examination of Malay and Javanese literature (his own field) to natural history descriptions of the Javanese landscape.[3] Synthesis would come later; the philologist C. F. Winter used an entire issue of the journal of the Batavian Society to publish a transcription, not even a translation, of the Kawi epic the Romojono.[4]

The Batavian Enlightenment was initially inspired by European models. The public dissemination of enlightenment began at the Batavian Society for Arts and Sciences, an eighteenth-century Enlightenment institution that had languished since the 1790s but had been revived in the mid-1830s.[5] The Batavian Society consolidated its position as the central scholarly institution in the early 1840s when it purchased its own press, only the third printing press in the colony. During the 1840s eleven different journals were printed in Batavia, most of them on the press of the Batavian Society.[6] For example, the journal *De Kopiïst* (Copyist), founded in 1842, began by reprinting noteworthy articles from European journals and papers. And it was also in the 1840s that Europe and the Indies became much closer in time and space; in 1844 a mail route opened up via the Isthmus of Suez, and Batavians could read newspaper articles, letters, or books written just two months earlier in Europe.[7] While the models for scholarship were imported from Europe, the Batavian scholars' ambitions went beyond copying European achievement. In 1843 *De Kopiïst* was redubbed the *Indische Magazijn* (Indies Magazine), and thereafter contributors wrote exclusively about issues directly pertinent to readers in Batavia.[8] It did not take long for the Batavian scholars to gain confidence in crafting their own scholarship for their own benefit. In 1842 in *De Kopiïst* the naturalist Franz Junghuhn asked all inhabitants of the colony, officials and nonofficials alike, "to offer their cooperation in collecting material for a history of the incidents and changes in the nature of the archipelago."[9] Generating knowledge of the colony's natural history was to be a collective endeavor.

To its adherents scholarship promised to be the foundation of an enlightened community in the tropics. In planning for the future, the apostles of enlightenment shaped their public scholarship similarly to contemporaneous intellectual currents in Europe. Their scholarship was to be public knowledge, written, presented, and vetted among the Europeans of Batavia. Like the contemporary developments in Central Europe studied by the historian Ian F. McNeely, these Batavian gentlemen emphasized knowledge of and about their community. The results of their research were to be spread widely by filling gazettes and journals with geographic, social, and economic data. Their knowledge might begin in an elite setting, but its destination was the colonial public. The ambition of the Batavian Enlightenment, like its counterpart in southwestern Germany, was the creation of a road map to civil society.[10] Public knowledge about the complexity of their new home would set the parameters for the debate about the future of the colony's society. It would establish knowledge about the still largely unknown territory but also create a community of scholars with the authority to debate the direction of the colony. They expected that scholarly organizations, along with their publications, would be the earliest institutions of the colony's civil society.

During the 1840s the apostles of enlightenment not only examined their surroundings but also began to think of themselves as fathers of a nascent society. For the first time in decades the colony was stable and prosperous for its imported inhabitants. The Dutch colonial state had consolidated its military grip on Java during the Java War (1825–30), and by the early 1840s the cultivation system, which forced Javanese peasants to grow crops for the export market, had extended Dutch administration across the island of Java. But the administration was still in its formative phases. It was expanding, and had begun to attract a new wave of recruits from Europe, but there were as of yet few formal requirements for joining the administration. Embarking on an assignment in the colony was still an adventure and not a career.[11] Nonetheless, as the Dutch colonial administration expanded it brought educated professionals from Europe, including doctors, ministers, and career soldiers, as well as administrators. The identity of these new arrivals was still in flux, not synonymous with the colonial administration. The historian of the cultivation system Robert van Niel has argued that during the 1840s the European bureaucrats' worldview was dominated by liberal economic ideals despite the draconian economic measures handed down from The Hague.[12] Europeans in Batavia, even those employed by the colonial administration, thought of themselves as more than state officials; they were the avant-garde of a new European society. And they imagined a leadership role for themselves, taking on the task of creating an identity for their community, which would advise the government. Throughout the 1840s, Batavian scholars printed numerous suggestions, projects, and ideas for the colonial administration. In 1842 a Batavian journal proposed a plan to build a railroad network across Java.[13] These colonial apostles of enlightenment saw it as their responsibility to guide policy for the future of the colony.

All of this was possible because during the 1840s the central bureaucratic apparatus in Buitenzorg was in transition. The administration had an unsure mandate now that it had been stripped of its economic duties. Fiscal guidelines about the workings of the cultivation system were set in The Hague and were not to be tampered with by even the governor general.[14] In the colony the administration was to maintain political and economic stability, but no one, in either The Hague or Buitenzorg, knew yet what that specifically meant. Furthermore, in the first half of the decade the administration was run by a string of temporary governor generals with brief tenures, most of whom had been appointed from among the local inhabitants and none of whom created a new administrative spirit.[15] Weak leadership at the top prevailed until the arrival of J. J. Rochussen in September of 1845.[16] Rochussen brought new life to the governor generalship, and he was unequivocal in his policy of forbidding the citation of any official government notice in any of the local journals.[17] But many of the Enlightenment initiatives overlapped with his ambition of establishing the colony as an "enlightened, strong and honorable autocracy."[18] He was not simply a tool of the colonial policy makers in the Netherlands, who considered Rochussen an overambitious "project maker,"[19] and he consciously attempted to harness the creativity of the Batavian apostles of enlightenment. He adopted the plan of installing a railroad between Batavia and Buitenzorg first proposed in 1842 (this plan was later canceled by Minister of Colonies Baud due to expense). Although by 1848 Rochussen had gained the upper hand in dealing with the political aspirations of the Batavian European community, during his tenure he permitted a wide range of nonpolitical scholarship.

## Java's Humboldt

Most pressing for the Batavian apostles of enlightenment was a full understanding of the colony's strange, and to these newcomers possibly daunting, nature. And hence natural history, ranging from travelogues to taxonomy, was a central preoccupation of the Batavian Enlightenment. Naturalists turned to European natural history writing for models and were in particular inspired by the taxonomy of Linnaeus and his followers, as well as the protoecological writing of Alexander von Humboldt. Natural history articles in all manner of genres dropped off the Batavian printing presses in the 1840s, all contributing to a fuller accounting of the tropical environment. Most had a taxonomic element as part of a wider ambition to identify and explain the components of the jungle landscape. Many analyzed the geographic distribution of the flora and fauna and, like other European biological studies (including those of the British naturalist Alfred Wallace, who traveled in the Malay archipelago between 1854 and 1862), were inspired especially by Humboldt.[20] While the tropical environment was daunting and strange, colonial naturalists endeavored to use European practices to classify, organize, and comprehend Javanese nature. In particular, Humboldtian science was seen as a way to demystify the unfamiliarity of the expansive rain forests. And in imitation of Humboldt the Batavian

Franz Junghuhn, self-portrait, 1860 (Courtesy of KITLV/Royal Netherlands
Institute of Southeast Asian and Caribbean Studies)

naturalists explained where and how the Europeans would belong in the Javanese landscape. Moreover, much like their contemporary European Humboldtian scientists, naturalists in Batavia saw the practice of natural history as a
means of transforming a colony into an enlightened society.[21]

Between 1835 and 1848 the Humboldt of Java, a military doctor named
Franz Junghuhn, spent every spare minute crisscrossing the volcanoes of Java.[22]
He regularly updated the European population of the Indies in the Batavian
journals. This research was ultimately collected in his thirteen-hundred-page

*Java: Its Form, Covering, and Inner Structure.*[23] The skeleton of Junghuhn's Java was the volcanoes, all thirty-three of which were described based on personal inspection. Like Humboldt, he emphasized the unity of flora within twelve altitudinal zones. Java's spiritual greatness came alive in his descriptions of soil, rock, and plants. Using thirteen years of familiarity with the mountains of Java, he showed how to be in touch with the mood of the land. Through mixtures of architectural, medical, and botanical languages, the island metamorphosed into a conscious being with a rigid body and a sensuous spirit. It was a clearly delineated island with absolute boundaries, a large mass of land but finite, manageable and at the same time diverse. Java's inner mass remained a constant, but its character revealed itself in the mountains. For Junghuhn, the soul of Java was manifested by the diverse set of flowering plants growing on the surface of its volcanoes.

Junghuhn was the first naturalist to give Java depth; his Java was vertical and spiked, not flat and two-dimensional. Instead of treating the volcanoes as blank and hence intimidating spots on the map, he depicted a landscape with a sentient and regular personality. This was possible, he explained proudly, because of his use of a calibrated barometer.[24] This quintessentially Humboldtian tool measured the insides of Java by recording the height of the volcano peaks.[25] Because normal climate change moved in a regular twenty-four-hour period, and because the temperature of the barometer could be kept constant, accurate elevations could be taken on all the Javanese volcanoes as long as measurements were taken early in the morning. Junghuhn easily shifts into his role as nature's physician; the measurements of Java's internal pressure, temperature, and skin condition record Java's vital signs. The barometer, in other words, gave life to Java's nature. Junghuhn projected a Javanese nature that was not only knowable but easy to get along with.

While Junghuhn was living in the Netherlands in the early 1850s, he took the next step and elaborated on how Europeans should interact with Java's landscape. No longer merely descriptive of Java's nature, he is prescriptive, making suggestions about European naturalists' roles as leaders of Javanese colonial society. Junghuhn's *Licht- en schaduwbeelden van de binnenlanden van Java* (Light and Shadow Images from the Interior of Java), first published in 1854, is a chronicle of the adventures and conversations of two men — "Day" and "Night" — on a trip through the mountainous regions of Java. Their debate about the merits of bringing Christianity to Java is interwoven with descriptions of the natural landscape and interactions with Javanese villagers. Night advocates Christianizing the Javanese people in order to raise them out of their primitive condition. Day — who speaks for Junghuhn — disputes the need for these Christian missions on Java and specifically takes issue with Night's claim that Christian learning had produced the "light of knowledge that spreads its bright rays across Europe."[26] Day disparages the superstitious belief in miracles that the church preaches. He then lectures his companion on the achievements of the natural sciences and the naturalist's passion for the "enduring immutability and consistency of the natural laws," which provide "truth, the best explanation." He praises "the constantly growing certainty of that which we discern in the heavens above and here below on earth, that has above all flowed from the discoveries done in the fields of geology, astronomy, chemistry, physics, and physiology. Principally through this the truth has been revealed and announced."[27] Night, who has fewer and fewer moments as the book continues, slowly gives way to Day's natural metaphysics. God, light, and truth were revealed without the intermediary of the church, but instead through careful study of nature. Nature itself was Day's religion: "One must not imagine God as a force divorced from nature, external to her. But instead as a force contained within her, as the universal spirit in nature—as the Worldspirit."[28]

Junghuhn suggests the building of a new kind of colonial society in Java, one in which members of the indigenous elite are taught the methods of European natural history. Junghuhn claimed that the colonizers, as a people spiritually further advanced because of the geography and climate of Europe and the superior intellectual capacity of the Caucasian race, would bring the tools and methods necessary to harmoniously and productively live on the land.[29] This would allow European and Javanese populations alike to properly harmonize with nature and the Worldspirit. At the end of the book Day and Night meet an imam, an Islamic holy man, and give him a "thermometer, a small compass, a magnet, a pocket telescope, a simple microscope, and other similar instruments. We instructed him of their proper uses. He was greatly pleased to be in possession of these tools. He promised us he would use all his power to spread the Gospel of nature among his compatriots, while we promised further written instruction."[30] The imam readily sees the drawbacks of organized Christianity and enthusiastically embraces the tools of Humboldtian science. Junghuhn and the Javanese are bound through their common love of the rich landscape and its universal spirit. This bond will be cemented through the tools of natural history and with further instructions from Junghuhn and his fellow naturalists. There is no hint of conflict between the imam's Islam and the Worldspirit.

*Licht- en schaduwbeelden* is not a contemplative book, nor was it intended as one. Night is a one-dimensional straw man, and all characters besides Day are rather crudely drawn; it is first and foremost a polemic meant to advance Junghuhn's argument that his natural history could harmoniously bring Java under enlightened Dutch leadership even going so far as to claim an easy co-opting of Islamic leaders who might otherwise put up a fight against Christianity. He proposed a colonial relationship modeled on scientific cooperation between the Europeans and the Javanese. Moreover, he posited competition between European knowledge and religion and suggested that the enlightened Europeans should seize the initiative by beating back the influence of the church. Junghuhn, of course, would be the high priest of this scientific naturalism but with backing from the colonial administration.

Junghuhn's big ideas sold well, in Europe at least, as *Licht- en schaduwbeelden* went through four editions in ten years.[31] The idea that scientific manliness could conquer nature and establish a true colonial presence was exciting to many European readers familiar with romanticism.[32] And it suggested to Junghuhn and other naturalists that their vision of a colonial civil society was coming to fruition. Junghuhn had become something of a star in the Netherlands and had cemented important relationships with the new liberal leadership, including the Liberal Party leader J. R. Thorbecke, and the minister of colonies, C. F. Pahud. He also won a debate with C. L. Blume, the head of the Rijksherbarium (now known as National Herbarium of the Netherlands, Leiden University branch), over who controlled colonial scholarship. After a nasty back-and-forth argument carried out publicly in rival publications in 1850, the old scientific association controlled by Blume and his cronies was disbanded by Thorbecke.[33] Junghuhn and other colonial scientists would henceforth be allowed to control their own scholarship. It began to look like important powers within the government were supporting apostles of enlightenment such as Junghuhn, including their political agenda.

In Batavia confidence in transforming civil society was high even after Rochussen's clampdown on political activity in 1848. In 1852 Governor General A. J. Duymaer van Twist—Rochussen left the colony in 1851—granted a license for the printing of the first regular colonial newspaper, known as the *Java-Bode*, and soon thereafter it proved able to do more than simply publish official and commercial announcements. It was run by a professional editorial staff and featured regular columns, including a serialized novel and a prominent section titled "Arts and Sciences, Agriculture and Industry."[34] The cultural birth of the colony seemed to be moving forward. The apostles of enlightenment were still busy building knowledge of the colony. The possibility of transforming the colony into an enlightened society reached its zenith with the staging of an exhibition in Batavia in 1853.

## The 1853 Exhibition

The culmination of the work of this era's apostles of enlightenment was an exhibition staged in Batavia in 1853. The exhibition showcased natural and manmade objects from all corners of the Indies and had been planned as an exhibit of what its organizers considered the popular and industrial arts of the Indies. Its planners, European colonial officials who had come to Batavia in the 1830s and 1840s, proceeded believing that a comprehensive exhibition of the resources of the colony would advance the colonial population's industrial and agricultural production. Once Batavians understood what raw materials were available, they would be able to plan future exploitation in a reasoned fashion. The spark for the exhibition was the Great Exhibition of 1851 in London, and the Batavian planners consciously attempted to imitate the European model, for "in almost all countries of that part of the world [Europe], these exhibits have repeatedly offered people the opportunity to learn the products of different regions, the needs, the tastes, and the desires of many nations, down to their smallest detail."[35] Transforming colonial society was front and center, and, as one of the organizers explained, its goals were "to decide which means are viable for increasing and improving the products from the fields, to awaken and dignify the industriousness and taste of the native, and also, to learn scientifically some of the little or completely unknown agricultural products and crafts, which in the future could diversify trade with more life."[36] All Batavian residents were invited to the exhibition, and summaries printed in Dutch, Javanese, and Malay were circulated in the months before it opened.[37] It opened in late 1853 to popular acclaim. In the first week, more than five thousand people paid their admission, with approximately half classifying themselves as natives or Chinese.[38] It was the social event of the year, and, for the inhabitants of Batavia at least, it was a popular excursion for the two months that it remained open.

The exhibition was conceived by members of the recently established Natuurkundige Vereeniging in Nederlandsch Indië (Naturalists' Association of the Netherlands Indies), which since its founding in 1850 had been looking for ways to institutionalize the achievements of the 1840s Enlightenment. The Naturalists' Association was the brainchild of one of Junghuhn's oldest friends in the colony, Piet Bleeker, a military doctor and amateur naturalist. It was imagined as a forum dedicated to promoting, discussing, and publishing the results of research about the colony's nature. The association quickly sought to build scholarly credibility by signing up corresponding members in the Netherlands.[39] In October of 1850 it launched its own journal.[40] The Naturalists' Association grew out of the Batavian Society of Arts and Sciences, and early copies of its journal were mailed to all the members of that society. Although the islands had been under the naturalist gaze for more than fifty years, this organization and its journal were the first institutions dedicated to natural research founded on local initiatives.

The new organization and its journal emerged quite naturally out of the Enlightenment initiatives of the 1840s. It planned to use European scientific traditions to reveal the workings of the colony, all for the purposes of progress. In early 1852 Bleeker said in his yearly presidential address, "Gentleman, we are in a region, where natural history, even more so than in other civilized countries, is called upon to locate natural riches, and on a large scale make useful to humanity, the properties of nature."[41] This was, however, a journal that would publish practically anything that could be construed as natural history.[42] The submission of articles and their acceptance were high enough to warrant almost one thousand pages of text in 1853. Like earlier attempts of the 1840s, the association was to advance knowledge of and for the colony. But the longer range goal was the making of lasting institutions capable of permanent scientific leadership, for "only a division of effort can lead to reaching the great scientific unity, on which all of our energy must be spent."[43] In 1852 Bleeker confidently announced at the annual meeting of the Naturalists' Association that their organization was beginning to reach the level of similar ones in Europe.[44]

The idea of an exhibition was raised in late 1851 by S. D. Schiff, a member of the Naturalists' Association, who proposed sending colonial materials to the Dutch national exhibition to be held in Arnhem the following year. Once time constraints made this plan unworkable, the association convened an exhibition committee charged with planning a local alternative: an exhibition staged in Batavia of "popular and industrial arts" in the Indies. The governor general's office gave its seal of approval and also offered to formally instruct local colonial officials to help in the gathering of goods.[45] By the middle of 1852, the commission had raised nearly f.20,000 (guilders) almost entirely from local committees.[46] An official program was published at that time with an inventory of permissible items. Almost anything was requested: minerals, leaves of plants, peanut oil, chairs, honey, machines, batiks, and so on. Only live animals, odiferous material, and explosives were prohibited.[47]

On October 10, 1853, the organizers led Governor General Duymaer van Twist on the first tour of the exhibition. The governor general spent two hours taking in the varied riches of the archipelago, which were arranged residency by residency. Every outpost of the colony had sent something. From Batavia came evidence of European ingenuity: a special centrifuge used in sugar fabrication, type settings from a lithography company, and a European plow modified for use on Java. Sprinkled in were the governor general's sizable collection of native boats and an ichthyologist's collection of rare fish taxidermy. Representative ceramic dishes, krisses (Javanese daggers), and silverwork rounded out the Batavian collection.[48] Rural areas, such as Ciribon, contributed limestone, oil, and a very hardy piece of sandstone, as well as proudly showing off prize specimens of coffee, indigo, and tea.[49] From Solo came leather puppets for *wayang* shadow plays, a model chariot from the Surakarta *kraton* (palace), and a specially constructed gamelan ensemble.[50] From the evidence available it appears that there was little interpretive material describing the collection. The captions in the catalog briefly described the item and its location and only provided the name of the person who sent it in.[51]

Schiff opened the exhibition on October 10, and Bleeker, the president of the Naturalists' Association, gave the keynote address. He praised the apostles of enlightenment who had organized the exhibition, hailing their service to the colony, in particular by supporting native industries with European knowledge and tools.[52] The exhibition was quite popular, and in the first forty days, more than twenty thousand people visited, which was an unprecedented accomplishment. Most of their reactions are unrecorded. The local newspaper the *Java-Bode* was congratulatory about the achievements of Dutch colonial society in staging such an informative and systematically rigorous exhibition.[53] Governor

General Duymaer van Twist was keen about the project from the beginning, and in a letter he wrote to the minister of colonies when the exhibition was in its planning phase he explained that although this was a private endeavor it nonetheless went a long way toward advancing his goal of improving private industry.[54] The colonial leader's first impression of the final result was guarded, however, stressing the need for an edited catalog if attendees were to make sense of the disparate objects.[55] Despite this, to Duymaer van Twist and others the program advocated by the exhibition looked eminently possible.

To its organizers the 1853 exhibition itself was proof that civil society already existed in the colony. The products sent in were not just dispassionate scholarship but the fruits of general involvement in taking the colony forward. The exhibition was designed to take advantage of the broad participation of colonial elites, including not just colonial gentleman in Batavia but also European and non-European officials outside of Batavia. The process of collecting material was decentralized and conceived with a democratic spirit, with circulars and appeals going out to all elites in the colony, European and non-European alike. An explicit goal was popular participation both in collecting material and in the exhibition itself. The organizers envisioned a movement led by Europeans but encompassing the entire population. Furthermore, the exhibition's epistemology was not hierarchical. The organizers issued only very general regulations for what objects would be allowed. Of course this also demonstrated that there was still weak central control over knowledge, but to the organizers of the exhibition this was positive. Collecting material locally meant the colony was generating its own knowledge without the need to defer to Europe. The role of the Naturalists' Association in organizing the exhibition was to facilitate progress by motivating the colony's inhabitants to find ways to promote prosperity and industriousness.

## Civil Society and Elite Leadership

When Bleeker left the colony in 1860, he noted that science there was on the decline and generations behind Europe. He was even pessimistic about the colony's future as a whole.[56] He expressed this despair despite his prominence within both scientific and official circles in Batavia throughout the 1850s. What had happened? The 1853 exhibition was to be the capstone of the Batavian Enlightenment. It was intended to decisively use natural history and scholarship to shape the future of the colony, imagined by Schiff and the other organizers as the triumphant arrival of a civil society. Their optimism was not borne out. The exhibition was soon forgotten; its artifacts were moved into storage, and nothing was ever done with them. The promises of the exhibition turned out to be empty. By the middle of the 1850s, shortly after the exuberance of the exhibition, the popular Enlightenment was moribund. Scholarship like that of the

1840s continued to be practiced but now as an amateur hobby. Civil society, if it existed at all, was hollow.

Why did the apostles of enlightenment fail to produce changes in the early 1850s? Mostly they failed because their Enlightenment initiatives did not constitute a popular movement. They were produced by newly arrived Dutch elites for the consumption of newly arrived Dutch elites. Moreover, most were colonial officials themselves. All the leaders of the Batavian Enlightenment, from Van Hoëvell to Bleeker, had come from Europe in the 1830s and 1840s in order to work for the colonial administration. These adventurers quickly cast themselves as the leaders of a dramatic cultural creation but did so without broad social support. Their scholarship's audience was narrowly based in Batavia. Their confidence in natural history's capacity to kick-start progress was misplaced. Their scholarship, and even the 1853 exhibition, had little or no real impact on other communities in the colony. While in May of 1848 Van Hoëvell had attempted to broaden his coalition to include the Eurasians of the colony, the older Batavian Society was by and large hostile to the cultural and social ideals of the nineteenth-century imports.[57] Furthermore, there is no evidence that Batavian scholarship reached Javanese elites. The Dutch scholars in Batavia were distant from the rest of the Batavian urban population and had little contact with the other inhabitants of Java.

For the Batavian apostles of enlightenment, the road to civil society ran through natural history and scholarship, not increased social participation. The Batavian scholars of the 1840s believed the latter would follow from the former. This was a serious miscalculation, as even contemporary observers noted. A. W. P. Weitzel, in an 1860 exposé of the European community in Batavia, concluded his book by taking the Batavian apostles of enlightenment to task for failing to push at the social and political boundaries erected by the bureaucracy. In analyzing the legacy of the 1853 exhibition, Weitzel noted that the 1850s had failed to encourage free economic activity in part because Batavian residents had little or no influence on colonial policy.[58] Weitzel argued that the Indies could become productive only by building a society founded on its own economic and intellectual foundation. He emphasized the need for shaping a civil society freed from metropolitan control, writing, "As long as the Indies is populated by Europeans, who reside there temporarily in order to change their fortunes, as long as it is *an overseas possession* and not a colony with an autonomous, political, and intellectual life, there shall not be any significant changes."[59] Weitzel suggested that the Batavian gentlemen found a school where the sons of Europeans could, gratis, pursue a technical education.[60] This would ensure the reproduction of an educated and freestanding European community in Batavia and be a catalyst for creating a productive, enlightened culture in the Indies. This, he pointed out, was exactly what the leaders of the Batavian Enlightenment had failed to accomplish. Though they clearly believed in the dissemination of information, it is apparent that they did not know how to educate those less educated than themselves.

The failure of an exhibition to launch a cultural and social transformation of the Netherlands East Indies is in retrospect not surprising considering that the purpose of nineteenth-century European exhibitions was to expand bourgeois economic power into the political realm. European exhibitions were staged as part of bourgeois cultural engineering and were intended to legitimate the right of the bourgeoisie to speak for all the urban classes.[61] Conditions in the Indies in the 1850s were vastly different from those in England or elsewhere in northwestern Europe. At midcentury no entity in the Dutch colony, not even the colonial administration and certainly not a small group of scholars, had the capacity or even the ambition to engage in widespread cultural engineering. Moreover, the apostles of enlightenment did not stand on a solid economic base. Their project was staged by idealists and romantics who actually believed the ideological literature of the 1851 London Great Exhibition, which proclaimed that exhibitions could promote economic and social progress.[62] The organizers of the 1853 exhibition imagined theirs would do the same. It did not. It did little to promote an autonomous and enlightened Indies civil society. The Batavian apostles' naïveté shows in their haphazard, unexplained displays, which were very different from those of the carefully constructed European exhibitions. With no contextualization, indeed only a title for each item, its geographic origin within the colony, and the name of the person who procured it, these items en masse could inspire awe, but could not give the audience any deep or useful knowledge. The organizers of the exhibition, though they may have intended to educate the populace, had no idea how to do so. A generation later the 1853 exhibition was forgotten.[63]

While the seriousness of the 1840s Batavian Enlightenment scholarship dimmed after 1853, the Enlightenment ideals of expanding civil society were still alive, now inside the colonial state. After 1850 the colonial administration began appointing leading apostles of enlightenment to new positions where they were charged with implementing Enlightenment goals. The confidence of these men was such that they harbored great hopes in bending the colonial administration to their will. For example, Piet Bleeker in 1851 was appointed the first director of a new two-year medical course for the sons of Javanese aristocrats, later known as the *dokter-djawa* school.[64] As shown in the next chapter, Junghuhn became the principle official in charge of acclimatizing the quinine tree. As the apostles of enlightenment received official appointments, they initially believed that they could build the railroads, schools, and museums they had previously dreamed of. There was much reason for this confidence, as the colonial administration had been, for example, an unabashed supporter of the 1853 exhibition, for which the f.15,000 from the governor general's office largely paid.[65] It appeared quite plausible to many of the apostles of enlightenment that the road to civil society could be pursued most naturally through the colonial state's adoption of its programs. At the opening of the 1853 exhibition, Bleeker publicly thanked the governor general for his generous monetary support.[66] To the Batavian scholars it seemed the state was lining up behind them.

As the colonial state absorbed the Enlightenment principles of the 1840s, the government bent scholarship to its will. The tangible legacy of the 1853 exhibition was the integration of scholarship into the colonial administration in the colony. The popular Enlightenment's road map to a civil society, which envisioned jump-starting popular participation in the colony's future by generating the data fit for discussion, was largely forgotten ten years later. By then the colonial administration was firmly in control of Batavia and Java, and was making inroads across the archipelago. Political and social change would be channeled through them, not through the scholarship and discussion of urban elites in Batavia. The Batavian Society of Arts and Sciences and its allied institutions still published hobby scholarship but with greatly truncated ambitions. To be sure, a few of the apostles of enlightenment found work within the colonial administration. Bleeker and his colleagues began their careers within it with high hopes for political change in the colony. And by the end of the century science was even a career option within the colonial administration. But their goals and possibilities were a far cry from the great social engineering ambitions harbored by the 1840s apostles of enlightenment.

## The Buitenzorg Botanical Gardens

The colonial ideology of peace and prosperity rapidly came to dominate the work of the apostles of enlightenment who became official scholars in the 1850s. The colonial government controlled the agenda of official scholarship very effectively. The apostle Junghuhn's very difficult transition to state science will be treated in the next chapter. To understand the roots of the clash between the popular Enlightenment and the state's administrative needs, however, we must go back to the 1840s, and examine the institutionalization of natural history within the Buitenzorg Botanical Gardens. The natural history scholarship of the botanical gardens in the 1840s had been far less vibrant than the Enlightenment and had generally been ignored by the apostles of enlightenment. But the arrangements worked out in the 1840s at the Buitenzorg Botanical Gardens between colonial officials and professional scientists established an important institutional precedent for bureaucratic science. And these arrangements remained in force when the apostles of enlightenment joined the colonial institutions of science in the 1850s.

Prior to the 1840s the colonial administration of the governors general had no role in generating and regulating knowledge about the colony. Expert knowledge of the colony was still under metropolitan control. After the end of the Napoleonic wars, during which time the Netherlands was under French control and Java was captured and administered by the British, the Netherlands East Indies became a crown colony. King Willem I and his advisers tried to develop administrative systems that exploited the colony's agricultural potential. With an eye toward this the king was keen to employ naturalists and their investigative skills in learning as much as possible about the natural potential of the colony. In 1815 one of the first decisions made by the newly crowned king was to dispatch the Dutch botanist C. G. C. Reinwardt to his new tropical possession. Because the Dutch East had been under private company management for two centuries while controlled by the VOC, the king knew very little about his newly acquired empire; even the physical geography of the archipelago was largely uncharted. Reinwardt was given wide powers to direct natural history and agricultural matters in the colony, with an emphasis on ascertaining the economic possibilities.[67] After Reinwardt had been in the colony four years, his mission was expanded in 1820 through the royally created Natuurkundige Commissie (Natural History Commission). With this commission, knowledge of the Netherlands East Indies was institutionalized under the control of the king.[68]

The Natural History Commission's monopoly on natural history research remained in place until the early 1840s, after which it became clear that natural history was of little importance in further exploiting the agricultural potential of the colony. King Willem II recommended in 1844 that the commission not be perpetuated "because natural history research has been going on for so many years, and . . . for basic issues, not so much can be expected from its continuation."[69] From the king's point of view the economic benefits derived from the Natural History Commission had been minimal. No profitable indigenous flora or fauna had been found. By then nonnaturalists among the king's men had devised a cultivation system that profitably extracted wealth from Java using non-native crops. The colonial administration oversaw the cultivation system, which forced Javanese peasants to use a portion of their land and time to grow cash crops, including indigo, sugar, and coffee, and facilitated their export to Europe.[70] Expert knowledge of agriculture and nature played no part in the "basic issues" of the cultivation system. It was in this context of the decline in the prestige and importance of the Natural History Commission that the governor general in Buitenzorg came to be responsible for official scholarship, especially that involving natural history.

Beginning in the 1840s the colonial administration in Buitenzorg first began to see the advantage of using outside scholars to gather and analyze knowledge, as well as to provide expert advice. In the 1840s, natural history institutions came under the control of the Buitenzorg colonial administration, outside the control of the king, the minister of colonies, or Parliament. This arrangement, which was arrived at quietly and largely unnoticed, created the mold of professional science until the very end of the Netherlands East Indies. The earliest scientific institution was the botanical gardens, located in the park surrounding the governor general's palace in Buitenzorg. The gardens had begun as a collecting station for the Natural History Commission, but by the 1840s it had reverted to the control of the governor general. Without the king, funds were scarce — the few remaining naturalists at the Buitenzorg Botanical Gardens waited in vain in the early 1840s for promised funds with which to print a new catalog of the plants in the gardens — but the monopoly of natural knowledge claimed earlier by the king was gone too. While the minister of colonies, J. C. Baud, carefully centralized economic decision making in The Hague, the governor general was given free rein to implement policies that did not cost money or threaten the now mandated profitability of the colony.[71] Sponsoring his own natural history research was just such an area.

An argument between metropole and colony about the rights to a Japanese ornamental tree in which the head gardener of the Buitenzorg Botanical Gardens bested the preeminent Professor Blume, the Dutch botanical head of the Rijksherbarium, serves to illustrate the sea change in scientific administration in the colony. Blume had been head of the Natural History Commission between 1824 and 1826. In 1840 Blume and P. F. von Siebold, a famous Japan specialist and iris grower, started a business importing exotic plants from Asia to Leiden. It had formal government backing, with Minister of Colonies Baud agreeing to their monopoly on importing Japanese plants. As the popularity of the Japanese iris had shown, this was prestigious for the Dutch colonial endeavor but above all lucrative economically.[72] In 1843 Von Siebold and Blume discovered that one of their "unique" Japanese plants, *Paulownia imperialis*, a hardy ornamental tree today known in the United States as the princess tree, was being cultivated in Paris, and they cried foul. They immediately blamed the Buitenzorg Botanical Gardens for diverting a tree to Paris, on the reasoning that only Buitenzorg could have had the opportunity to divert a specimen, and demanded that Baud instruct the botanical gardens to stop sending Japanese plants anywhere besides the Rijksherbarium in Leiden.[73] Baud declined to intervene directly and passed the demand, and hence the responsibility over matters of natural knowledge, to the governor general. When the chief gardener of the botanical gardens, J. E. Teysmann (the position of director had lapsed with Blume's departure from the colony almost two decades earlier), heard the complaint, he responded that he had not mailed the seeds to Paris. Moreover, according to a French botanical journal the plant had been imported into France in 1834.[74] For the governor general this effectively ended the matter. Teysmann was to keep to the rules about passing material on to the Natural History Commission but was otherwise free to administer its scientific affairs, now under the supervision of the governor general.[75] Blume was in no position to demand otherwise.

It was also during the 1840s that the Buitenzorg gardens acquired the bare essentials of an institution with European scholarly credentials. All of this was paid for by the governor general's office and was done under the supervision of the military intendant who commanded the Buitenzorg palace. In 1844 the governor general paid for the publication of assistant gardener J. K. Hasskarl's catalog of the plant collections at the Buitenzorg gardens.[76] In the same year, Teysmann persuaded the governor general to build a small herbarium, made of bamboo, for the modest cost of f.1,855. A library was attached to the herbarium, and by the mid-1840s Teysmann was in regular contact with botanists in the Netherlands and elsewhere.[77] Buitenzorg became an outpost of the already existing international network of botany, sending rare and interesting specimens across the globe. In 1845 the palace's intendant authorized Teysmann to exchange material with institutions in the Netherlands, within the archipelago, and in the Cape Colonies in South Africa.[78] In 1847 W. H. de Vriese at Leiden University applied for permission from the minister of colonies to trade plants with Teysmann.[79] In the 1850s this grew into a special relationship between the Buitenzorg Botanical Gardens and Leiden University.[80] This all took place outside the arrangement with the Natural History Commission. By 1849 Teysmann was regularly shipping boxes full of plants to the universities in Utrecht, Leiden, Amsterdam, and Groningen.[81] The exchange of live material, seeds, and herbarium specimens expanded rapidly, with Buitenzorg also receiving material from non-Dutch botanical gardens.[82] Teysmann collected extensively outside of Java in the 1850s with the goal of the "enrichment of the Botanical Gardens and the expansion of botanical knowledge."[83]

For his work in establishing the gardens as a scientific institution, Teysmann has been hailed as the major pioneer in the annals of the Buitenzorg Botanical Gardens. To his successors in Buitenzorg, 1844, when the herbarium was built, is the date when botanical science became possible in the Indies.[84] His achievements contrast distinctly with those of Junghuhn and the contemporaneous Enlightenment in Batavia.[85] The apostles of enlightenment had tried to spur the creation of an undefined civil society through the rarified power of natural history; Teysmann's efforts created a professional institution with European contacts and official backing. Teysmann's basic accomplishment was to bring the authority of European expert natural knowledge into the colonial administration. He made the plants and specimens he collected legible to European scholars and organized the material in Buitenzorg according to European systems. This explains the singular effort to reorganize, and replant, the gardens so that each of its sections contained plants from one family, according to the Austrian botanist Stephan Endlicher's classification system, recently published as *Genera Plantarum Secundum Ordines Naturales Disposita*.[86] Hasskarl's catalog of 1844 did the same.[87] After Teysmann and Hasskarl, botanists in Europe working on plants from the Indonesian archipelago were obliged to note the work done by the Buitenzorg Botanical Gardens. Still, under Teysmann the botanical gardens remained poorly funded (f.1,500 in 1847, two-thirds of which went to Teysmann's salary).[88] For all practical purposes the governor general saw the botanical gardens as an extension of his palace.[89] Thus, while the botanical gardens had institutional authority, it had virtually no connection to the people of the colony. And its framework for understanding nature was imported from outside. Teysmann's ambition of establishing the botanical gardens as a regular administrative budget item guaranteed its survival but also limited its horizon.

By the mid-1840s Teysmann had become an articulate advocate for government-financed botanical science, as is evidenced in letters in the colonial archive.[90] Still, under Teysmann's direction the botanical gardens remained a modest institution. When Alfred Wallace visited in the early 1860s, he was largely unimpressed and devoted only a half page to it in *The Malay Archipelago*.[91] And later governors general tended to neglect the botanical gardens as a scientific institution, treating it again as their private domain. Teysmann was never able to do much more with the herbarium than dispense occasional advice to the bureaucracy. Yet his botanical gardens remained the most robust scientific institute in the colony and became the model for institutionalized and professional colonial science.

## Conclusion

The apostles of enlightenment believed their message would engage colonial society broadly, expecting that civil society would allow them to be heard beyond the rarefied halls of the Batavian Society. Instead it was the state that listened most intently and absorbed the values and ideas of the apostles. E. Douwes Dekker's famous 1860 novel *Max Havelaar* exemplifies this dynamic. Douwes Dekker ends Max Havelaar's famous speech to the Javanese chiefs of Lebak with a spirit reminiscent of the 1853 exhibition: "We have noble work to do. If Allah preserves our lives, we shall see to it that prosperity comes. . . . Once more I ask you to look upon me as a friend who will help you when he can, especially where injustice has to be fought. And in this I shall be most grateful for your co-operation. In due course I shall return to you the Reports on Agriculture, Cattle-raising, Police and Justice, with my decisions."[92] This speech, whether based on a real one or not, was heard not by the Javanese chiefs of Lebak, and not even by the Batavian population, but by the future colonial officials of the Netherlands East Indies who read *Max Havelaar* in the Netherlands. After the 1850s the colonial state would pick and choose aspects of the Enlightenment to make their own. By the middle of the 1850s the projects of the Batavian apostles of enlightenment were a part of the colonial bureaucracy.

Following the May 1848 meeting organized by Van Hoëvell, the colonial bureaucracy grew adept at absorbing the potential of the apostles of enlightenment. Natural history, for example, was henceforth a valuable tool in both the social and physical realms. Ann Laura Stoler, writing about colonial educational policy, has argued that in the wake of the 1848 movement colonial officials believed that "local knowledge should never be too local and that familial attachments were to be mediated and reworked through concerns of the state, filtered through a fine sieve, through the ears of Dutch categories, distilled into European typologies, reconfigured as sanctioned knowledge in a usable form."[93] Much the same was true of other legacies of the 1840s Enlightenment of Batavia, including that of natural history. The bureaucracy did not allow a movement that was too popular, too closely allied to local traditions of knowledge. It quickly swallowed the most compelling ideas and talents of the apostles of enlightenment while favoring the "universal" taxonomy instituted by Teysmann at the Buitenzorg Botanical Gardens. There was a brief resurgence of the Batavian Enlightenment under Rochussen's more liberal successor, A. J. Duymaer van Twist, who patronized the 1853 exhibition. But as a whole after 1848 the colonial bureaucracy was very cautious about promoting general interest in Enlightenment knowledge.

Until the very end of the empire, most Dutch colonial elites arrogantly believed in the superiority of European knowledge and the inadequacy of indigenous knowledge. This common ground was the opportunity state scientists such as Teysmann exploited to maintain scientific authority. Starting in the 1840s, colonial bureaucrats and officials turned to local institutes for guidance in demystifying nature. In part they did this because these institutions were cheap and far easier to control than metropolitan institutions. And when naturalists such as Teysmann, Hasskarl, and Bleeker, all with unexceptional scientific backgrounds in Europe, became identified as professional scientists in their own right by maintaining collections and corresponding regularly with professors in Europe, their authority to speak expertly about nature grew. They became the first professional scholar-officials of the Netherlands East Indies, the first generation of floracrats. And their allegiance was to the colonial bureaucracy, not colonial society.

# 2

# Quinine Science

The 1848 convulsions in the Netherlands led to far-reaching political changes, which by 1870 would have greater consequences for colonial Indonesia than the 1848 Batavian uprising. Over the course of the 1850s and 1860s, liberalism dismantled the ancien régime of the Dutch conservative oligarchy through a series of constitutional changes. This began in March of 1848 when King Willem II ceded considerable power to the liberals in order to forestall a social revolution such as was under way in France and Germany. Under the leadership of J. R. Thorbecke, liberals pushed through a series of political and social reforms, beginning with expanding the electoral rolls, holding elections with secret ballots, and devolving power to provinces and counties. In economic policy it ended protectionist tariffs and laws and hence promoted trade more free from government control. The capstone of the liberal achievement came with the education law of 1879, which created and financed mass primary education in the Netherlands. Three decades after 1848 the Dutch government was a liberal state.[1]

The ripples of these political changes were not felt in the colony until after 1870, and even then it was only economic liberalism, with its goal of expanding private ownership and investment, that would be implemented. In the 1850s and 1860s the Netherlands East Indies remained distant from the Netherlands, with the governors general still acting as autocrats. The economic system of government-run export-crop agriculture, known as the cultivation system, was still in place. Those coming from the Netherlands saw it romantically as a land of adventure where hard work, a little luck, and good connections might land them a fortune. The government administration had only a rudimentary bureaucracy. For example, in the late 1850s Governor General Pahud retained on his staff the charismatic naturalist Franz Junghuhn, who was to create quinine plantations that reflected Pahud's glory and power. Pahud would not stand for any kind of criticism of Junghuhn, not from other natural experts in the Indies and not from officials and politicians in the Netherlands. Professionalism, whether in science or administration, had as yet no place in the colony.

Nonetheless, after 1848, liberals in the Netherlands took a keen interest in ending economic and political protectionism in the Indies. This was possible due to a new constitution enacted that year, which gave Parliament, for the first time, power over its overseas possessions. The abolishment of the cultivation system and the opening of Java's economy to Dutch entrepreneurs became an important component of the Liberal Party's platform during the 1850s.[2] In the 1850s real changes were few, though, in part because of caution and worry over destroying the monetary remittances to the Netherlands generated by the cultivation system, the so-called *batig slot*, and in part because liberals knew very little about how the colony worked. Moreover, holdovers from the conservative 1840s such as Pahud—at that time still minister of colonies—remained important in making colonial policy. Even so, by 1854, liberals had decided that the colonial question would need to be settled by creating an economy of freed labor no longer reliant on coercing and exploiting Javanese peasants. They searched for policies that would lift the heavy work burden off of the Javanese peasant, break the system of forced cultivation, expand Dutch business interests in Java, weaken the old guard's grip on colonial power, and increase colonial profits, all in a system that was fueled by free Javanese labor.[3]

How to do this remained unclear. Liberal politicians turned to their scientists for answers, asking them to find ways to administer colonial nature more effectively, similar to what had happened in the British Empire.[4] In the mid-1850s, in a letter to the minister of colonies, the Utrecht chemist G. J. Mulder urged the government acclimatize a well-selected range of plants: "It is my belief that the goal should be to quickly make Java cultivate everything which can be produced profitably. And only those products that are at a mature stage of development in warm areas and have trade potential, should be considered for transfer to the colony."[5] And it was a Dutch academic scientist, the botanist W. H. de Vriese, to whom the liberals in 1857 assigned the job of ascertaining the state of agriculture in the Indies and to travel to the colony with the charge of answering the question of how to best free the export economy from the government monopoly.[6] At the culmination of a three-year trip, a report to Parliament—which promised to be the most comprehensive survey of the archipelago yet—was to be the basis of a new policy about colonial agriculture.[7] From the start De Vriese advised that any change in colonial agriculture needed to be built on a scientific foundation: "Improvements should be designed from a basis of real theoretical and practical knowledge, and accomplished through comprehensive cooperation of all appropriate people and elements."[8] De Vriese, like Mulder, envisioned a leadership role for Dutch scientists such as themselves, who would coordinate colonial agriculture. De Vriese idealistically believed that if scientists provided enlightened knowledge about tropical nature the right economic and political relationships would organically follow. In fact the colonial system repelled De Vriese and his advice very easily. Despite De Vriese's parliamentary mandate, the first thing Governor General Pahud did was to explicitly forbid De Vriese from offering political advice to him or any other colonial official. Pahud intended to utilize De Vriese for his own purposes and directed the professor to find out how well scientific knowledge was being used by local colonial officials. He was to submit a plan to bolster cooperation between colonial officials and scientists before leaving the colony.[9] De Vriese never fully complied, but even so, this command was quite different from that set out by liberal politicians from the motherland at the beginning of the trip.

Once Dutch politicians began to plan colonial reform, the central stumbling block was colonial profits. Throughout the 1850s, entrenched interests among sugar contractors made it impossible to satisfy the basic liberals' desire to open up the rewarding of sugar contracts to public scrutiny.[10] When the liberals finally took the reins of colonial policy in 1861, their reforms never challenged the tenet of profitability. Instead, throughout this entire period, they looked for ways to expand the ranks of private Europeans as investors, entrepreneurs, and planters who could safely and profitably live in the Indies. This entailed more than just economic reform. Thus it was in the 1860s that the colonial administration underwent organizational reform, creating a professional bureaucracy in which civil servants trained in the Netherlands implemented Parliament's colonial policy. And it also encouraged the Dutch government to find ways to make life in the Indies safe from disease. It is in this context that in the 1860s the acclimatization on Java of cinchona, the tree whose bark produces quinine, became an obsession for the Dutch. Creating cinchona plantations promised to be profitable economically, as cinchona could be sold as an export crop. And it was also a means to turn the Indies from a colony only fit for adventurers into a land safe from malaria.

## The Cinchona Adventurers

As Philip Curtin noted fifty years ago, the 1840s was a major turning point in the history of malaria prevention. Regular quinine prophylaxis against malaria became standard in British West Africa, and elsewhere as well, during the 1840s. After 1840, preventing the fever among soldiers and bureaucrats was a critical requirement for establishing European colonial rule in Africa and Asia.[11] By then few disputed the efficacy of the drug extracted from cinchona bark. Quinine alkaloid, with the aid of contemporary developments in pharmaceutical chemistry, could be easily extracted from the bark of the cinchona tree. And it was the wonder drug quinine that prevented and even cured malaria, the deadly tropical disease. The difficulty was the bark supply, as cinchona only grew wild in the Andes. While Europeans had known about quinine's healing properties for centuries, and the bark of the tree had long been available to those who could afford it, the European imperial powers had greater need of it, fairly suddenly, in the middle of the nineteenth century. As the worldwide demand for quinine grew, the governments of Peru and Bolivia increasingly controlled its collection and sale, holding hostage European imperial expansion. After the 1840s, Europeans looked for a steady, reliable, and economical supply of the cinchona bark, grown under European control, which would cut the South Americans out of the loop.

In the 1840s the acclimatization of cinchona in the European colonies took on a sudden urgency. The acclimatization of new world crops in Europe and Asia had been attempted with varying success since the end of the eighteenth century.[12] Alexander von Humboldt and others had since the 1820s suggested the acclimatization of cinchona by the European powers, but because of the difficulties and expenses involved nothing was done prior to 1840. Only when quinine prophylaxis became commonly accepted as a necessity among all Europeans in the tropics did the Europeans expend considerable effort transferring cinchona from South America to plantations in the European colonies in Asia and Africa. The challenges were considerable. The cinchona tree grew high in the Andes, and its natural history was poorly understood. There was very little expertise about cinchona in Europe, and it was not clear who would acclimatize the quinine tree. Botanical expertise was needed, but as cinchona was endemic to South America and needed to be transferred to Asia or Africa, the acclimatizer would require an adventurous spirit, as well as an involvement in the European colonies. Hence the challenge attracted apostles of enlightenment with colonial and scientific experience, who saw an opportunity to use scientific knowledge to expand European empires. These self-fashioned experts would initially lead the effort to acclimatize cinchona in the soils of the European colonies. Sir Clements Markham, the best known of the mid-nineteenth-century apostles of enlightenment who took up cinchona acclimatization, played an important role in propagating into the twentieth century the argument of a dependent link between science and empire.[13] While none of the Dutch quinine adventurers has remained as well known, they were cut from the same cloth as Markham.

In 1851 C. F. Pahud, the Dutch minister of colonies, contracted the botanist J. K. Hasskarl, who had first made his name as an assistant at the Buitenzorg Botanical Gardens in the 1840s, to collect live specimens and seeds of the cinchona tree in South America. He was to scour the highlands of Peru in search of cinchona trees and bring seeds and live plants to Java, where the mountain environment was thought to be suitable for creating cinchona plantations. Hasskarl's trip was the first earnest Dutch entry in the race to acclimatize cinchona. The French had already left the starting gate, but their cinchona trees did not grow to maturity on plantations in Algeria, probably because of the hot climate.[14] Mountainous Java, with topography similar to that of the Andes, was a far more promising location. Moreover, by sending Hasskarl, who knew Java from the 1840s, the Dutch government was tapping an expert in colonial nature as well. After three years of surreptitiously crisscrossing Peru, Hasskarl, carrying seventy-five young cinchona plants, was picked up by a specially sent Dutch warship, the *Prins Frederik*, in 1854. Most died on the Pacific crossing to Java. Hasskarl planted the two surviving saplings in the Cibodas garden above the Puncak Pass in West Java. More trees were cultivated at Cibodas from seeds Hasskarl had sent to Holland and the Indies before he left South America.[15]

When Hasskarl arrived on Java with his two live cinchona saplings in 1854, he became the state-appointed director of quinine production. The governor general now had responsibility for the acclimatization project, as well as colonial science more generally, and from then on The Hague had little control over the direction of the quinine initiatives. This shift in scientific authority was completed when Pahud moved from his job as the minister of colonies to the governor generalship in early 1856. When Hasskarl resigned due to illness that year, Governor General Pahud handed the reins of the cinchona initiative over to Junghuhn.[16] Junghuhn had been back in Java for about a year but under completely different circumstances from his earlier stay. The publication of his *Java*—and his subsequent assignment to draw a new, definitive map of the island—had made him a superstar. Plus he was married to a niece of former governor general Rochussen and was close to Pahud. The successful acclimatization of cinchona was a major priority for Pahud, and he thought highly of Junghuhn's ability, having gotten to know him in the Netherlands. Junghuhn came to the job not only with more direct field experience in Java than anyone else but with a vision for creating a European society there. And for the next eight years he labored ceaselessly—planting cuttings, testing bark, and writing guides—all in the name of producing more cinchona trees. His acclimatization efforts failed and with it the dreams of influence and power of his generation of apostles of enlightenment. That failure would lead the way toward a formal integration of science into the bureaucracy.

## Pahud's Cinchona

As the reformed colonial bureaucracy began to take shape in the 1850s and 1860s, scientists stood on firmer ground in the Indies. Colonial officials in Buitenzorg appreciated scientists' ability to open up tropical nature as long as they stayed out of politics. The antecedents for the institutionalization of science in the colony go back to the early 1840s when the governor general gained control over the Buitenzorg Botanical Gardens. It continued with Governor General Rochussen's informal projects, which included hiring Junghuhn outside of the then mainstay of colonial science, the Natural History Commission, to conduct surveys in southwestern Java. By the 1850s, scientists were integrated straight into the governor general's office. This was not like the practice in European scientific institutions, as Minister of Colonies P. Mijer warned the aspiring colonial chemist J. E. de Vrij; in the Indies scientists were subservient to the political hierarchy.[17] The acclimatization of cinchona after 1854, though a much larger operation than any previous scientific project, was still run as a pet project of the governor general.

It was as governor general that Pahud came to personally control the first stage of the scientific acclimatization of cinchona.[18] Pahud was a transitional leader poised between the king's cronies, who had set colonial policy in the 1830s and 1840s, and the liberal regimes of the 1860s and 1870s. As minister of colonies in the early 1850s, as liberal parliamentarians began fitful attempts to exert control over the Netherlands East Indies, Pahud was responsible for reforming colonial policy. He took small steps, and his directives retained much of the colonial architecture of the 1830s and 1840s, drawing on his experience as an Indies official.[19] The liberal statesman J. R. Thorbecke once characterized him as "a diligent, hard-working administrator, but like all administrators, wedded to tradition."[20] Pahud was primarily J. C. Baud's man even in the liberal cabinets of the 1850s. And he continued some of the patronage-style colonial politics that existed under Baud's reign as minister of colonies in the 1840s. Once he was chief of the colony, Pahud saw himself as a patron of the scientists he employed, managing them directly without deference to Dutch politicians or ministers.

Pahud's relationship with Junghuhn epitomized this vision of harnessing science to the colonial state. And Junghuhn was happy to play his part. Finally, after multiple setbacks at the beginning of his career, he had a powerful patron in Pahud. Pahud gave him great power, autonomy, and funds, not just to acclimatize cinchona but to direct all science in the colony. His title was inspector of scientific research in the Netherlands Indies. An 1859 official inventory of his office's books lists 117 titles on a wide variety of natural history topics.[21] It has not been possible to re-create Junghuhn's full budget, but between 1856 and 1863 he commanded a yearly budget of f.4,200 for the purchase of books and instruments alone.[22] Acclimatizing cinchona, though, remained Junghuhn's chief charge, and therein he answered only to Pahud. Pahud, for example, explicitly forbade De Vriese from expressing an opinion about Junghuhn's handling of the cinchona plantations.[23] Junghuhn was even outside of the reach of The Hague. In 1855 the Dutch government, on the recommendation of the Utrecht chemistry professor G. J. Mulder, appointed Mulder's student K. W. van Gorkom as chemist in service to the quinine acclimatization project. After Van Gorkom arrived in the colony, Junghuhn managed to have the Rotterdam chemist J. E. de Vrij appointed instead, leaving Van Gorkom stranded.[24] Junghuhn well understood why this was possible. In 1857, with only three hundred trees in the ground, of which the quinine content was not yet known, he wrote in his first official report, "[T]he transfer of the quinine trees from South

America to Java, an accomplishment which successive ministerial and colonial governments have pursued in vain for more than twenty years, has finally been accomplished and brought to fruition under our current Governor General Pahud, previously minister of colonies."[25] In exchange for this autonomy, Junghuhn did more than give Pahud credit; Junghuhn had Pahud's name placed on a new species of cinchona.[26]

Junghuhn quickly set out to put his individual stamp on the Javanese cinchona. He changed the location of the plantations from Cibodas (above the Puncak Pass on the slopes of Gunung Gedeh) to Pengalengan in the Malawar range. Both sites were in the highlands of West Java, but Junghuhn did not like the shallow layer of topsoil at Hasskarl's site (it was actually chosen by the head of the Buitenzorg Botanical Gardens, J. E. Teysmann, in the early 1850s). Hasskarl had furthermore planted the trees in a cleared field, and Junghuhn believed that the trees would grow more naturally in forests similar to the ones in South America. In preparing his site, he cleared away underbrush but kept the old-growth forest intact.[27] He stopped actively cultivating the species *Cinchona calisaya*, at that time considered the best of the cinchona species, and concentrated on trees that had been raised from seeds Hasskarl sent to the colony via the Netherlands before he left South America. Hasskarl called this tree *C. ovata* while Junghuhn first thought it was a variety of *C. lucumaefolia*. In 1860, however, on Junghuhn's urging, the British quinine chemist J. E. Howard declared it a new species, and called it *C. pahudiana* after Junghuhn's patron, Governor General Pahud.[28]

Junghuhn saw the cinchona acclimatization as a continuation of the work he had begun in the late 1830s while exploring Java's volcanoes. After 1855 he had realized not only his dream job but his vision of shaping the colony through science. Political protection was provided by his patron Governor General Pahud. He did not believe he answered to or was even a part of a bureaucratic chain of command. But to many he was the official in charge of creating tangible quinine alkaloids in the cinchona bark. As the head of cinchona cultivation with an appointment at the rank of resident, he was part of the hierarchy of the Netherlands East Indies state, although in practice he reported only to the governor general. Pahud continued to protect him, but his job — to produce quinine-rich bark on Java — was easily scrutinized. Pahud asked for regular reports; they were circulated inside the government, and some were made public. Initially his reports were generally sanguine: trees were growing, and quinine would be available soon.[29] Nobody doubted Junghuhn's skill in growing trees; after only three years more than one hundred thousand cinchona trees in all stages of development were growing in the forests of the Malawar range.[30] Still, some quietly began to question why the official in charge of cinchona acclimatization had not produced trees rich in quinine. Governor General Pahud, Junghuhn's patron, patiently accepted the explanation that only mature trees

Mature *Cinchona pahudiana* on Java (Reprinted from J. C. Bernelot Moens, *De Kinacultuur in Azië, 1854 t/m 1882* [Batavia: Ernst, 1882])

contained bark with quinine. With Pahud as governor general of the Netherlands East Indies, no public disapproval of Junghuhn or his work was permissible in the Indies, and, as Junghuhn remained outside of any bureaucratic hierarchy, there was no channel for internal government criticism. In 1861 J. E. Teysmann complained in a private letter that Pahud was blind to the faults of Junghuhn. Off the record Teysmann had told the governor general that Junghuhn had purposely destroyed everything that was not his own by ignoring all trees planted before he was put in charge, but Teysmann could do nothing publicly with Pahud still in the colony.[31] When Pahud left the colony in September of 1861 and Junghuhn's trees were still without quinine, however, Junghuhn came under much closer scrutiny.

Local criticism came first from the man who had helped plant the first cinchona tree on Java in Cibodas in 1852, the head of the Buitenzorg Botanical Gardens, J. E. Teysmann.[32] Teysmann had never admired Junghuhn's Enlightenment idealism, and in 1843 he had printed his criticism of Junghuhn's opposition to botanical nomenclature.[33] Finally, in an 1863 article, with Pahud having departed, Teysmann attacked almost all of Junghuhn's decisions, pointing to Junghuhn's tendency to value his theories as best, which had led Junghuhn to throw all previous knowledge about cinchona away and to start the acclimatization process from scratch. By planting a new species of cinchona, in the shade and at a new location, Junghuhn had overlooked the horticultural experience of Teysmann, who had been cultivating foreign trees since 1830. Junghuhn had also ignored the firsthand knowledge of Hasskarl, who had seen the trees in South America. Finally, he had neglected the botanical expertise of H. A. Weddell, who judged *C. calisaya* to be the best species.[34] At about the same time Clements Markham, by the early 1860s the chief cinchona acclimatizer in the British Empire, offered a scathing critique of Dutch efforts to domesticate cinchona on Java in his 1862 book.[35]

Junghuhn's eventual failure suggests that his contemporary critics were right. His successors overturned almost all of his choices.[36] What stands out from his work as chief of the cinchona cultivation project, however, is the method by which he tried to bend nature to the will of knowledge. Junghuhn initially believed that a large number of healthy cinchona trees would mean large quantities of quinine. Because there were more seeds from *C. pahudiana* than the other trees, this species received preference. Considering the material available, this was perhaps the only road open to him in 1856. But when his new species did not pay off, he was not flexible. Both his honor and the honor of his patron (whose name was on the species) were at stake. Junghuhn did not portray himself as a cog in the system or as one of many experts, hiding behind reports. Moreover, he had little interest in trial and error experimentation to determine which group of trees produced the most quinine. He took all the responsibility on himself. He saw himself not as a servant of the colonial administration but as the scientific savior of the Netherlands East Indies.

## Quinine in Parliament

As I have pointed out, in the 1850s The Hague had little direct control over its Dutch citizens in the colony, and the colonial administration still managed affairs as its governor general saw fit. But to the Liberal Party this was highly unsatisfactory, and beginning with the De Vriese expedition in 1857 it asked naturalists to provide systematic solutions to the colonial question. In 1862 the liberal government appointed Piet Bleeker, one of the apostles of enlightenment who had just returned from Batavia, to head a commission investigating the future of Dutch pedagogy in colonial rule. Specifically he was instructed to recommend what kind of curriculum would best prepare administrators for a career in the colonial bureaucracy. Bleeker—friend to E. Douwes Dekker, F. Junghuhn, and W. R. van Hoëvell—had almost twenty years of experience as a naturalist in Java, where he had also founded the Naturalists' Association and had a part in staging the 1853 exhibition. Not surprisingly, then, the commission recommended that the cornerstone of the colonial bureaucrats' education should be natural history, with an emphasis on geology, botany, and zoology. By the time the commission's recommendations became law in 1864, however, all of the natural history had been dropped. The minister of colonies and Parliament instead proposed that the curriculum for educating future colonial officials be indigenous language and cultures. Improving the efficiency, administration, and tractability of the colonial bureaucracy would not require scientific training.[37] Bleeker was understandably disappointed that his proposals for creating a prototechnocracy were ignored, and he wrote in his autobiography "that Dutch society stood without the aid of science; that in general it had no interest in fostering scientific knowledge, and simply left it to intellectuals."[38] Bitterness aside, Bleeker was right that Dutch politicians planning colonial policy had little interest in fostering scientific knowledge in the early 1860s. In part this was because science had been delegated to the governor general's desk nearly two decades earlier and had remained annoyingly outside The Hague's sphere of control. And in part it was because of the debacle of Junghuhn's quinine initiative, which came to be debated in Parliament in 1863.

After Pahud's departure in September of 1861, Junghuhn lost his protecting patron. Pahud continued to correspond with Junghuhn about "our cinchonas" and offered advice and suggestions for interacting with the senior colonial officials.[39] With Pahud a private citizen, Junghuhn's critics, many of whom had been forced into silence for years, were free to attack Junghuhn and his methods in public. And while these criticisms began with Teysmann and Markham, they found their way quickly to the Netherlands, beginning with a report, commissioned by the office of the minister of colonies, by the Utrecht University professors Mulder and F. A. W. Miquel, which analyzed the quinine content of *C. pahudiana*. This extensive report became the basis for the Dutch debate about cinchona and Junghuhn over the next few years. Based on chemical tests on one sample and a wide overview of the cinchona acclimatization literature, Mulder and Miquel recommended that the government immediately cease cultivating *C. pahudiana* and instead experiment widely with different species. To this end they encouraged exchanges with the British, as well as another expedition to South America to gather seeds and saplings. For the failures of the last few years, they blamed Junghuhn personally.[40]

Although this report had been commissioned under earlier administrations, and Junghuhn had cooperated in sending a specimen of the *C. pahudiana*, the final report came out just as a newly invigorated liberal cabinet came to power in Parliament. G. H. Uhlenbeck, Van Hoëvell's protégé and a participant in the 1848 Batavian uprising, became the minister of colonies in Thorbecke's cabinet in 1862. Uhlenbeck had himself been associated with the 1840s apostles of enlightenment while he was director of public works in the colony.[41] Uhlenbeck's liberal idealism animated his short year as the minister of colonies. He expected that if the Dutch moved swiftly to end the cultivation system and the government cultivations it would soon be replaced by private industry and free labor. For Uhlenbeck the quinine initiative was emblematic of the problems with the government cultivations. Armed with Mulder and Miquel's report, he ordered the governor general to stop Junghuhn from cultivating or propagating the *C. pahudiana* species. He further suggested that it was time to end the government's oversight of the acclimatization process, leaving it to private entrepreneurs, as there was no dispute that cinchona grew easily in Javanese soil.[42] This suggestion likely was part of his liberal agenda to reform the colonial economy, as it actually contradicted Mulder and Miquel's report, which had suggested new ways in which the government would stay involved in the quinine initiative.

After receiving his orders, Governor General L. A. J. W. Baron Sloet van de Beele duly forbade Junghuhn from planting any more *C. pahudiana*. This undoubtedly sat badly with Junghuhn, who thereafter apparently did not like the new governor general.[43] Nonetheless, Sloet van de Beele did not lose confidence in his scientist and continued to back Junghuhn as the leading authority on cinchona. He wrote that it was difficult for him to take the word of two academics in Utrecht, who had looked at just one sample of *C. pahudiana*, while Junghuhn had reason to believe some of his still young trees would produce large quantities of quinine.[44] Junghuhn's adamant rejection of privatizing the acclimatization was also seconded by the governor general.[45] A lengthier response to Mulder and Miquel concluded with the argument that the two Utrecht scientists had neither the authority nor the know-how to judge the cinchona acclimatization. Junghuhn and De Vrij's authority, on the other hand, was based on their experience applying scientific knowledge in the colony itself.[46] Among colonial officials many believed it when Junghuhn argued that eventually his trees would supply the world with quinine and that this important work, which could open up the tropical world to increased European settlement, would be pioneered by colonial scientists such as himself.[47] It was only illness, which in early 1864 led to a medical leave, that led to Junghuhn's replacement. In a March 1864 letter to the minister of colonies, Governor General Sloet van de Beele proposed that K. W. van Gorkom become the new head of cinchona acclimatization, with the rank of *controleur*, and that when and if Junghuhn returned (Junghuhn actually died a month later) he would retain only his job as head of scientific research.[48] Van Gorkom would lead the cinchona initiative for the next eleven years.

The 1862 and 1863 exchange among Uhlenbeck, Sloet van de Beele, and Junghuhn demonstrates that the then almost twenty-year dynamic of managing science was still in place; while the minister of colonies could issue specific directives—stop cultivating *C. pahudiana*!—the governor general managed his science and scientists. Uhlenbeck had challenged but not dislodged this arrangement. His reform proposals, in particular his desire to abolish all government cultivations in the colony, were out of step with the views of most of his liberal colleagues, and after he failed to pass the colonial budget in late 1862 he was replaced in early 1863. Still, it was Uhlenbeck who raised the issue of quinine in Parliament and brought to light the highly irregular and informal manner in which colonial science had been institutionalized in the colony. After Uhlenbeck was gone, the inadequate oversight of colonial science remained a political issue that many parliamentarians believed required reform. In his budgetary request of 1862, Uhlenbeck had informed Parliament that he had ordered the governor general to stop cultivating *C. pahudiana* and suggested turning the acclimatization over to private hands. This stark challenge to Junghuhn was noticed, and in May of 1862 there was a detailed questioning of Uhlenbeck in the Tweede Kamer, the lower house of Parliament. What gave him the confidence to discredit Junghuhn? Uhlenbeck admitted that his appraisal of the disastrous state of the cinchona initiative was solely based on the report of Mulder and Miquel.[49] Parliament's questioning forced Uhlenbeck to recant his privatization idea, but the directive about *C. pahudiana* stood.[50] And now the continued failure of the *C. pahudianas* to produce quinine was part of the public and parliamentary record. Within the next year, a parliamentary committee, which included P. Mijer, the former minister of colonies and himself an apostle of enlightenment from 1840s Batavia, concluded that the questions around the cinchona acclimatization project were sufficient to warrant further questioning of the new minister of colonies, including a fuller accounting of Mulder and Miquel's report.[51] The new minister of colonies, I. D. Fransen van de Putte, responded that he thought the old status quo, in which the governor general managed science in the colonies, was the most practical application, and hence he recommended no immediate change. But he did concur that further investigation was prudent.[52] The matter was alive, though, and was again debated in the lower house of Parliament in June of 1863.

The June 2, 1863, debate was the first time cinchona science was debated extensively in Parliament, and the principal conclusion, that colonial scientists needed to be brought into line, had long-term repercussions. By the time of this debate many parliamentarians had read or heard criticisms of Junghuhn, including those in Markham's book of the previous year. In 1862 the now retired but still influential W. R. van Hoëvell used Markham's critique of Dutch cinchona cultivation to express his dismay with a system that kept such an important endeavor in the hands of one man. He argued that political liberalism had answers for this problem and that the infusion of fresh blood in the form of Dutch entrepreneurs would guarantee success.[53] The June 1863 debate reveals a keen awareness on the part of the Dutch politicians of the complex scientific and political issues surrounding cinchona acclimatization. No one shared Uhlenbeck's antagonism for Junghuhn or the *C. pahudiana*, but there was agreement that this important matter should be brought under systematic control. Science, the consensus was, needed to be reined in by the political process. Lower house member W.H. Idzerda spoke in favor of the minister of colonies' decision to inquire further about the state of the cinchona cultivations:

> [I] believe it is of great importance to especially plant a great number of different sorts of cinchona species, in order to have access to a wide array of choices. And from time to time, especially as the trees approach maturity, it is necessary to do a careful and impartial chemical investigation of the quinine content in those different sorts. This study will then decide which sorts are most appropriate for cultivation in our colony. Then we will have a foundation of certainty. The quinine cultivations, which have already cost so much trouble, efforts, and fortune, and shall cost more yet, will have, with success, incalculable advantages for the motherland and the colony. We must prevent the quinine cultivations from being sacrificed to personal ambition and stubbornness to preconceived ideas.[54]

The central realization of the Dutch policy makers was that colonial science could not be left to find its own way as the colonial scientists now needed to be told how to do their jobs. In the future it would be necessary to organize science with institutional controls. As the colonial administration was reinvented in the 1860s, in large measure by Fransen van de Putte, science was brought under firm bureaucratic control.

The new minister of the colonies, Fransen van de Putte—millionaire and former colonial sugar planter—had little use for Uhlenbeck's idealism. He, too, wanted to open Java to free labor and believed capitalism in the colony was only possible through the expansion of Dutch private property rights to land currently under Javanese ownership. His comprehensive bill of 1865 divided the liberals, though, many of whom were shocked by its potentially far-reaching social implications for Java, and Parliament postponed transforming Javanese land tenure laws until 1870.[55] Fransen van de Putte had greater success centralizing the colonial administration, which, as H. W. van den Doel argues, effectively ended the cultivation system in 1866 when he abolished the practice of paying colonial officials through agricultural commissions. A new professional bureaucracy, with salaried officials, was organized into four centralized departments.[56] The quinine initiatives, along with other former components of the government cultivations, came under the Department of Civil Administration.[57] And with this the governor general's patronage of scientists ended as well.

Natural scientists were brought into the colonial administration in the 1850s because of their perceived unique abilities. Many still glowed with the aura of idealism that had come out of the 1840s, all the more so because many of the administrative leaders in the 1850s and 1860s had themselves been associated with the apostles of enlightenment. The notion that science would advance colonial society remained generally felt in the Netherlands and the Indies. All this was changed by what many came to see as Junghuhn's failure, made public after 1862. In the future science would be just one more part of the colonial bureaucracy. Just as the cinchona scientists were subjected to hierarchical oversight in 1866, so, too, was the Buitenzorg Botanical Gardens, which came under control of the new Department of Education, Religion, and Industry. But even as the era of Batavian Enlightenment closed, scientists continued to play a special role in the colonial administration. They would need to prove their value to colonial capitalism by creating economically useful knowledge.

## Liberals, Planters, and Scientists

The liberally envisioned colonial system was to operate on the twin pillars of expertly trained bureaucrats and autonomous Dutch entrepreneurs. By design, bureaucrats did not have direct economic functions, instead serving as the infrastructure on which private capital could build a prosperous colony. Liberal idealism expected the bureaucracy to manage society but keep its hand out of the economy to the point of not even planning economic life. This ideal was difficult to put into practice. Most complex was the transformation of the Javanese countryside following passage of the 1870 Agrarian Law, which allowed Europeans to lease land on Java. How this was to be managed, with an eye to not allowing Javanese society to be disrupted but still creating revenue, came to be the colonial state's problem as it was not spelled out in the Agrarian Law itself. In the 1870s it was difficult to figure out how to encourage private capital investment. For example, private companies were unwilling to risk investing in building the Javanese rail network even though this was a major priority of the colonial state. This meant the colonial state would need to build it using surplus money from export crop revenue, the batig slot, which was remitted to the Netherlands and came to form a sizable portion of the Dutch budget (one-third in the early 1860s). The privatization reforms, however, meant denying the government this revenue; which was difficult for Dutch policy makers. It was with the last of the batig slot money that the Dutch funded the start of the Javanese rail network and other infrastructure projects. By the late 1870s the remittances had ended. Notwithstanding Parliament's extensive attempts to solve the colonial question in the 1860s, their tentative answers meant the details of implementing liberal society were left to the colonial leadership.[58] This gave scientists a new opportunity, in particular for the new head of the cinchona initiatives, K. W. van Gorkom.

As the Dutch political regime had not evolved out of a preexisting social order but had been imposed from the outside, it needed to invent systems that managed society. Its job was to contain civil society—by preventing, for example, indigenous political action—while at the same time generating policies that made society run to the civil specifications of the Dutch government. As the colonial regime in Buitenzorg was presented with problems—of economics or security usually—it turned to experts and specialists who could work out solutions. The sociologist J. A. A. van Doorn, by examining in particular colonial engineers, argues that this created an unfulfilled technocracy in which technocrats were unaware of their political power and hence did not wield it.[59] My own sense is that using technocracy as a category of analysis here is a distraction, as the state policies were not about creating technical solutions per se but about generating systems that they could effectively administer. The colonial state pointedly ignored the advice of Junghuhn, Bleeker, and De Vriese, whose proposals had all suggested creating something like a technocracy in which scientists directed colonial policy. The end result was quite different. The colonial state absorbed individuals who had the capacity to shape civil society while keeping control over the goals it pursued. For the duration of the Netherlands East Indies, scientists suggested technocratic reforms, but little came of it. It was not until the 1940s that a technocracy came into existence in the Indies, a process analyzed in chapter 6.

Between 1870 and 1900, the dominant liberal directive was for the government to create systems that encouraged private capital investment. The most profitable and successful export crop, sugar, already had a successful system, which it had inherited from the cultivation system. But after 1870 the colonial state was expected by all to create further sectors of the export economy based on private initiative and using free labor. This proved very difficult in the 1870s, when no one wanted to invest in either coffee or tobacco endeavors.[60] The colonial state needed more than new policies, as there was nothing there from which to start. The state would need to invent an entire system to generate an export crop. This is just what happened with quinine. In the late 1870s and early 1880s, the cinchona initiative would become the bedrock of Indies science in large part because it was useful in managing the gap between the colonial bureaucracy and the entrepreneurial planters. And it would be the autonomy, idealism, and nonofficialdom of the scientists that allowed them to create a viable system.

## Inventing Commercial Quinine

In some ways K. W. van Gorkom, the official in charge of acclimatizing quinine after the death of Junghuhn in 1864, was the perfect official and professional scientist; he followed Idzerda's 1863 advice to Parliament, quoted above, closely. He planted every type of cinchona seed he laid his hands on, did not hesitate to experiment broadly, and used chemical tests of the quinine content in the bark to decide which species were best suited to the Javanese soil. He did not let into his work preconceived ideas about how nature operated and did not believe that nature could be bent by the will of science. He respected bureaucratic authority, produced regular reports, and did not publicly argue with others. And eventually he did acclimatize high-yield cinchona. Yet this understates his achievement. His lasting legacy was a system that allowed cinchona cultivation to be carried out commercially on Java. By the 1870s, although he had yet to find the magical species of cinchona that would provide high percentages of quinine, Van Gorkom and his colleagues shaped the new cinchona system so that it comported to the Liberal Party's colonial policy, deftly navigating the split between colonial administrators and capitalist agricultural production. Thanks to Van Gorkom's books and nursery, Dutch private planters could grow quinine, and hence profits, in the Javanese soil. And in the process the colonial state found a way to harness scientists for political purposes that greatly extended the scientific ciphers for which Idzerda had called.

When K. W. van Gorkom, a student of the Utrecht University chemist Mulder, began as the director of cinchona cultivations, he inherited 1,151,810 trees, of which there were 1,139,376 *C. pahudiana*, more than 99 percent of the total number of trees.[61] Van Gorkom had little practical knowledge of acclimatization, but, as he wrote at the end of his life, "[M]y exertions were made lighter by experience insofar as we had learned at least not to follow in the footsteps of Junghuhn."[62] He left the *C. pahudiana* alone, moved other species out of the forest and shade, and closed down many of Junghuhn's plantations. He did not keep the government or the public in the dark about setbacks. Tree illness, misjudgments concerning the hardiness of young saplings, or unsatisfactory quinine content were not elided, hidden, or blamed on others.[63] Instead in government reports he defended the endeavor as a difficult one and cautioned "that the development could not be forced on, that the young plants demanded time to grow into trees, and that the necessary experiments could take up many years, before all debated points were solved."[64] Debated points required outside review and input, which he actively sought. He wrote to Miquel, "In order to avoid Junghuhn's mistake, I repeatedly requested that the government invite an expert to review my method of cultivation."[65] There were no basic truths to start with, and in order to solve this multifaceted problem Van Gorkom enlisted allies in Java and Europe. Teysmann visited the plantations in 1866 to review Van Gorkom's method of cultivating;[66] Van Gorkom sent a cinchona herbarium to

Miquel in Utrecht for help in classifying the varieties and species.[67] And R. H. C. C. Scheffer, the director of the Buitenzorg Botanical Gardens after 1868, advised that insects had caused a disease that had afflicted a number of cinchona plants.[68] Van Gorkom was not attached to any particular species, and he propagated the trees richest in quinine content regardless of its name or the ease of its cultivation. Although Junghuhn had concentrated on *C. pahudiana*, there were many other stocks available, procured during the 1850s and early 1860s. Van Gorkom was careful to experiment with everything he had. He continued to receive seeds from South America and India, the latter through exchanges with the British cinchona endeavors. Although Van Gorkom had, through chemical tests done by his colleague J. C. Bernelot Moens, come to believe that true *C. calisaya* produced the most quinine, he did not cling to this as dogmatic fact. In his handbook of cinchona cultivation he wrote about his efforts (in the third person):

> The continued debate as to the intrinsic value of the numerous cinchona barks, obliged [Van Gorkom] to be especially cautious in the choice of stock-plants for propagation, and it was about that point, in which he from the first sought as much enlightenment as possible from chemical analyses of different material. In these attempts he found his chief support in J. C. Bernelot Moens at Weltevreden, whose help since 1865 has been incessant and disinterested. Samples of bark were submitted to examination, from trees considered to be of good quality, which promised shortly to produce seed. If analysis proved the superiority of this or that tree, then seeds were gathered exclusively from it, and sown.[69]

Van Gorkom continued not on the basis of his convictions but driven by the goal of selecting for trees richest in quinine content.

Van Gorkom's system was achieved with an independent scientific community in which its members pooled resources, ideas, and expertise and worked toward a common goal. Under Van Gorkom's predecessor cinchona cultivation had meant the achievement of fame and honor for Junghuhn and his patron. This allowed no cooperation with rival scientists or collaboration with rival patrons. For all of Junghuhn's outrage over the earlier mentioned power-hoarding schemes of the king's botanist, Blume, he ended up being just as exclusionary of others' input once he reached a higher rung on the ladder of success. Van Gorkom's bureaucratic practice cordoned off a place for colonial science free from political oversight. Under Van Gorkom, his colleagues all believed, in the words of Bernelot Moens, that the "final goal of the quinine-cultivation is the production of the chinine [quinine] alkaloid."[70] But this community could not be held together with scientific goals alone, and Van Gorkom worked hard to create a professional identity for the cinchona work. In the year or two after Van Gorkom took over, he had little authority, as many still believed Junghuhn had been right and quinine would abound when the trees grew older.[71] Van

Sampling cinchona bark from one of the original Ledger trees in the government plantation in Cinyiruan (Courtesy of the National Herbarium of the Netherlands, Leiden University branch)

Gorkom parried criticism not with personal invective or scientific opinion but with a professional shell. He always presented himself as a government official first and a scientist second. And, while individual components of his program were technical and scientific—especially Bernelot Moens's chemical experiments—his overall method was administrative. In the early 1870s he officially created a professional entity when he proposed that an official Indies scientific commission, with Teysmann, Scheffer, and Bernelot Moens as members, review the state of cinchona cultivations.[72] This independent board could certify his scientific achievements and would show the government that he was not above administrative review. At the same time it kept the bulk of the colonial state out of his way.

All the same, the late 1860s were hard years, as complaints were raised in Parliament almost every year about the slow progress of the quinine cultivations. The British seemed to be far ahead, even though they had started later.[73] Reflecting on this years later, Van Gorkom wrote, "In the East Indies, the public ridiculed the expensive hobby of the Government in maintaining a cinchona plantation which after so many years had not shown any prospects of results in hard cash and of which only negative results could be expected. . . . Private individuals or companies could not be induced to make experiments with the cultivation of cinchona."[74] After a decade of failure under Junghuhn, every year without success seemed particularly painful. But because quinine content did in fact increase with age, and only trees older than five years had bark containing over 3 percent quinine alkaloid, there was no other choice than to wait. But there was little consistency; high quantities were only found in certain exceptional trees (and this followed no logical rule).[75] In 1870 Van Gorkom had been in his job for six years and was only then finding sufficient quinine content for commercial cultivation to be started. Pessimism continued into the 1870s even after 876 kilos of cinchona bark were auctioned in Amsterdam in 1870. This amount was hardly enough to claim victory. In 1871 in London, 6,244 kilos of cinchona bark from India and Ceylon were auctioned. And a few hundred kilos paled in comparison with the hundreds of thousands of kilos of cinchona bark coming from South America every year.[76]

Van Gorkom's mix of professionalism, administration, and careful experiment succeeded where Junghuhn's patronage science failed. But this was still unclear as late as the early 1870s. During his first eight years on the job, Van Gorkom's trees only produced very small quantities of quinine. It was not until 1872 that his fortunes changed. In that year, the trees grown from seeds sold years earlier to the Dutch government by the British animal trader Charles Ledger blossomed. And, much to the surprise of Van Gorkom, the flowers were creamy white, not pink, which would indicate that the trees were not the well-known species *C. calisaya*.

In 1865 Ledger had fourteen pounds of what he believed were *C. calisaya* seeds smuggled out of Peru to his brother in London (by then virtually all

South American states had enacted laws against the export of cinchona seeds and plants). In London his brother offered it to British government officials first, but they refused to buy from an unknown.[77] Ledger's brother then took the seeds to Dutch officials, who consulted with their local quinine expert, Professor Miquel at Utrecht University.[78] Miquel recommended that the government buy, although he, like others, thought it was unlikely that seeds found by an unknown animal trader would result in quinine-yielding trees. The government bought one pound for f.100. Miquel sent the bulk of these seeds on to Van Gorkom.[79] The Ledger seeds turned out to be quite a bargain for descendents of the trees grown from that seed have produced much of the world's quinine since—even more than the trees that grew wild in South America.[80] By the beginning of the twentieth century, almost all the world's quinine would be extracted from trees descended from two of the twenty thousand Ledger plants that germinated in Javanese soil in late 1865 and early 1866. At the time Van Gorkom believed them to be just more of the *C. calisaya* species, but he tended them all the same. Ten years later this diligence would pay off.

What allowed Van Gorkom to survive difficult years with little good news from the cinchona trees was his crafting of a new scientific identity. As I argued above, he was an excellent scientific manager and official who managed to insert professional science into the colonial bureaucracy. But he also came to see that he needed to create something outside of the bureaucracy, not just a productive cinchona tree but a system for producing quinine commercially. Junghuhn had planted a million trees in the belief that the government, led by himself of course, would run the cinchona plantations. Van Gorkom understood, probably very early on, that private planters would run the eventual plantations. In 1866 he started sending young cinchona plants to individual planters and thus began transferring his knowledge to private industry.[81] It is unclear whether this was on his own initiative or if he was directed to do so. All the same, a relationship began in which eventually planters would use Van Gorkom's methods and plants in their plantations. And his close contacts with the planters, especially the tea planters of West Java, gained him valuable allies. Whether Van Gorkom was following liberal policies or not, by 1870 it was clear to him and other interested parties that ascertaining the comparative success of the Dutch and British cinchona initiatives would be done by examining the amounts and prices fetched for cinchona bark on the Amsterdam and London quinine markets. By then British commercial operations of cinchona were already on their way. And thus Van Gorkom needed to do more than find a suitable type of cinchona tree for Java; he needed to help found a Dutch commercial operation. He did not so much service the quinine agribusiness as create it.

In 1872, when the trees grown from Ledger's seeds blossomed and the creamy white flower was not the same color as the blossoms of *C. calisaya*, Bernelot Moens analyzed the quinine content in samples of bark from the five-and-a-half-year-old trees. He and Van Gorkom were surprised to find an exceptionally high quantity of quinine. They promptly harvested a few hundred kilos of bark and sent them to Amsterdam for auction, where in May of 1873 they sold for f.9.13 per kilo, almost four times what *C. calisaya* bark sold for at the same time.[82] This new bark's reputation spread quickly. In 1877 auctioneers paid f. 17.58 for a kilo of Ledger bark.[83] This turned the twenty-year enterprise around and made it a public success. With quinine percentages over 5 percent and the bark fetching excellent prices in Amsterdam, it looked like commercial operations on Java could succeed. Although a number of tea planters in West Java had already started growing cinchona from seeds provided by Van Gorkom, it was only after 1873 that individual planters made sincere efforts to devote extensive tracts of land to cultivating cinchona. The government continued to encourage commercial cinchona planters by providing seeds and plants free of charge until 1883.[84]

It would be a mistake, however, to chalk up the success of Van Gorkom and Bernelot Moens to serendipity and their lucky discovery of the perfect cinchona species. Even discounting Van Gorkom's management of the Ledger stock, it was his scientific network that brought Ledger's cinchona into commercial production. Miquel forwarded the seeds, Van Gorkom did not hesitate to experiment on another long shot, and Bernelot Moens ascertained the high quinine alkaloid content, all with the cooperation of the Dutch colonial government. Van Gorkom certainly believed that success had not come from Ledger's seeds alone. He later claimed that another variety of *C. calisaya*, in fact stemming from Hasskarl's collecting, had sufficiently high quinine content to have become the basis of a commercial operation.[85] But in any case it would be the continued care of the Ledger stock by Bernelot Moens especially, who took over as head of the government quinine operation in 1875, that made cinchona cultivation a commercial success.

Based on pharmacological evidence, Bernelot Moens thought the Ledger tree a new species, *C. ledgeriana*.[86] The British quinologist Eliot Howard agreed in 1879, in fact going so far as to suggest that *C. ledgeriana* was the original form and *C. calisaya* one of its hybrids.[87] This kind of Victorian idealism—to assume quinine content was the central feature of the original cinchona, from which subsequent generations had decayed—was not shared by Bernelot Moens. Still, Bernelot Moens by then knew that cinchona hybridizes easily, that is, it reproduces with other species, and this was the greatest challenge to keeping the Ledger stock pure. The original Ledger trees could easily be pollinated by other species of cinchona, as there were numerous trees of various known and unknown species growing in government plantations. Thus not all seeds from known cinchona studs would germinate into trees with high quinine potential. Only chemical analysis of mature trees' bark could determine whether the cinchona was still a pure Ledger tree. This meant considerable care had to be taken in producing seeds. In order to maintain a reliable supply, Van Gorkom and Bernelot Moens took cuttings from Ledger trees with the

Herbarium sheet of *Cinchona ledgeriana* (Reprinted from J. C. Bernelot Moens, *De Kinacultuur in Azië, 1854 t/m 1882* [Batavia: Ernst, 1882])

highest quinine content and replanted them in separate areas, preventing contamination by *C. pahudiana* or other species. The trees in these seed gardens would fertilize each other and hence produce pure *C. ledgeriana* seeds. This was done with a number of different varieties using different source trees. After the mid-1880s most varieties were descendents of either "mother" tree no. 23 or no. 38, descendents of which were cultivated in a separate area after 1878.[88] In a test from 1888, ten-year-old descendents of tree no. 23 contained between 10.31 and 12.35 percent quinine, while one descendent of tree no. 38 contained 12.30 percent quinine. These descendents were the source of future grafting scions.[89] Although it was possible for the planters, using reliable seeds from the government, to start their own seed gardens, initially this was rare, and the government plantations remained the primary source of dependable seeds.

Under Bernelot Moens's care, the *C. ledgeriana* stock improved drastically. Shortly after the discovery of *C. ledgeriana* in 1872, quinine officials set the target for bark quinine content at 5 percent. By the end of the decade, they optimistically predicted that only trees with more than 10 percent would be harvested.[90] Bernelot Moens's careful selection process, however, was not leading quickly to an abundance of quinine on the Amsterdam market. The reliable seed gardens could not be trusted to produce the large number of trees needed for propagating by broadcasting (only about fifty full-grown trees resulted from the twenty thousand original Ledger seeds). And cinchona took at least five or six years to flower for the first time. Moreover, the Ledger trees would only grow vigorously after germinating in excellent, fertile soil, basically limited to cleared virgin forest sites. Before plantations could be built on a large scale, Bernelot Moens had to solve the problem of propagation. In 1879, apparently following the suggestion of the Dutch visiting horticulturalist J. W. van Loon, Bernelot Moens's staff started experimenting with grafting Ledger scions, cuttings from the existing supply of *C. ledgeriana* trees, onto the stems of *C. succirubra* rooted in the soil.[91] *C. succirubra*, a species from the northern Andes obtained in an exchange with the British, was not high in quinine content but would grow practically anywhere from seed.[92] Plate grafting was a well-known nursery trick but previously unused in the Indies plantations. The scions, making use of the *C. succirubra* root system, quickly grew into proper trees with high quinine content. This grafting process would become the critical operation for maintaining a large plantation because the *C. ledgeriana* scions could also be grafted onto stumps of practically any cinchona species.

The 1880s were the beginning of the cinchona boom. In 1882 Bernelot Moens published an extensive guidebook with detailed instructions on all aspects of cultivating cinchona. The work of the quinine scientists was now public knowledge. Dutch entrepreneurs with no botanical, chemical, or agricultural knowledge could follow his instructions, which included everything: preparing the soil, planting seeds, grafting, fertilizing, and harvesting.[93] Entrepreneurs flocked to West Java to ride the cinchona wave. Whereas in 1880 only

Visual instructions for grafting *Cinchona ledgeriana* scions onto *Cinchona succirubra* (Reprinted from K. W. van Gorkom, *A Handbook of Cinchona Culture* [Amsterdam: J. H. Bussy, 1883])

124,000 kilograms of Javanese cinchona bark were auctioned in Amsterdam (most of it coming from government plantations), in ten years' time this multiplied twentyfold to 2,901,000 kilograms in 1890.[94] This was less bark than was coming from India, but because of its high quinine content it fetched far better prices. By 1900 Java was producing more than 80 percent of the world's quinine. In 1930, 97 percent of the world's supply of the bark came from Java.[95]

Although Javanese cinchona dominated the market, it was not a bonanza for long. The depression of the 1880s and early 1890s lowered prices and need. And when the British cinchona planters decided to start growing tea instead, they flooded the market with all the bark they had, further lowering the price in the early 1890s. In 1895 the unit-price of processed quinine was ten times less than it had been a decade earlier.[96] By 1895 even the Javanese cinchona plantations were not profitable anymore. Although prices went up in the next few years, the cinchona planter led a precarious existence after 1890. Still, because of the very high quinine content and the excellent technical support from the government cinchona estates, it was possible for planters to eke out a living until the end of Dutch colonialism in 1942.

## Conclusion

It was by no means inevitable that colonial science would become the domain of floracrats and other government scientists. Other models were being discussed as late as 1861. In that year, De Vriese, after returning to the Netherlands following three years in the Indies, lectured about "science and civilization." He encouraged scientists to do more than write pretty natural histories and to pursue research that would turn the colony into an enlightened, modern, and thriving European society, declaring that "science and civilization must be the foundation of the prosperity of the lands and peoples of the Indies archipelago." He optimistically concluded, much like the organizers of the 1853 exhibition had done, that natural science could promote moral and social development and would advance the "instillation of work, industriousness, and trade amongst the population."[97] To be sure, this speech has long since been forgotten. Contemporary Dutch politicians, to whom it was addressed, pointedly ignored his advice. In the 1860s colonial leadership was placed in the hands of professional administrators, who took care to subject science and scientists to their control. Romantics were not again trusted with leading colonial society.

Van Gorkom and Bernelot Moens's legacy, in addition to the cinchona acclimatization project, was the creation of a professional identity for Dutch colonial scientists. It meant a secure place in the hierarchy of the Dutch colonial state. Van Gorkom and Bernelot Moens were afforded autonomy to do their work as long as they participated in generating and maintaining Dutch colonialism. Other scientists followed. In 1868 R. H. C. C. Scheffer, a twenty-three-year-old student of Utrecht University botany professor Miquel, took a job as director of the Buitenzorg Botanical Gardens. Initially he set himself the goal of writing a new topographical flora of Java, and he began work on the kind of taxonomy and natural history he had learned in Utrecht.[98] In letters to his mentor, though, he despaired of the disarray and difficulty of taxonomic work.[99] After the early 1870s he gave up on natural history. From the mid-1870s until his death in 1880, Scheffer concentrated on acclimatization endeavors, including educating colonial officials at an agricultural school and working on importing and spreading the best varieties of Liberian coffee.[100] Scheffer modeled his career on those of Van Gorkom and Bernelot Moens even though his work did not attract the same kind of acclaim.

*C. ledgeriana* seed was never a black-box technology given in completion to the planters and never needing further maintenance. It continued to require professional expertise to work right. Most obviously, the quinine scientists remained in firm control of distributing seeds and cuttings. Because they administered the cinchona stock, they were indispensable. But they also served as technical experts, advising on what soil to plant in, how to thin young groves, and when and how to harvest the bark. In this unsure business they remained the voice of reason among government bureaucrats and planters, and that is exactly how they saw themselves as well. Van Gorkom, Bernelot Moens, and their successors never doubted the critical role they played in making cinchona cultivation a success on Java. For them it was not just that the Javanese location was perfect. It was the whole system, which centered on the government cinchona plantations. In 1919 Arnold Groothoff, recently retired as director of the government cinchona plantations, defended a thesis with the title *Rational Exploitation of Quinine-Plantations*.[101] The quinine scientists' web of knowledge about cinchona kept them at the nexus of government officialdom, scientific research, and liberal capitalism.

In the decades following the widespread cultivation of C. *ledgeriana*, a multitude of scientific experiment stations sprung up in the Netherlands East Indies, where scientists catered to the needs of sugar, coffee, tea, indigo, tobacco, and rubber planters. At these stations biologists, chemists, and others researched weather patterns, soil types, harmful weeds, and higher-yielding strains.[102] Some experiment stations were established by planters' associations, although the government stepped in to save many of them during the lean years in the 1880s and 1890s. Within the hierarchy of a planter-sponsored or government-run institute, scientists were helping to maintain the export-crop economy. Experiment stations, including the government cinchona operation, did not permanently join planter society to the bureaucracy, but this form of science, in which knowledge was created for the benefit of agribusiness, did create a template for future cooperation. These experiment stations remained the major employer of scientists in the Indies until the demise of Dutch colonialism in 1942. To some scientists, though, it only partially fulfilled the promises of science; the possibility of generating a civilization in harmony with the environment, Junghuhn's dream, was never entirely extinguished. Some scientists, at least, remained committed to the possibility and promise of professional science to infuse the colony with the hope of creating something more meaningful than mere profit.

# 3

# Treub's Beautiful Science

By the 1860s the apostles of enlightenment's optimism about transforming the colony through science was in the distant past. The colonial state supported acclimatization research and to a lesser degree scientific education, including medical and agricultural schools. But other sciences had little government support and were only practiced in the amateur societies, which met less and less regularly and published fewer and fewer journals. Most of their scientific work was shared only within a small European community. On Piet Bleeker's retirement in 1860 as head of the Batavian medical college, he predicted in a public speech to the amateur Naturalists' Association that their science still had a long way to go to become the equal of European science: "Is further specialization in our work useful? Will we one day have separate associations in mathematics and astronomy, physics and chemistry, geology and botany, zoology and anthropology? Will, in a far off future, just as in Europe, associations exist, dedicated to only one branch of science? . . . We wish it, because it will be the reflection of the highest flowering of society in these parts. But that time, if indeed it comes, shall be that of our distant descendants."[1] Bleeker was one of the last apostles of enlightenment of the 1840s generation to leave the colony, and there was no new generation to replace them. He was right to be pessimistic, as the next decades bore out Bleeker's prediction. For thirty years, professional science did not flourish. Nor did Indies society flower in the sense Bleeker meant it, that is, like enlightened Europe. A professional career in science was only possible through a handful of state appointments in the medical, agricultural, and engineering fields.

Science, though, did have a rebirth at the end of the nineteenth century at the Buitenzorg Botanical Gardens. By 1900 it was the most famous tropical botanical gardens in the world, known as 's Lands Plantentuin, an archaic yet memorable name literally meaning "Plant Garden of the Land." By then the institute was famous all over scientific Europe and was believed to be exemplary of Dutch colonialism's commitment to science. The German Darwinist A. F. W. Schimper, who visited in 1890, repeatedly praised in print the biology laboratory and its role in advancing the study of plant adaptations.[2] Two years after Schimper's visit, Gottlieb Haberlandt made the journey to Java, and in his account of that trip, first published in 1893, he predicted that the "rich fruits achieved in the general morphology of plants, the anatomy, physiology and biology, can also be expected in the future of the Botanical Gardens."[3] How did the vigor and reputation of science return in the Netherlands East Indies, not as an extension of acclimatization research but as a research institute?

The agent of this change was the new director, Melchior Treub, who arrived in 1880. Since 1900 few observers of Indies and Indonesian science have doubted Treub's importance to science in the Netherlands East Indies. Still, how he turned biology at the botanical gardens, a backwater institution made to serve at the pleasure of the governor general, into an icon of Dutch colonialism has not been adequately explained. This is largely because the work he did during the last years of his colonial career, highlighted by his directorship of the Department of Agriculture between 1905 and 1909, has overshadowed his earlier work. Recent scholarship has sought to probe how Treub's science at the Department of Agriculture was connected to the colonial project of administering and developing peasant society.[4] In the rush to deconstruct his colonial project, questions about how he built scientific authority in the first place have been neglected, taking for granted that science was generally admired and respected inside the colony. Interpreters of Treub have glided over important questions of how he created scientific credibility in a society that had lost its appetite for apostles of enlightenment decades earlier. Before examining the Department of Agriculture, which I will do in the next chapter, it is necessary to understand what Treub's regime of science was.

Since the publication of H. H. Zeijlstra's biography fifty years ago, no systematic study has looked at the scientific practices of Treub and his staff at the botanical gardens. Nor has an adequate explanation been advanced of what secured Treub's status as the premier Dutch colonial expert on nature by the end of the nineteenth century.[5] Zeijlstra's contention that Treub's success as the first director of agriculture came from his pioneering research in plant physiology has stood for fifty years. Still, this argument should give us pause. How was a plant physiologist famous for laboratory research, with no training in agricultural science and little experience as a bureaucrat, able to turn the botanical gardens into the administrative center of agriculture? Zeijlstra never explains how laboratory work positioned Treub as the chief scientific official in the colony. As I argue below, Treub's great success was not as a researcher, in fact, but as the fashioner of a beautiful image of international-level scientific research being done in the Indies. The botanical gardens flourished not because it reflected a flowering of colonial society but because it created an enduring fantasy that the colony was on the road to civilization because the gardens was now a center of professional science.

## International Science

In 1880 the botanical gardens in Buitenzorg was little more than a riding park for the governor general despite its sixty-year connection to natural history. In the 1820s it had served as Reinwardt and Blume's headquarters for the royal Natural History Commission, but this association vanished with Blume's return to the Netherlands in 1826. And, although the gardens revived in mid-century under the chief gardener's care, after 1830 it was under the direct control of the governor general's palace guard. There was no scientific director, and the chief curator and gardener, J. E. Teysmann, was the nominal head. A Dutch visitor in 1858 referred to the gardens as the governor general's park, noting further that there was an excellent collection of "botanical treasures of the tropics."[6] As late as 1865 Teysmann's superior, the intendant of the palace of Buitenzorg, instructed that "when felling or pruning a tree, do not forget that the Botanical Gardens are almost entirely a part of the palace of Buitenzorg, and serve as a walking area for the Governor General."[7] The botanical gardens was a place where the colonial elite could enjoy the tropical qualities of the Indies safely. It was part of the world of nineteenth-century *mooi Indië* (beautiful Indies), sun-drenched and lush nature mellowed to suggest a tranquil, sentimental, and pretty landscape.[8] Let us not forget that it was surrounded by tropical rain forests. On the cover of my paperback copy of H. W. van den Doel's book about the Indies civil administration is a reproduction of an 1842 lithograph of the governor general's palace viewed across a manicured pond and flanked by the botanical gardens. Two European gentlemen ride horseback, a woman watches swans attended by her chivalrous husband, and three dark-skinned gardeners are weeding.[9] This image of the beautiful Indies has endured for almost 170 years, and tourists, Indonesians and foreigners both, can still, with some imagination, view this image of the tropical sublime in Bogor (formerly Buitenzorg).

Nonetheless, Teysmann had established science as a part of the gardens' mission. To be sure, science at the gardens remained secondary to its aesthetic purpose despite Teysmann's success in building scientific credentials in the 1840s and 1850s. In 1851 he proposed new regulations for the botanical gardens and in 1858 recommended turning it into an independent scientific entity. These and other suggestions about professionalizing the gardens were ignored.[10] Only in 1861, with the arrival in Buitenzorg of the reform-minded governor general L. A. J. W. Baron Sloet van de Beele, did Teysmann find a willing listener. Teysmann was made interim director, and they started a serious search for a botanist with scientific credentials.[11] C. A. J. A. Oudemans, botanist at the University of Amsterdam, was offered the job but turned it down.[12] R. H. C. C. Scheffer, a student of Utrecht University botanist F. A. W. Miquel, finally took the job in 1868. In the same year the gardens was transferred into the care of the Department of Education, Religion, and Industry. As detailed in the previous chapter, Scheffer, too, came to concentrate on acclimatization efforts. Scientific contacts with Europe diminished, and he stopped maintaining either the herbarium or the library.[13] He died in 1880 after a lackluster twelve-year tenure as the scientific chief of the botanical gardens.

When the still young Treub arrived in the Netherlands East Indies, he found little to suggest that this was a scientific research station. And in fact his first few years were inauspicious, as neither he nor the botanical gardens had any real authority either inside or outside the Netherlands East Indies. Even the gardens' ability to run a small agricultural school was doubted, and the school was shut down in 1884.[14] Treub, who had been trained as an experimental biologist, struggled to continue his physiological research.[15] The gardens were disheveled, the few scattered buildings were not built to permanently weather their tropical presence, and there were no research tools or facilities. He also had little leverage with other colonial leaders. It is likely that initially Treub's ambitions still were with the world of European science.[16] Still, it appears that from the start he believed that his reputation was linked to the reputation of the institution he controlled. Consequently, the first few years were dedicated to making the botanical gardens look like a scientific institute to other European scientists. It took him nearly a decade to create something he could present proudly. In an 1888 letter to F. A. F. C. Went, who he was recruiting for a job as a staff researcher, he wrote, "As you well know, it is no longer necessary to fear the scientific isolation for someone attached to the Buitenzorg Gardens."[17] Treub could write with such confidence because by then word of Buitenzorg's scientific credentials had reached Europe.

In his first decade in Buitenzorg, Treub created authority, legitimacy, and power for the botanical gardens and himself through the shaping of a physical space where tropical nature could be investigated in a comparative fashion. Before he left the Netherlands for Buitenzorg in 1880, Treub had written in the intellectual monthly *De Gids* (The Guide) that if Dutch researchers could be spurred to greater research accomplishments the Netherlands might return to its glorious stature, regaining its position as an intellectual leader of European science.[18] His ambitions were great, but they were oriented toward showing European science what the Dutch could do in their colony. Consequently, he addressed his science, certainly in the first decade, to European scientists, not Dutch colonial officials. The institution he built came to have all the tools needed to make sense of the rich flora and fauna of Java: a library, work space, chemicals, microscopes, floras, live specimens, and travel grant money. His laboratories became a convenient research station for studying tropical nature, which by the late nineteenth century, thanks to Darwin, had become crucial for understanding biological processes more generally.[19] As the botanical gardens was linked to the world of European science, it created knowledge that had little apparent connection to the Dutch colonial mission. In the process the botanical gardens came to register in the world of science not as a colonial institution but as a tropical one. This image began in Europe but eventually carried itself back to the colony, where its detachment from the vulgarities of colonial knowledge brought great admiration.

Treub began his career in 1880 by instituting European practice in regard to the few assets the gardens had: the herbarium, the library, and the gardens proper. The botanical gardens had maintained a herbarium since 1844, but for most of its existence the sheets of dried plants had lain unsorted in a temporary wooden shack. Insects were eating the paper. Nobody knew what collections the herbarium contained. Most sheets had no determination, that is, no determination of the specimen's taxonomy. Because the library had not been maintained, labeling on already determined material was out of date.[20] And all the material gathered before 1844 was in the Netherlands. This included the type-specimens for the Javanese plant species discovered and named by Blume in the early nineteenth century. While it would not be possible to write the flora of Java, or even a definitive monograph about a large plant family, the herbarium was still the reference point for the whole gardens. At the end of 1881 W. Burck joined the gardens as assistant director and chief of the herbarium. In an attempt to make sense of the largely unordered and undetermined sheets, he started by cataloging the ferns.[21] For the first time ever, sheets were placed in tin canisters after having been fumigated. Burck and Treub arranged exchanges of herbarium material with the botanical gardens in Calcutta, as well as with herbaria in the Netherlands. A start was made at properly determining the more recent collections of Teysmann, O. Beccari, and H. O. Forbes.[22] Only in 1884 did proper cataloging begin.[23] Treub established guidelines for concentrating on species that grew in the colony, but separate herbaria were maintained for plants growing in the gardens (including the trial cultivation gardens) and a local herbarium of Buitenzorg. The library was cataloged in 1881 and began subscribing to European scientific journals. The gardens proper were cleaned up, with the goal of "expanding the public's interest in plants and their products," and throughout Treub's directorship great efforts were made to shape an aesthetic "beautiful Indies" experience for the casual visitor.[24]

Treub's best asset in establishing the gardens' reputation as a research institute was print media. He quickly created a scientific mouthpiece for the Buitenzorg Botanical Gardens, the *Annales du Jardin Botanique de Buitenzorg*. A lone issue of a scientific periodical had appeared under his predecessor Scheffer, but it was Treub who turned it into a European-style scientific journal. Treub's *Annales* was printed in Leiden by the publisher Brill to facilitate better quality and easier distribution in Europe and also to guarantee that it looked

Palm quarter in the Buitenzorg Botanical Gardens, 1900 (Courtesy of KITLV/Royal Netherlands Institute of Southeast Asian and Caribbean Studies)

like a scientific journal. Treub and Burck wrote most of the articles for the first few issues, but by the end of the 1880s foreign visitors were writing up some of their research in it. Treub did not stop with a single journal. By 1900 the gardens published six different journals, including the *Annales*.

At first Treub worked to create a research institute of the kind he knew in Europe. Initially, his own research set the scientific standard. He began by continuing his embryological research, engaging with the scientific debates in European biology. His early research about *Lycopodium* was typical. Initially these were examinations of the life stages of the well-known Lycopods *L. cernuum* and *L. phlegmaria* in which he reproduced work done in Europe and published the corroborating results in his own journal.[25] The point of these first studies was to establish his credentials as a biologist following standard scientific practices even though he was in the tropics.[26] Once his credibility was on a solid foundation, he was able to branch out to study Lycopods that only he, as a researcher in the tropics, could access. He concentrated on species found on Java, cultivating them in the laboratory, hoping to discover the complete embryology of one of the species. He had, in the end, only limited success, but he did discover some new material growing in the eastern part of the archipelago.[27] This in itself was an important result. Embryological research could be carried out anywhere, including Buitenzorg, but new species were not to be found in Europe. Not only did he prove that Buitenzorg could enter into scientific dialogue and debate with scientists in Europe, but he showed the advantages of a tropical location with its many live available specimens for comparative analysis.

Neither Treub nor Burck were gifted field biologists; both were more comfortable with experimental techniques and comparative anatomy. Burck's theoretical treatment of the taxonomy of the Rubiaceae family, which engaged with recent writings by Darwin, strove for a quality associated with European scholarship.[28] Burck, more so than Treub, targeted economic botany, although this research area was likely selected with the aid of Treub, a notorious control freak. In his first few years he taxonomically revised the economically important tropical lumber family Dipterocarpaceae, as well as the Sapotoceae family, which includes the genus from which the latex gutta-percha is made.[29] By concentrating on commercially important species that grew in the Dutch colony, Burck followed in the footsteps of K. W. van Gorkom. The gutta-percha research had resulted from a special trip made to Sumatra in 1882 to collect specimens and observe the local cultivation of the crop, not to botanize indiscriminately.[30] But the trend in Buitenzorg was away from acclimatization research. During the early 1880s, the poor state of the herbarium collections in Buitenzorg could not support more general taxonomic research. Still, in order to attain the status of a botanical research station, breadth was needed. At the end of the 1880s, Treub expanded the collecting goals of the gardens and turned it into a depository of Indies nature, where any and all of the colony's nature could be found in storage on coded herbarium sheets. As research about economically important species declined during the 1890s, Treub transformed the beautiful gardens into an institute for the study of all of the colony's tropical nature rather than a specialized institute for the comparative study of economic botany.

Treub meticulously crafted the image of the botanical gardens, especially among scientists in Europe. In the 1880s, word of the tropical research station spread via journals, books, and pamphlets printed by Treub. Treub went to great effort to impress on his peers in Europe what he was doing. In 1887 he printed a catalog of the Buitenzorg library and mailed it to all the botanical institutes in Europe; not only was the strength of the book collection (five thousand volumes) a mark of distinction but Treub proved to scientists that Buitenzorg was up to date with the latest research.[31] Treub perhaps cemented his European image through a description of the gardens in an 1890 article in the French international culture revue *Revue des Deux Mondes*, where he touted the attributes of his tropical research station. An English translation was printed by the Smithsonian a year later.[32] Through a lively description of the botanical gardens, he argued that Buitenzorg was the ideal place to study and research tropical biology. Herein he also begins to sketch an argument for the political meaning of colonial science: "The more civilization advances the more it is demanded of nations which possess great kingdoms in faraway countries blessed by heaven that they should not forget that royalty has its responsibilities and that it can not be allowed to withdraw itself from the noble task of adding to our knowledge of nature, independent of any direct advantage, either present or future."[33] Treub's principal audience was European science, but, as the quote shows, by 1890 he had worked out a connection between tropical science and colonial politics. By then he was actively burnishing the image of his work among Dutch colonial and metropolitan elites. This was not an overnight endeavor. A year after he arrived in Buitenzorg he started producing yearly reports, in Dutch, of the work done at the gardens. They actively highlight the year's accomplishments, and in the absence of archives from the gardens, which are no longer extant, and the scant interaction Treub had with the colonial bureaucracy, these reports are the best way for the historian to access the gardens in the 1880s. This chapter relies heavily on them as well. The reports from the 1880s, especially, make exciting reading, vividly showing how Treub created the scientific infrastructure of a research station in a few short years. Later reports, even while Treub was still director, were more representative of a typical and routine institutional report. To be sure, these early reports were calculated to impress, but the verifiable information therein has proven accurate, and in any case they serve as excellent evidence of Treub's crafting of the image of the botanical gardens.

The yearly reports were followed by two important institutional histories, which settled the reputation of the botanical gardens. The first was a short institutional history written by Treub in 1889, which created an origin myth for the science at the gardens and covered its history from 1817 until 1844. Using long forgotten printed material, as well as letters from the governor general's archive, he unearthed and reconstructed the gardens' scientific achievements under Reinwardt, Blume, and Teysmann. In particular, Treub argued that the gardens attained an intellectual and scientific autonomy under Teysmann in the 1840s. Treub subtitled his book suggesting it was the first of two volumes and that he would finish the history between 1844 and 1880.[34] The second volume never appeared. The lasting impression is that Treub picked up where Teysmann left off in 1844, when Teysmann won scientific status for the botanical gardens from the governor general if not financial independence. Treub's institutional history likely served as a dress rehearsal for his 1892 volume celebrating the seventy-fifth anniversary of the botanical gardens.[35] This oversized volume appeared in conjunction with a celebration on May 18, 1892, and depicted a research institution at maturation. Treub penned a short history of its seventy-five-year existence, but it included little of interest not in his earlier institutional history. Burck and others chronicled contemporary achievements. Over five hundred pages showed the Dutch reading public something of which the colony could be proud.[36]

What did Treub get out of this? There is little doubt that he was satisfied with his career in Buitenzorg and chose not to use it as a stepping-stone to European academia as he may have initially planned. By the early 1890s Treub felt more than satisfied with his opportunities. At the end of his life he wrote that he passed up five attractive job offers while in Buitenzorg, including three from Dutch universities, one from a German university, and one from a foreign botanical institute. He cited in particular his pride about the international scientific reputation of the botanical gardens.[37] By the 1890s he had won wide scientific acclaim, not for his research or publishing but for his talents leading an internationally acclaimed research institute. He also won respect and credibility inside the colony. In 1898 the government made him, symbolically at least, the leading intellectual in the colony by granting him the title of professor; at that time he was the only person in the Netherlands East Indies to hold that rank.[38]

## Biology's Mecca

A scientific reputation could not be built on the printed word alone. Much of the prestige of the botanical gardens was generated by scientists who visited it, used its research facilities, and spread the word about the scientific opportunities afforded in Buitenzorg. Treub catered to visiting scientists' needs. In 1884 he established, with the blessings of the government, a lab open to visiting scientists. It was this lab that raised the international stature of the botanical gardens. And it was this transnational authority that in later years gave Treub so much authority inside the colony. H. Graf zu Solms Laubach, a professor of botany at Göttingen, visited for a few months during the winter of 1883 and 1884 and published a positive review of the gardens in the *Botanische Zeitung* with the recommendation that others follow his lead in visiting Java.[39] Even before the lab was founded in 1884, Treub was openly stating his wish to make Buitenzorg a center for tropical botany by luring visiting scientists: "With today's numerous and convenient possibilities for transportation to Java, we can imagine that other botanists will wish to spend some time at the Botanical Gardens. It is clear that with this development our institution shall expand its scientific meaning in a new direction. Often it is believed, entirely erroneously, that scientists in the tropics can do nothing else besides making extensive trips, in order to bring together large collections. For at least the next few years, one will have the right to claim, that where an institute such as the Botanical Gardens in Buitenzorg exists, a stay of three or four months will provide the opportunity, partly for doing, partly for preparing, interesting scientific research."[40] In November of 1884 the government relinquished the old military hospital in Buitenzorg to Treub. He had it converted into a simple botanical research station, the first one, he was careful to note, in the tropics.[41] The large hall of the visitors' lab contained five windows with a desk at each one of them. A herbarium of material cultivated in the gardens was kept on the premises so there would be no need to walk to the herbarium for identification of commonly used material.[42] During the first decade of the lab's existence, it hosted 46 visitors, two-thirds of whom were Dutch or German. Between its founding and the beginning of World War I, 154 foreign scientists visited and used this research station.[43] In 1891 Treub expanded visitors' research opportunities by having a small research lab built in the satellite botanical gardens at Cibodas on the slope of the Gedeh volcano east of Buitenzorg.[44] More than any other part of the botanical gardens, the visitors' labs tied the gardens to a larger scientific endeavor, though one closely aligned with the science practiced in Germany and the Netherlands. The constant stream of scientific visitors, and the positive word they spread back in Europe, allowed Treub to reinvent the governor general's riding park as a respected tropical research station, all in less than twenty years.

Along with the visitors' laboratory, Treub established formal scientific relations with institutions in Europe. In exchange for receiving new herbarium material, the Leiden herbarium in 1885 sent duplicates of some type-specimens of Dipterocarpeae—an important family of tropical hardwoods—from earlier nineteenth-century collectors.[45] This established a level of parity between Leiden and Buitenzorg. The maintenance of regular institutional correspondence with scientists all over the world also increased the scientific reputation of the gardens. Requests for seeds or herbarium material were promptly granted. The majority of this correspondence concerned mundane matters, but it was the multiplication of the contacts that made the gardens more and more a tropical resource center. In 1888 the director's office of the gardens sent 1,061 letters answering inquiries from inside and outside the colony, almost twice as many as two years earlier.[46] Of course, Teysmann, Binnendijk, and Scheffer had

Botanical laboratory at the Buitenzorg Botanical Gardens with M. Treub seated, left, facing the camera, 1900 (Courtesy of KITLV/Royal Netherlands Institute of Southeast Asian and Caribbean Studies)

maintained contacts with other botanical gardens in the decades before Treub arrived, but by the end of the 1880s Treub was in contact with biologists of all specialties. In 1886 he started a personal correspondence with Adolf Engler, the director of the Berlin Botanical Gardens, after receiving a letter from him asking about plant material. Although both had different biological research interests, over the next two decades Treub and Engler continued to write, exchanging publications about plant taxonomy, informing each other about people, and discussing the state of science in Buitenzorg.[47] Engler was in Buitenzorg from December 1905 until February 1906, exploring the tropical flora on excursions and discovering in the Buitenzorg herbarium new species of Araceae, a large family of tropical monocots related to palms and Engler's taxonomic specialty. Significantly, Treub communicated with Europe's most important plant systematist as a colleague. Not only was Engler the editor of the important *Botanischen Jahrbücher*, but he was also the coeditor of *Die Naturlichen Planzenfamilien*, then and for many more decades the most widely used classification scheme. Engler's system, a "natural" one in which plants were arranged evolutionarily from primitive to advanced, was the first attempt at a comprehensive world flora at the genus level.[48] In fact Engler was emblematic of the direction plant taxonomy would take in the twentieth century when experts of large tropical families (such as Engler's Araceae) dominated the field. For Engler, his relationship with Treub was an excellent way to remain tied into a network of tropical botany and helped the Berlin Botanical Gardens maintain its preeminence as a center of European botany, especially considering the dearth of German colonies with tropical botanical gardens. Treub was one link in Engler's Central Botanical Bureau for the German Colonies, an office founded in 1891 to serve as a clearinghouse for tropical botanical information.[49] For Treub, maintaining a personal correspondence with Engler meant that the Buitenzorg Botanical Gardens as an institution had scientific authority in Europe as more than a collecting station. Treub and Engler discussed biology as equals without Buitenzorg being subservient to Berlin.

On his first sabbatical to Europe in 1887 and 1888, Treub created the Commissie tot Bevordering van het Natuurkundig Onderzoek der Nederlandsche Koloniën (Commission for the Promotion of the Physical Research of the Dutch Colonies) and on returning to Buitenzorg its sister organization, the Indische Comité van Wetenschappelijk Onderzoek (Netherlands-Indies Committee for Scientific Research).[50] Both were intended to sponsor large-scale scientific expeditions. Although both organizations were officially recognized by the government, at first neither had any funds. But his control of them allowed Treub to exert influence over colonial science, even outside Buitenzorg. In 1888 he encouraged the early explorations of the paleontologist Eugène Dubois through the offices of the Netherlands-Indies Committee.[51] In 1890 the head of the state ethnology museum in Leiden managed to reorganize the committees into the Maatschappij ter bevordering van het Natuurkundig Onderzoek der Nederlandsche Koloniën (Society for the Advancement of Physical Research in the Dutch Colonies), with funding from local sources for exploratory expeditions. Treub remained the head in the Indies; in fact the organization usually was called simply the Treub Society. The society sponsored a number of large expeditions, including an expedition to Borneo in 1893-94, the Sibolga marine-biological expedition in 1899-1900, and various expeditions to New Guinea in the first decade of the twentieth century. All these trips coordinated their botanical research via the Buitenzorg Botanical Gardens.[52]

Buitenzorg, and Java more generally, was moving closer to Europe and the rest of the world. By the end of the nineteenth century it was already a famous tropical destination, "the garden of the East," a month's steam away from the northern hemisphere winter. The American tourist Eliza Scidmore visited in the 1890s and picked up on the possibilities for science in "the most picturesque and satisfactory bit of tropics anywhere near the world's great routes of travel," calling it the "the scientist's greatest storehouse."[53] Transportation was convenient in Java as well. Buitenzorg had been linked to Batavia via rail since 1867.[54] During the 1880s, with its cooler climate and two hotels—the Hotel Chemin de Fer and the Hotel Bellevue—it was filled with Europeans escaping the heat of Batavia. During their stay scientists had the opportunity to mingle in the "pleasant [ *gesellige*] social life of the European" at either the hotels or the European club.[55] Trips into the tropical rain forest were possible too. Roads and railroads linked all of Java together, and most visitors combined a stay in Buitenzorg with excursions into West Java or even an overland trip via Central Java to Surabaya. But it was not just the ease of travel and accommodation in Java but also the steamers operating between Europe and the Netherlands East Indies via the Suez Canal, which were able to make the journey in under a month, that brought European scientists to Java. This made a trip to Java during the European winter a possibility. These visitors might only be gone from Europe for six months, nothing like Wallace's trip to the Malay archipelago in the 1850s and 1860s, which lasted eight years.

Treub's visitors heaped praise on the botanical gardens and its research facilities. A number of biologists who took a few months' sabbatical to visit Treub's gardens, including Gottlieb Haberlandt, Andreas Schimper, Jean Massart, Elmer Merrill, and David Fairchild, wrote glowing reviews of both the gardens and the research opportunities. It was at this time that the idea arose that Buitenzorg was a botanical mecca that all botanists should try to visit before they died. In 1898 the German plant morphologist Karl von Goebel argued that the botanical gardens in Buitenzorg, with its modern laboratory facilities, had opened up the tropics to serious research into the lives of plants. He suggested that plant biology would stagnate unless botanists turned to studying the myriad adaptation techniques of tropical plants.[56] Gottlieb Haberlandt, who stayed in Buitenzorg in 1890 and 1891, praised the gardens for being organized around plant groups.[57] His later wrote that this organization taught visiting biologists about the widely differing adaptation techniques of tropical plants.[58] During his stay Haberlandt worked at his own desk in the visitors' lab, and it was in this lab that he discovered the leaf pores that enable plants to evaporate moisture in humid, tropical climates.[59] To the visitors the work needed to generate such convenience was largely hidden. But even so, they knew this was the result of a political decision, and they consistently thanked the Netherlands East Indies state for its progressive policies in funding the gardens and allowing visitors from all over free access to it.[60]

At the turn of the twentieth century, visiting scientists were introduced to the Dutch colonial lifestyle, including the use of native servants. Apparently every visitor received a servant to help in the lab, as well as at the hotel. David Fairchild, a twenty-five-year-old American agriculturalist who visited in 1895 and 1896, got eight months' use of one of Treub's servants, Mario: "From mounting microtone sections on microscopic slides to managing a caravan across the mountains, he took care of everything."[61] Fairchild's research into the ecological connection between termites and fungi would have been impossible without the aid of "Papa Iidan," who hunted termite nests in the gardens.[62] Even his trip up to Cibodas, to the small laboratory nested in the virgin forest above the Puncak Pass, was carefree: "Loaded down with a hundred pounds of baggage on each end of a bamboo pole, the coolies trotted off over the pass while Mario and I followed in a dos-à-dos."[63]

The ease with which visitors such a David Fairchild moved around and did research was possible because of a well-established colonial infrastructure, including the large labor force of Sundanese and Javanese working at the botanical gardens under Treub. As is clear from photographs, native workers kept the paths through the gardens clean, worked in the nursery, and assisted in all aspects of the administrative and research work. In addition to holding a variety of clerical positions, native employees of the gardens did highly skilled but anonymous work such as illustrating the botanical images used in the gardens' journals and books. There were also native plant collectors, some of whom were sent on their own collecting trips. And men such as Iidan and Mario were indispensable for their assistance in research. But the Dutch scientists never systematically trained these research assistants, certainly not in autonomous research and only haphazardly in the routine jobs around the laboratory and in the herbarium. Moreover, native labor is very rarely mentioned in correspondence from the Dutch floracrats at the botanical gardens. To my knowledge Fairchild's account of working with Iidan and Mario is the most extensive discussion of native assistants at the colonial era Buitenzorg gardens. Still, as Fairchild suggests, effective collectors and research assistants were highly prized. Most learned on the job, as presumably Mario had done while working for many years next to Treub. During the entire colonial period, no native employees of the gardens advanced past the position of servant to Dutch science, although they obviously helped in much of the research conducted by the scientists, both the permanent staff and the visitors.

Under Treub's guidance, the Buitenzorg Botanical Gardens became a hub in the circuit of botanical knowledge. Its solid institutional base and wide-ranging access to tropical knowledge made it the envy of botanists the world over. Treub's institute became a modular component of Dutch colonialism, one that was subject to duplication by other, non-Dutch colonial botanists.[64] Most spectacularly, the new American colony of the Philippines sent a botanical emissary shortly after a government was established in Manila. In 1902 the young American botanist Elmer Merrill (later famous as the director of the New York Botanical Gardens and then head of the Harvard herbarium) visited with an eye to establishing a similar scientific discipline in Manila, where he was the staff botanist in the newly established Philippine Bureau of Forestry. Merrill knew next to nothing about the flora of the region because the Spanish era herbarium in Manila had burned down in 1897 and herbaria in the United States had scant Philippine collections. He brought with him six hundred specimens recently collected in the American colony, and using the library and herbarium in Buitenzorg he classified them. But in his two months in Buitenzorg he did much more; he took photos of the institute, arranged for an exchange of publications between Manila and Buitenzorg, ordered herbarium specimens of trees thought to grow in the Philippines, and made careful observations of the methods of the Java Forest Flora.[65] His report is only one part taxonomy. He introduces the final chapter, about the workings of "this admirable institution" under Treub, with the statement that it was his specific intent "to record some of the methods of investigation and call attention to the equipment of the 's Lands Plantentuin."[66] Perhaps with an eye to a future institute in Manila, he lists the 1897 budget for the gardens (a total government appropriation of f.215,140), as well as salaries of the professional staff (in 1897 Treub's salary was f.14,400). A footnote helpfully explains that "one guilder = $0.40 United States currency."[67] A separate chapter details the workings of the Java Forest Flora. Merrill starts this excursion by arguing that foresters in the Philippines could not use methods developed in the United States but needed a special system suitable for tropical forests such as the one developed in Buitenzorg. Principally this meant the detailed surveying of different areas designated as "typical." Before he left he had the chief of the Forest Flora carefully examine his essay for any possible errors.[68] What interested Merrill was not the results of the Java forest survey or the lab research on tobacco diseases but the method employed for organizing government-sponsored botany. And Merrill learned from Treub's system. He spent the next twenty-two years building up the Manila collections, becoming director of the Bureau of Science and a professor of botany at the University of the Philippines before leaving for a career as the "American Linnaeus" in the United States.[69]

By the turn of the century the Buitenzorg Botanical Gardens had become an established center of tropical biology. It was an ideal location for a young European or American scientist to make his mark in biology, conducting research in the tropics under controlled and convenient conditions. Maintaining this institution meant that Treub had to enforce strict discipline even among its visitors. Such discipline did not appeal to every visitor, as is illustrated by the visit of the German biologist Ernst Haeckel, probably the most famous biologist to visit Buitenzorg under Treub. Haeckel arrived in late 1900 with plans to research freshwater plankton (marine invertebrates had been his first love), paint watercolors of interesting forms for a new picture book, and collect material for the zoological museum at Jena. By then the biology discipline at the University of Jena had been under Haeckel's sway for almost forty years, and his institute played an important part in the development of evolutionary morphology.[70] While in Buitenzorg, Haeckel and Treub freely discussed the scientists and scientific ideas of the age and interacted as equals. But Haeckel, perhaps hoping to duplicate his six-week stay on the south coast of Ceylon twenty years earlier, where he had the freedom to create his own zoological laboratory in a bungalow, was disappointed by "enervating Buitenzorg."[71] He wrote that nature in the gardens did not appeal to him, and the climate had a deleterious effect.[72] He chafed under Treub's requirement that every night he

Melchior Treub, ca. 1900 (Courtesy of the National Herbarium of the Netherlands, Leiden University branch)

wear formal European dress to his host's dinner.[73] Haeckel apparently saw Treub's formality as an obstacle to gaining unadulterated access to the tropics.

Treub was by all accounts the perfect colonial gentleman. Although he was a bachelor for most of his years in Buitenzorg, he maintained a scrupulous decorum everywhere in the botanical gardens. When he eventually did marry in 1905, it was to a young woman from a well-known Rotterdam bourgeois family. His unblemished image, never tainted by scandal or lowness, was one part of his success as a colonial scientist. And this sacred image has remained an important part of the legacy Treub left to the botanical gardens. Later leaders of the gardens have been careful to keep this image on a pedestal. In 1925 the director of the gardens, W. H. Docters van Leeuwen, found a package of letters in the director's archives that included love letters from Mrs. Treub to Mr. Treub and correspondence about Treub's contentious relationships with other colonial officials. Docters van Leeuwen unfortunately but predictably destroyed these letters immediately after reading them.[74] In 1914 Treub's protégé and successor, J. C. Koningsberger, said of him, quoting the ancient historian Livy, "Treub too was remarkable, even in his short-comings."[75] Because of Treub's careful regulation of his own image—I have yet to find an extant letter by him that reveals a personal side—and his successors' protection, we have remarkably little insight into what those shortcomings were. We are left solely with the impression of a superior gentleman dedicated in his pursuit of science.

## Conclusion

By the late nineteenth century the mantra of the Dutch state was peace and tranquility. The relationship between the native inhabitants and Europeans flared violently at times, for example, in the peasant revolt in Banten, West Java, in 1888.[76] Still, after 1830 and the end of the Java War, the Dutch administration quickly suppressed most armed resistance. By 1900 many of the outer islands had been peacefully integrated into the Netherlands East Indies. The major exception was Aceh, where the Dutch were unable to defeat opposition to their occupation. J. B. van Heutsz became a hero in the colony and the Netherlands after he was credited with breaking the Acehnese resistance in 1901. For the Dutch he epitomized the kind of official who calmly pacified the Netherlands East Indies and only used violent means when absolutely necessary.[77] Even if violence lurked just below the surface, the surface was not often broken, and for many it appeared quite real.[78] And by the late nineteenth century the colonial state was taking an active interest in crafting a society that looked like it was more than just subjugated to state power.

In the last generation, historians from Leiden have made a persuasive case that the calm surface was a result of the efforts of bureaucrats and other officials who manned the Dutch administration. Their calm and quiet exercise of power, as in the title of H. W. van den Doel's book about the civil administration, kept violence at bay.[79] Clearly the methods of administration were well developed by the end of the nineteenth century, and the Dutch colonial state utilized a broad array of techniques to rule the native population. Still, many officials and colonial inhabitants wished for more, and this probably fueled interest in the beautiful Indies paintings of the late nineteenth century.[80] Science, as well as art, could contribute to the sheen of modernity. A biological science institute, Treub discovered, was ideally suited to suggest that the colony was a modern and enlightened society. It was cosmopolitan in that it was tied into the European network of science, yet it was specifically about and for the tropical setting of the Indies. Moreover, it was not in the least political; unlike the scientific agendas of the apostles of enlightenment in the 1840s, the botanical gardens of the 1890s did not attempt to influence colonial policy. It created the fantasy of an enlightened society in the Netherlands East Indies by erecting an institution where scientists created beautiful and enlightened knowledge. Senior colonial officials, in particular, now became interested in associating themselves with the world-famous Treub. In so doing, the fantasy of mooi Indië garnered an official endorsement that was sealed when the colonial government extended to Treub the title of professor. The allure of framing the colony behind the mediating glass of science remained potent. Rudolf Mrázek has argued that for colonial inhabitants in the 1920s this fantasy of the beautiful Indies was "[well] known and often more significant than the colony itself."[81] By then it had been official state policy for decades. Beginning in the 1890s the colonial state began to consciously cultivate an official culture of enlightenment, in particular by supporting institutions such as Treub's. As I will show in the next chapter, after the turn of the century Treub took advantage of this to further expand the reach of the botanical gardens, as well as his own reputation as an enlightened leader worthy of power.

# 4

# Ethical Professionals

At the end of the nineteenth century, new reform proposals began coming out of the Netherlands and the Indies calling for greater political and economic autonomy for the colony. This reflected a sense among colonial elites generally that their lives had been tethered to the Netherlands long enough and that a modern Indies would need to shape its own future outside of the whims and interest of the motherland. These arguments were not new and in any case might have been ignored by Dutch politicians. The difference was that some of the reformers were able to reframe the political debate about colonialism in moral terms, arguing that the Netherlands had obligations to its colony and that the population, both native and European, had rights, which entailed their participation in discussions about the social, political, and economic future of the colony. These reform ideas were built around a missionary impulse in which European culture, variously Christian, scientific, or liberal, would uplift the native population. To generate progress in the colony would require expanding cultural, educational, and economic opportunities for all colonial inhabitants. As Robert Cribb has pointed out, these reforms, which came to be known collectively as the ethical policy, were always a conglomeration of ideas and goals, not all of which were reconciled to each other. Still, three areas of reform came to define the ethical policy initiatives: protecting native society, expanding educational opportunities for native inhabitants, and improving the welfare of native society through development initiatives.[1] As I argue in this chapter, this last initiative had long-lasting effects on the character and composition of the colonial state, greatly expanding the presence of state-run science.

While the initial ethical policy directive came from the Netherlands, the colonial bureaucracy would have to see to its implementation, much like the liberal reforms thirty-five years earlier. The metropolitan directive to reform came in 1901 when, to most people's surprise, Queen Wilhelmina declared in her speech at the opening of Parliament that the Netherlands would pursue a colonial ethical policy, which would begin with a scientific investigation of native welfare. The Hague empowered the governor general to find new ways to respond to the economic and political needs of colonial society. Although the queen's speech directed officials to examine native welfare, there were few suggestions on how to accomplish this. And an early initiative, the European Pauperism Commission, demonstrates that the new responsibilities were not gladly or easily borne by current officials.[2] It was not until 1905 that the colonial government was able to implement real policy changes.[3] In large part this was because the established sections of the colonial government were not equipped to carry out the various policy initiatives. The ethical policy required a change in the colonial state.

In 1901 the governor general exercised centralized power through the colonial military and four different departments—Civil Administration; Finances; Public Works; and Education, Religion, and Industry—in addition to his executive staff, the Algemeene Secretarie.[4] The officials of the Civil Administration Department were the face of the colonial state for most inhabitants outside of Batavia. In the second half of the nineteenth century, this department anchored colonial power, and it was only in civil administration that one could make a career and work one's way up to a prestigious senior position. The colonial state's finances were overseen through a dedicated department in Batavia. The Public Works Department built and maintained the public infrastructure. And, finally, the Department of Education, Religion, and Industry contained a hodgepodge of state institutions intended to manage colonial society, principally on Java. Science, education, and economic planning were lodged in this department. It was this department and its units that would invent a number of the new systems of the ethical policy. And the result was a branch of the colonial state whose officials came to rival officials of the Department of Civil Administration in terms of influence and power.

With the inauguration of the ethical policy, the colonial state looked for ways to more actively engage with native society in the Netherlands East Indies. After 1905, as the colonial state added new officials, institutions, and departments, power was granted to a host of technical experts, including natural scientists, social scientists, and engineers, charged with leading the colony's population, especially the native inhabitants, toward modernity. This resulted in professional science becoming a formal and permanent part of the colonial administration in the first decade of the twentieth century. It did so because scientists could generate knowledge useful to ethical policy goals. In the context of an official reform movement, Melchior Treub, as the leading scientist in the colony, was in a perfect position to expand the power of scientists in the Netherlands East Indies. In the first decade of the twentieth century he created a powerful new institution, the Department of Agriculture, on the promise that the botanical gardens and its research facilities would improve the colonial economy for natives and Europeans alike. Treub was backed by a professional scientific institute that had widespread credibility and prestige. In the context of reform, Treub convinced other officials that scientists should have the power and authority to lead colonial agriculture.

Treub envisioned that his department would not just generate useful knowledge but would also lead society with science. He worked to establish his department and its scientists as the locus of modern economic reforms by creating mechanisms to spread scientific knowledge directly to the native population. He failed in this endeavor in part because he did not find a way to spread science popularly. More important, though, he failed because the state did not judge him by how well he led through science but by how well he implemented ethical policy goals. Other colonial officials expected the department to improve the native economy. And while his science was beyond reproach, his ability to affect economic progress was widely criticized. When Treub left four years after the department's founding, he was replaced by the agricultural expert H. J. Lovink, who had proven in the Netherlands that he could create economic uplift for farmers. By then scientific institutions were much more secure professionally, and much greater in number, than they had been twenty years earlier. But notwithstanding the growth of professional science, Treub was under the same pressure that Junghuhn had been under decades earlier; professional scientists, even revered ones, were expected to implement policies created elsewhere in the colonial regime.

The long-term result of the ethical policy was a larger, and more powerful, colonial state. After 1900 the colonial state had steadily increasing revenues from a new income tax levied on Europeans, as well as continued income from taxes on export crops, which had strong international demand for most of the period from 1895 to 1929. Government income increased, and state expenditures in 1920 were over one billion guilders, five times what they had been two decades earlier.[5] The Dutch empire gained at least nominal control over the rest of the archipelago not already colonized by other colonial powers, thereby butting up against them in British Malaya, the American Philippines, Portuguese Timor, and German and British (and later Australian) East New Guinea.[6] And as the "pacification" of the archipelago was completed, less money went to military expenses and more to ethical policy initiatives. The branches of colonial authority reached deep into the landscape, spreading horizontally across the archipelago and vertically into local society. On Java the state train system reached most large towns, and by 1940 the density of rail and tram lines was comparable to that in much of Europe.[7] The result was a colonial state much more intertwined with the population's economic and political life. And for scientists it meant far more opportunities for positions within the colonial government.

## Science Power

Beginning in the 1890s, Treub transformed the botanical gardens a second time, turning it into a center of agricultural knowledge. As analyzed in the previous chapter, the botanical gardens had risen from obscurity to become what was considered by many inside and outside the colony to be the finest tropical research station in the world, all in less than two decades. Treub's great prestige, among naturalists and scientists internationally but especially among Dutch inhabitants of the colony, gave him an opportunity to do what the apostles of enlightenment had dreamed of fifty years earlier: to place scientists at the heart of colonial leadership, where science would guide the future of the colony. Mirroring the hopes of Junghuhn, Bleeker, and other earlier colonial scientists, Treub promised the use of universal knowledge for practical manipulations of nature. But he stood on much firmer ground when interacting with the colonial state. He was the consummate insider, with a strong professional institution behind him, who was in a position to launch scientists, especially himself, into the colonial administration.

Treub's continued opportunities lay not in the realm of research but in colonial administration and politics. He came to see colonial agriculture as an obvious outlet for his expertise. In an 1890 article in a French journal, he wrote that in an era of competition among colonial powers in exploiting tropical products, empirically verifiable information about plants would be needed to guarantee colonial profits: "We are already far from the period when the grossest empiricism was usually sufficient, permitting the acquirement of wealth by those destitute of education and often even of intelligence. To ensure solid results, tropical agriculture—no less than that of temperate countries—demands judgment and special knowledge, and the need is felt of establishing it also on a firm scientific basis."[8] In the decade that followed he took a number of steps to put tropical agriculture on his scientific basis, and, as his 1899 instructions *About the Responsibility and Work of the Botanical Gardens in Buitenzorg* make clear, he had by then refined his ideas about linking professional science and agriculture.[9] He explained that the massive expansion of the botanical gardens in the last two decades was the result of the European biologists' wish for a tropical research station. He predicted that in the future agricultural responsibilities at the gardens would grow while at the same time more scientists and nonscientists would come to rely on the gardens for scientific knowledge. He urged his staff to muffle personal desires and work for the greater good of the whole. Only a well-oiled professional science would keep the gardens close to the agricultural interests of the colony.[10] He further urged everyone to make visitors welcome, as they provided free labor and, more important, raised the prestige of the Dutch colonies within important intellectual circles in Europe. The interaction between the gardens' staff and foreigners had a "political meaning," and promoting it was, he urged, a patriotic duty.[11] Stronger still, the gardens' world rank had to be thought of at all times. Treub exhorted "*that nothing go undone in keeping the Gardens in first place.*"[12] The maintenance of the institute — library, collections, and research space — would serve both agriculture and visitors.[13]

Treub's international fame gave him a head start as a government scientist, just as Junghuhn's Humboldtian science of Java had endeared him to some colonial officials. But it should not be assumed that the diffusion of European scientific practices to Buitenzorg under Treub automatically secured authority in the Indies. For many years Treub's immediate supervisor, the director of the Department of Education, Religion, and Industry, checked the botanical gardens' growth. The archives of the governor general are silent on the botanical gardens in the 1880s, suggesting that in his first few years in the colony Treub was treated as a simple subordinate of the director of education, religion, and industry. Yet the botanical gardens' opportunities lay within the government, as Treub understood from the late 1880s onward. With the exception of plantation owners, who he also courted (see below), there was no other institutional patron. In his first few years in the Netherlands East Indies, Treub did not consciously court colonial officials beyond publishing the yearly reports of the botanical gardens. It is possible that he only came to understand the importance of state support for science after 1889 when he did some archival digging for his history of the Buitenzorg Botanical Gardens. The main conclusion of that book, that the local stature of the botanical gardens had risen and fallen in parallel with Teysmann's authority with the governor general, showed Treub the importance of adopting bureaucratic methods.[14] Researching and writing this book taught Treub a further lesson: the state would not be swayed by Enlightenment ideals. For Treub to use the professional scientific network to improve colonial society, he would need to present tangible evidence of his ability to mold tropical nature to the Dutch colonial will. This proof would need to pass muster with other colonial officials.

He made a start by strengthening the position of the gardens within the colonial administration. Since 1868 it had been part of the Department of Education, Religion, and Industry. The original statutes, in fact written just before the large-scale bureaucratic reorganization of 1868, were fit for a minor institute like the botanical gardens. They recognized the scientific existence of the gardens with a mandate to expand the knowledge of the colony's nature. But there were no guidelines for further integration into the government beyond its capacity to advise. Only three officials of the gardens were named: the director, the chief gardener, and the assistant gardener.[15] By 1890 the gardens had far outgrown these early guidelines, as even the government noticed. In that year the governor general's office asked Treub to clarify the position of assistant director. Treub took this opportunity to completely rewrite the 1868 statutes and strengthen his position as scientific leader of the gardens free from scientific oversight by the head of the Department of Education, Religion, and Industry.

Treub's recommendations were eventually adopted nearly verbatim, even over the criticisms of the department's director, P. H. van der Kemp.[16] The latter, a career official with the Department of Civil Administration, complained that Treub's draft statutes placed the gardens under the department only administratively and let Treub keep final authority on all scientific and technical matters. He saw no reason to except the gardens from standard bureaucratic procedures.[17] But the governor general, his staff, and the Council of the Netherlands Indies sided with Treub; one adviser to the governor general speculated that Van der Kemp's rewrites stemmed from jealousy.[18] Treub had expanded his professional autonomy largely because of his scientific credibility.

Treub built his regime of science, and his authority to use universal reason to affect practical change, by making the plant and animal collections at the botanical gardens legible to bureaucrats. Starting at the end of the 1880s, Treub started building an archive of nature. He had this archive fashioned in such a way that it would be an item-by-item accounting of the plants and animals in the colony. The collections would link the professional world of science with the practical demands of colonial officials. Each individual specimen in the collection—for example, the dried plants of the herbarium—would refer to a specific item present in the colony yet be named and known in universal terms. This meant that any specific question about a plant or animal would start with the specimen in the collections. That specimen would then point, with its name, to general answers. It was thus important to have good labels on the specimens, and he concentrated on quality not quantity. He started with the plant collections. In 1888 P. de Monchy, a volunteer, was added to Burck's staff, and he started sorting through the undetermined material in the herbarium. This meant that the herbarium specimens would become useful comparative material and also allowed De Monchy to exchange duplicates with other botanical institutes. In the 1888 yearly report, Treub wrote that he imagined this effort to be the beginning of the composition of a new flora of the Netherlands East Indies.[19] During the course of the next year, Burck and De Monchy started numbering and redetermining the plants in the gardens proper. Three collections were made of all the plants in the gardens—one kept separate as the garden herbarium, one added to the general herbarium, and one placed for easy reference in the visitors' lab.[20] Treub also started on his quest for a general flora of all the plants of the archipelago, although he soon realized that a complete flora was many years away. Instead he sponsored a flora of the Buitenzorg region, spearheaded by J. G. Boerlage, the curator of the Rijksherbarium in Leiden, who in 1888 and 1889 spent four months in Buitenzorg as the first recipient of a Buitenzorg fellowship.[21] He spent some of his time collecting in the Buitenzorg region and prepared lists of his initial findings.[22] Treub also participated in a joint project between the gardens and the Forestry Department to produce a forest flora of the colony. All these collecting endeavors led to the rapid growth of the herbarium. By 1892 the herbarium had twelve hundred canisters full of about one hundred dried plants each, sorted alphabetically by family.[23]

Treub was by no means the first Dutch colonial scientist to become interested in the question of how to use science for the advancement of the colony. His solution, to make nature transparent to bureaucratic practices, had in many ways been anticipated by the cinchona engineers. He went further than Van Gorkom and Bernelot Moens, though, in suggesting that his professional science could comprehend all the plants and animals of the tropical colony. In an 1893 letter to his superior about the new herbarium chief, he wrote, "It cannot be emphasized enough, that at root maintaining the meaning and good name of the Botanical Gardens will largely depend on the state of the Herbarium. . . . I only need to remind you of the previously made comparison between the Herbarium and the archive of a large office, in order to show you how indispensable a good and properly ordered collection of dried plants is for a large Botanical Gardens."[24] The herbarium as archive hooked the botanical gardens, and the natural knowledge contained within it, into the colonial bureaucracy. The office archive was a model other colonial officials could understand. By archiving natural knowledge, Treub could administer nature authoritatively by administering the collections.

In the 1890s Treub halted the practice of staff scientists at the gardens writing monographs of interest to only a narrow group of European biologists and instead had them produce bureaucratic reports with scientific content. Boerlage, chief of the herbarium from 1896 to 1899 (and before that an informal collaborator of Treub's while still at the Rijksherbarium in Leiden), played an important role. He wrote a genus-level flora of the Indies. This flora was not a revision, and in fact did not include plants new to science, but was meant to serve as an introduction for curious colonial officials. Treub marketed the flora as an indispensable reference book, writing, "Many amongst the doctors, pharmacists, foresters, administrators, and staff at national enterprises, and certainly also under the officials in the civil service and other branches of the bureaucracy, long to expand their knowledge of the surrounding kingdom of varied plant life."[25] Unlike a flora of newly described species, which would be of interest to European scientists, this introductory flora was created in order to introduce other allied professionals to the tropical herbarium in Buitenzorg. This, though, was only the beginning. Treub would spend the rest of his professional life positioning the botanical gardens as an essential component of Dutch colonialism.

One of the first places Treub put into action his wish to place natural exploitation on a scientific footing was in forestry. Since the early nineteenth century, forest exploitation had been a Dutch state monopoly. In the 1850s and 1860s the Dutch government imported German forestry techniques and passed a number of laws regulating and controlling forest usage. After the liberal reorganization of the colonial administration in the late 1860s, the colonial forest

Josephina Koster and native employees at the Buitenzorg herbarium, ca. 1925
(Courtesy of the National Herbarium of the Netherlands, Leiden University
branch)

service was part of the Department of Civil Administration. Control over the forests increased following passage of the Agrarian Law of 1870, with the state laying claim as owner to all forested areas. The sale of teak was also a large source of state revenue.[26] Treub had been involved with the forestry service since 1888, when the young forester S. H. Koorders (who at the age of twenty-one had been the second user of the visitors' lab in Buitenzorg during 1885) started work on identifying the nonteak trees of Java. In the early 1890s this investigation expanded and gained government funding. Koorders—still a member of the forest service—moved to Buitenzorg. Over the next two decades he amassed a huge arboreal archive of Java. In Buitenzorg he and Th. Valeton (a botanical gardens employee paid specifically to help with the determinations) translated this archive into a thirteen-volume description of the forest flora of Java. It was written in Dutch and meant for a bureaucratic audience.[27] Books like this would win both Treub and his botanical gardens authority as centers of agricultural science. Despite the often difficult working relationship between Treub and Koorders, the forester remained an integral component of Treub's emerging scientific discipline.

## Botany and Agriculture

For Treub, shifting his expertise toward agriculture was a way to expand the reach of science, as well as his own personal and institution power. There were few other ways to accomplish this, as science remained weakly institutionalized. To be sure, in and near Buitenzorg Treub consolidated his rule over biology effectively. But moving beyond Buitenzorg was only possible through the tentacles of the colonial state. During the 1890s, when the scope of the biology practiced in Buitenzorg grew rapidly, it did so along routes already established by the government. For example, Treub built on Teysmann's success in exploring the outer islands, occasionally sending out a European botanist or collector with an assistant, starting with Burck's 1882 trip to Sumatra. The botanical expedition, the staple of mid-nineteenth-century tropical biology throughout the tropics and subtropics, could come at a high price. In 1899 Boerlage, then second in command at the botanical gardens, died on the trip back from exploring the Moluccan islands. Instead of the uncertainty and danger of jungle expeditions, Treub looked for simpler ways to extend his reach over colonial nature, which included an eventually unsuccessful attempt to create a field station in New Guinea.[28]

Treub certainly had the ability to amass natural knowledge outside of his gardens, but his power was centered in Buitenzorg. Expanding his influence meant expanding the institutions within Buitenzorg. One way he accomplished this feat was by attracting export-crop experiment stations to Buitenzorg, paid for by the now booming export-crop economy. These experiment stations were staffed by scientists picked by Treub who worked under his leadership. For

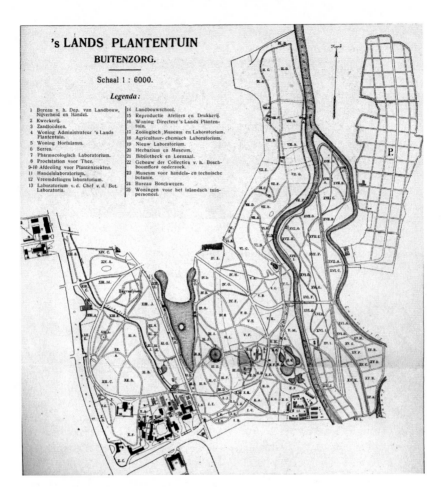

Map of the 's Lands Plantentuin, 1917 (Reprinted from J. J. Smith, *Geillustreerde Gids voor "'s Lands Plantentuin" te Buitenzorg* [Buitenzorg: Departement van LNH, 1917])

example, the Zoological and Plantpathology Museum, founded in 1894, concentrated on research about agricultural insect pests. On the side, it, too, spent a large amount of time collecting and naming specimens, many sent from other parts of the colonial government, with an eye to building an authoritative fauna collection in Buitenzorg to complement the herbarium. Treub was very skillful at procuring private support for new institutes such as the zoological museum within the botanical gardens family. The Netherlands-Indies Agricultural Company and Netherlands Indies Commercial Bank initially underwrote the maintenance of the zoological museum, including the first four years' of salary for its chief, J. C. Koningsberger, and the bulk of the funds needed to construct a museum building, finished in 1900.[29] This private money was usually finite and only available in flush years. But once an institute or laboratory was founded and had proven a success, it was not difficult for Treub to convince the government to take over paying the bills. Koningberger became a civil servant in 1898 and went on to become Treub's right-hand man and chosen successor.

Treub convinced other planters, including those in tobacco, tea, and eventually coffee and indigo, that funding a small laboratory in Buitenzorg dedicated to their crops would be beneficial and cost effective. By the end of the 1890s, export-crop markets had recovered after a twenty-year global slump, and planters' organizations had the ability to fund Treub's experiment stations. Treub would hire and oversee the research, which was done according to his professional standards, with the purpose of supporting the export-crop planters. All the experiment stations were in Buitenzorg, and he would take no responsibility for institutes not located near the botanical gardens.[30] By 1900 Treub held a virtual monopoly on biological research in the colony; not only did he control the best research resources and employ the best biologists, but even the sugar experiment stations, organized outside of Treub's direct control, remained mindful of his scientific authority.[31] By 1902 the botanical gardens encompassed the gardens proper, the herbarium, the library, the photographic laboratory, the botanical laboratory (including the visitors' area), the experiment garden at Cikeumeuh with attached chemistry laboratory, the gutta-percha experiment station at Cipetir, the pharmacological research laboratory, the mountain botanical gardens at Cibodas with laboratory, the tea experiment station, the forest flora project, the coffee research institute, the indigo experiment station, and the zoological museum. Only the sugar planters refused to deal with Treub, and in the 1880s three (and later two) experiment stations dedicated to sugar were instead controlled directly by a board of directors of sugar planters. Although these experiment stations were out of the grasp of Treub, they largely followed the same model of professional science supporting export-crop planters.[32]

## The Department of Agriculture

Treub was not an ideologue of the ethical policy. Nor did he have much previous experience with, or even interest in, developing the native peasant economy. Still, the botanical gardens fit nicely both the ideology and the practices of the ethical policy largely because Treub had already done the work of positioning it as a center of agricultural expertise. After 1901 he impressed a number of the early ethical policy leaders, including the socialist leader Henri van Kol.[33] Van Kol and the Labor Party positioned themselves as proponents of a path to independence for the Netherlands East Indies following a period of political and economic tutelage of the colonial population. In his 1903 book *Uit onze Koloniën* (Out of Our Colonies), Van Kol was critical of many colonial institutions, but he was greatly impressed by the botanical gardens. Even though he missed meeting the director, he represented the gardens in a manner that would have pleased Treub; he noted the history of foreign scientific visits, its systematic approach to studying colonial nature, and its potential for further improving agriculture. He concluded that the botanical gardens showed "evidence of serious study and the loyal service of science for agriculture. It is an institution we can rightfully be proud of!"[34] And Van Kol was not the only impressed onlooker. After 1900, Treub's vision of expanding nature's archive to control all agriculture struck many officials and politicians as an obvious way to rationalize the cultivation of crops in the colony.

The debate about colonial agriculture started with an export-crop planters' organization. In 1899, members of the Malang Coffee Planters Association wrote to the governor general, lamenting the poorly trained workforce available to them. They suggested that the resources of the Buitenzorg gardens would be ideal for creating a school, where young, local European men could increase their knowledge and gain real experience, which would lead to a "more rational practice of agriculture."[35] Everyone, planters and officials alike, loved the idea, including Treub.[36] The Department of Education, Religion, and Industry endorsed the plan, adding the caveat that the school should be for "Europeans without means," perhaps referring to the poor Eurasian class in the Indies.[37] As the proposal gained steam over the next few years, Treub expanded on the initial idea, using it to launch the botanical gardens into a much stronger position.

In the context of the ethical policy, the plans for the agricultural school, an unthinkable expense just five years earlier, quickly gained momentum. Treub readily agreed to run the school on his terms. His proposal focused on educating native elites and training officials in a formal scientific curriculum, including botany, zoology, chemistry, meteorology, bacteriology, and water engineering. Treub argued that after two years of agricultural education—instruction in how to administer the facets of the agricultural economy, not vocational training in running one plantation—these officials could fan across the colony, "advancing native agriculture through word of mouth, influence, and expert instruction." He envisioned students from aristocratic Javanese and Sundanese backgrounds, already in the possession of political authority. He dismissed the idea of training Europeans of limited means. This new class of native floracrats would bring science to the indigenous public. Much like the previous short-lived iteration of an agricultural school, as well as the contemporary medical school in Batavia, Treub's vision would create a class of scientifically educated native administrators that would became a useful interface between the needs of the European rulers and the native population.[38]

The proposal moved quickly, and the school opened in the middle of 1903, largely as Treub wanted it. Treub's superior, the reform-minded director of education, religion, and industry, J. H. Abendanon, overruled him only on the question of poor whites and came up with five government stipends a year, four for European students.[39] The final statutes made a place each year for fifteen European students (following the three-year course) and twenty indigenous students (for two years), the latter drawn from a pool of native official candidates who were proficient in Dutch. Native students would pay tuition only if they were from sufficiently wealthy families. The education would be split between the botanical gardens in Buitenzorg and the demonstration fields at the agricultural garden in Cikeumeuh.[40] J. C. Koningsberger, who was also head of the zoology museum, was appointed administrator of the school, and in mid-1903 it opened with five European and ten native students.[41]

At the same time that the agricultural school was being planned, the government directed other agricultural issues toward Melchior Treub. In 1899 he was asked to locate and prepare special fields throughout Java that were to serve as sites for demonstrating new rice-farming techniques to Javanese peasants. Later indigenous graduates of the agricultural school would become teachers at these fields.[42] The immediate results were not conclusive, as some fields were abandoned, but through 1904, colonial officials were under the impression that Treub's foray into native agriculture had been a success.[43] And it built nicely on his earlier achievements in helping the export-crop planters. In late 1901 or early 1902 Governor General Rooseboom asked Treub to look into the idea of forming an agricultural department encompassing everything that was related to agriculture.[44] This started the process that three years later led to what has become Treub's best-known achievement, the founding of the Department of Agriculture.

In January of 1902 Treub sent a highly detailed "Schematic Memo" to the governor general, proposing the expansion and transformation of the botanical gardens into a Department of Agriculture.[45] His tone in the memo fit within the ethical policy's parameters of cultural and economic reform, but its direct inspiration was elsewhere. The U.S. Department of Agriculture, he argued, was the model for the state taking proper responsibility for agriculture. In particular, he admired how in the U.S. Department of Agriculture scientists decided all administrative, public policy, and scientific matters pertaining to agriculture. In surveying the administration of agriculture in the Netherlands East Indies, Treub pointed to the lack of centralized agricultural planning. But his more serious critique was that department leaders were inefficient and arbitrary users of science; they did not "do scientific research, synthesize the desired knowledge, or spread that knowledge in a practical way." Treub's solution was to create a "technical department," a central institute for scientific research serving agricultural needs. Its tasks would be "observation, experimentation, and collection of data, and technical management of the institutes, organizations, and departments of agriculture in their entirety." With scientific research setting the agenda for agriculture, only a scientist could lead this endeavor, becoming the conduit between science and the government. Administrative work would fade to the background, only serving to "guarantee the implementation of general statutes of state administration" and "maintain an administrative link between the various sections."[46]

Treub's idea of centralizing agricultural research units in one department was greeted enthusiastically in the colony.[47] The contentious issue was its leadership.[48] Among senior officials there was skepticism about Treub's claim that with science in charge administration would be routine. Not only did this depart from the basic legal structure of the colonial state, but even sympathetic reformers could not understand the concept of a colony without an administrative hierarchy. There was also the suggestion that a scientist would make a bad director of agriculture, being too distant from practical matters.[49] Treub responded, explaining that in his department scientific research would be more than a task; it would be the means to create practical and useful results for agriculture. Moreover, Treub argued, there was no opposition between science and practicality: "Trustworthy, practical results *can only be reached* BY APPLYING SCIENTIFIC RESULTS."[50] A. W. F. Idenburg, a reformer frequently credited with being the first ethical minister of colonies, agreed, even going against advice from the Dutch Raad van State (State Council), arguing that the unique, technical functions of the new department would be carried out far more efficiently if the department were run along Treub's lines.[51] Idenburg brought the department proposal to Dutch Parliament largely following Treub's "Schematic Memo."[52]

Starting on June 2, 1904, the Dutch Parliament debated colonial science for the first time since it had taken up Junghuhn's handling of the quinine project forty years earlier. At issue was not the question of expanding the colonial state's responsibility over agriculture, which was widely supported. The debate was about who was most qualified to provide agricultural leadership. Minister of Colonies Idenburg ushered the proposal through the two houses of Parliament. In the lower house, socialist member Henri van Kol, who just a year earlier had celebrated Treub's botanical gardens, spoke first, railing against Treub's bid for increased power. Van Kol wanted agricultural expertise installed within the Department of Civil Administration, whose officials had their feet firmly planted in the Javanese soil and would do more for the Javanese peasants. For Van Kol, the plan for the "department-Treub," formulated in semisecret by Treub and circulated only between the governor general and the minister of colonies, did not have the interests of native peasants at heart. According to Van Kol, Treub's lust for power would lead to greater scientific consolidation and would not help the millions of peasants in need of practical leadership.[53] Antipathy to Treub, though, was not the rule, and the next day three parliament members and colonial experts defended Treub, the botanical gardens, and the proposal for a department of agriculture.[54] A week later Idenburg and Van Kol debated the matter once more, with Treub again the point of contention. Idenburg was here well-served by the extensive vetting of this issue in the colony and had ready talking points available that supported Treub's dedication to practical matters such as agriculture.[55] The proposal passed easily with seventy votes for and ten against. In the upper house only former governor general C. H. A. van der Wijck spoke against the proposal, arguing that Treub would not be proficient at handling economic matters. Idenburg ceded this point but felt that members of the civil administration would continue to have input in implementing agricultural policy.[56] The resolution passed with only two votes against. Later in 1904 Treub was appointed the new director of agriculture, starting January 1, 1905.

## Agriculture and the Economy

Treub ascended to a powerful colonial perch in 1905. Science and government were more closely intertwined than they had ever been before in the Netherlands East Indies. To both Treub and his critics, it appeared that professional scientists were now in control of the colonial state's agricultural administration. Treub was optimistic. His vision, expressed in the "Schematic Memo" and subsequent documents, underlay the department's design, which put scientists in real leadership roles. His scientific authority among colonial elites was at its peak, and he believed it would now spread to native society. And, indeed, for the remaining decades of colonial rule the Department of Agriculture, and the professional experts who staffed it, became a powerful new arm of the colonial administration. But even before Treub left the colony in 1909 the scientific authority of the botanical gardens and its research units waned. Scientific research itself was meaningful only when it satisfied other colonial policies, in this case the improvement of the native economy.

Treub started strong. The administrative transition from the botanical gardens to the Department of Agriculture went smoothly, with Treub taking over management of administrative units previously in other departments and then over the next year consolidating their tasks into scientific divisions within the department.[57] And Treub moved quickly to soothe concerns about the department's new responsibility for indigenous agriculture; for 1907 he reported cooperating with members of the civil administration in spreading information about new and improved cultivation techniques.[58] By then he could point to the expanding agricultural school. And his plan of spreading indigenous government officials trained at the botanical gardens around the colony was becoming a reality. Eight students had graduated and gone on to become *mantris* (technical supervisors), at demonstration fields scattered around Java. And seventeen new students had started the two-year course.[59] First-year students received instruction in botany, zoology, microscopy, chemistry, physics, general agriculture, plant cultivation, veterinary science, and geography (totaling eighteen hours per week of class time), as well as two hours a day of practical lessons in the experimental fields.[60]

Treub's challenge, one he ultimately failed to surmount, was connecting the department's professional science to native peasants. He employed a top-down approach, planning on bringing scientific thinking to the peasant via native elites. The agricultural school only recruited from the native aristocracy and bureaucracy. The education of the native elites as floracrats, and their dissemination across Java, was Treub's primary vector for spreading agricultural science to the native public.[61] Properly educated native experts would demonstrate the utility of rational, scientific methods of cultivation to communities across the colony. He rested this policy on his theory that Javanese peasants as individual cultivators did not then conform to any generalized system of rice production, and that in any case, Javanese peasants had "a total lack of economic awareness."[62] There was thus no reason to query how Javanese peasants grew rice. He trusted that only traditional leaders could properly persuade peasants without economic awareness to overhaul their farming practices.[63] In essence, he trusted that native elites trained in the sciences would succeed in persuading native farmers to change where the Dutch had not.

Treub's disparaging views of peasant economic abilities quickly ran him afoul of other officials. After 1908 he was almost continuously on the defensive, especially about his plans for developing native agriculture.[64] His opponents blamed him for moving too slowly and accused him of pursuing idealistic goals rather than implementing pragmatic policies. A 1908 report by H. E. Steinmetz about the state of the indigenous economy strongly criticized Treub for keeping the scope of native education so limited.[65] Steinmetz further objected that Treub's idealism had not provided immediate improvements to farming. Steinmetz and others believed there was a need for close contact between agricultural consultants and Javanese peasants. The Department of Agriculture should build on native farming practices rather than teaching new, foreign techniques.[66] Unlike Treub, his critics believed in a native economy, one that could be systematically engaged. Agricultural experts in the Netherlands and the colony complained that Treub by his own admission ignored economic issues and thus had failed to formulate meaningful policies to improve the native economy.[67] In 1909 H. J. Lovink, then the director general of the Dutch Ministry of Agriculture, contended that department officials needed to familiarize themselves with Javanese farming practices and "to increase together with the Javanese farmer his rice yields economically, taking into account his development, workforce and his capital."[68] Writing much later, Koningsberger admitted that in 1909 Treub had failed to anticipate that in addition to its technical-scientific work the Department of Agriculture needed to engage with economic problems.[69]

In the first half of 1909 Treub retired and returned to Europe to concentrate on botanical research.[70] At the end of 1909 incoming governor general Idenburg appointed Lovink, who had strongly criticized Treub's policies just months earlier, as the second director of agriculture.[71] The transition from Treub to

Lovink ended the strong link between professional biology and agricultural science at the botanical gardens. The scientific world of Buitenzorg was not dismantled, but no longer did biological science engage directly with the colonial bureaucracy. Practical-minded agronomists, many of them graduates not of the department's agricultural school but of the Wageningen Agricultural School in the Netherlands, would be the official government interpreters of Javanese nature and economy. After 1910, various economic sections transferred from other areas of the government into the department, which was renamed the Department of Agriculture, Industry, and Trade.[72] Lovink established an agricultural extension service through which Dutch agronomists taught peasants on demonstration fields.[73] Under Lovink the department was far more successful at building an extension service that created close contact between the colonial state and rural Java. It did so largely by creating an institutional dialogue between the department's experts and Javanese farmers.[74]

After 1910 Treub's hubris—leading agriculture with research science!—seemed so misplaced that commentators on the Department of Agriculture, including many of Treub's own protégés, found it hard to imagine that Treub had believed it himself. Surely Treub had done all this simply to defend the bastion of scientific research at Buitenzorg. But evidence from the last few months of Treub's life (he died shortly after returning to Europe) suggests that he remained wedded to the idea of professional biology leading agriculture. A letter written on September 22, 1909, the day after he heard that Lovink had been selected to replace him, succinctly (and angrily) laid out his ambitions of the last few years: "The entire organization of the Agriculture-Department was founded on the notion: that rational improvement in agricultural production can only be reached on the basis of applied scientific research."[75] While steaming between the colony and Europe in November of 1909, Treub wrote a 150-page tract defending the department's contribution to practical agricultural and economic matters. Treub held it as axiomatic that in a colony with no universities or other research institutes any department of agriculture must itself be a scientific institute. Treub wrote that Lovink, a stranger to colonial mentalities and not an academically trained scientist, would be incapable of providing effective leadership either to the scientists in Buitenzorg or to the native population. He ended his book with the prediction that with Lovink in charge the international scientific reputation of the botanical gardens would sink.[76]

The appointment of Lovink was a blow to Treub, who was understandably bitter about what he saw as the governor general's betrayal. But it did not lead to the downfall of science in Buitenzorg as Treub predicted. Government science, including science at the botanical gardens, thrived in the 1910s and 1920s, much of it inside the Department of Agriculture, with both applied and pure scientists coming from the Netherlands for a scientific career in the Indies. Moreover, the ethical policy, by setting a benchmark of success—in this case the effective development of the native economy—was able to nurture a more efficient and credible role for the bureaucracy. From the colonial administration's perspective, Treub's policies and direction had not been satisfactory, and they decided to go in a new direction with his successor. And the fact that Idenburg was able to lure Lovink, at that time the director general of the Dutch Ministry of Agriculture, to the colony shows the prestige and credibility of the ethical policy. The colonial bureaucracy, because of its defined mandate to exercise political leadership, had been able to replace Treub, who was not seen as connecting effectively with the population, with a leader who could connect economically and politically to peasant society.

### Conclusion

In this chapter I analyze the process by which the ethical policy animated the change and growth of the Dutch colonial bureaucracy, providing it with an argument for broad political leadership. After 1901 the ethical policy led not just to the growth of state power but to a qualitative change. Most scholars have seen the growth of the colonial state after the adoption of the ethical policy as evidence for the official implementation of the Dutch civilizing mission. Scholars have disagreed about whether to treat the Dutch civilizing mission as a serious attempt by Dutch officials to uplift the people of the Indies or, cynically, as a means for the Dutch to gain greater power over the colony.[77] That both aspects of the ethical policy could exist in one individual is evidenced by Melchior Treub. Through his creation of the Department of Agriculture, he believed sincerely in the positive influence of scientific knowledge on the Javanese peasants, yet he was also desirous of expanding the power and influence of scientists, himself foremost. Both the missionary impulse of improving native welfare, education, and prosperity and the creation of new colonial policies and institutions were part of the history of the ethical policy.

For the historian the wider significance of the ethical policy was that with a mandate for exercising political leadership the colonial state expanded its ranks of colonial officials, who did more than just administer the land and population of the Indies. New professionals joined the state, including scientists, lawyers, engineers, teachers, and social scientists, staffing new departments within the colonial bureaucracy. The long-term result was that the residents of the civil administration were no longer the demigods of the colonial state, wielding near autocratic power in their districts. Other experts, officials, and institutions infringed on their prerogatives. Because of the ethical policy, the governor general and other central officials had more and more levers of power.[78] In particular they now commanded a whole host of new technical, educational, and welfare agencies anchored by the Department of Agriculture. In these technical departments of the expanding state, scientists and other experts became handmaidens of state policy as set by senior officials in the colonial bureaucracy.[79] These men were not held down by the orientalist tradition of their colleagues in the civil administration and sought to administer colonial society in line with the European professions with which they were familiar. As a result, although the officials of the civil administration remained the face of the colonial government for much of the colony's population, especially outside of Java, in Batavia and Buitenzorg a new professional bureaucracy established the colonial state's broader political direction.

The state initially added European experts and professionals, but simultaneously it began to train and co-opt younger native elites who came to work alongside the European professionals. No longer would the relationship between the assistant resident and the Javanese regent, so memorably described by Multatuli in 1860, anchor colonial power. Graduates from the Department of Agriculture school, for example, entered the agricultural extension service, where they served as an interface between Javanese peasants and other European officials. This was duplicated in other areas, such as medicine, but eventually also spread to political administration. It was in this fashion that the ethical policy nurtured the first generation of native, professional administrators, most of whom had begun their colonial careers in one of the colony's professional schools. Their experience, and in particular the tension between their Enlightenment ideals and their professional service, forms the subject of the next chapter.

# 5

# The Nationalists' Enlightenment

Since the 1840s, generations of apostles of enlightenment in Indonesia have dreamed of using science as a way to provide knowledge that would help the land and its people. In the nineteenth century these apostles of enlightenment were all Europeans. In the twentieth century, Western education for native inhabitants created a small group of intellectuals who came to believe that science was not only useful but also had the power to allow natives to catch up with the West. At the same time, as part of the ethical policy, new professional careers opened up for these native apostles of enlightenment, initially as colonial officials but increasingly as entrepreneurial and political independent intellectuals. For the children of the native elites, no longer was a position in the native bureaucracy the only professional career option. Starting with a cohort of students who attended the medical college in Batavia in the first decade of the twentieth century, student apostles of enlightenment received scientific training and took up careers as doctors, teachers, and engineers. Even when political differences emerged after World War I, these intellectuals shared a common history in their systematic commitment to promoting science as a means of progress. Graduates of the medical college, including Soewardi Soerjaningrat, Tjipto Mangoenkoesomo, and G. S. S. J. Ratoe Langie, believed before and after the war that expanding Enlightenment methods and knowledge would lead to progress for the Indonesian people. This progress would in turn strengthen their position vis-à-vis the Dutch. In fact the apostles of enlightenment failed to lead Indonesia to independence, which was achieved in 1949 after a military power contest between Indonesians and the Dutch. Nonetheless, the dream of Indonesian professional science originated in this prewar movement and remained significant for the organization of science under the independent Indonesian Republic.

Until 1918 there was an expectation among the emerging native professional class that the Dutch colonial state would aid them in expanding scientific education to more Indonesians and hence broaden the authority and meaning of Western scientific knowledge. By the end of World War I, however, the extent to which the colonial government had manipulated and exploited Western-educated native elites became widely apparent to the native apostles of enlightenment. The naïveté they had held about their seeming benefactors, the colonial state, dissolved. They saw that their goal of leadership and enlightenment for the people must inevitably be a contest for power. At this low point of demoralization for the apostles of enlightenment, they lost momentum in their quest to foment Enlightenment knowledge. The political initiative passed to the social revolutionaries, including those influenced by communist and Marxist ideas, whose ranks had been growing over the previous years.[1] Still, even as the nationalists who saw the conflict with the Dutch as a contest of power became the voice of Indonesian nationalism, by the mid-1920s the apostles of enlightenment had regrouped as professionals who had lost their naïveté about colonial power but remained wedded to an enduring idealism. Enlightenment as a means of empowerment now would need to be achieved without government support. Professionals turned to teaching and publishing about science as a way to spread Enlightenment knowledge to the Indonesian masses.

The achievements in the 1920s and 1930s in spreading Enlightenment thought were modest. Two generations after native intellectuals began spreading their Enlightenment dreams, science was still an elite pursuit with professional science holding little more authority among the Indonesian population than it had at the beginning of the twentieth century. I will show in this chapter that while many of the educated and professional apostles of enlightenment believed science was important to the Indonesian people, they never successfully changed the identity of science as an elite pursuit. Consequently, whether as traditional authorities or colonial officials, the apostles failed to spark a popular Enlightenment. Moreover, in order to survive in this corner of colonial society, outside of radicalism yet also outside the goals of the state itself, the Indonesian science advocates stripped themselves of all politicalization. But removing the political goals from their program caused the overall meaning of and authority for scientific knowledge to become vague. The Indonesian nationalist movements that opposed the Dutch during the 1930s and 1940s, whatever their inspiration, all had clearly defined political agendas that led to complete independence from colonialism. As a result apostles of enlightenment played only a small role in the decolonization of Indonesia.

## Science and Leadership

Let us go back to 1908, to Boedi Oetomo, an advocacy group meant to promote the leadership of Western-educated Javanese elites. This organization of native civil society, the first conceived in a European institutional form, is often credited with being the first nationalist organization in Indonesia.[2] Boedi Oetomo was founded in 1908 after a handful of students at the School for the Training of Native Physicians in Batavia, known as the STOVIA, were influenced by the ideas of Wahidin Soedirohoesodo, who had been publicly arguing that scientifically trained Javanese elites could bridge the cultural worlds of the Dutch and the Javanese. He saw a future in which they would lead the colony and its native inhabitants toward greater participation in the modern world. Knowledge of science was central to Wahidin and his followers' arguments for the existence of Boedi Oetomo. And most of Boedi Oetomo's early members were either students at or graduates of the various colonial schools, including the STOVIA and the agricultural school run by the Department of Agriculture. R. A. A. Tirtokoesoemo, a graduate of the earlier iteration of the Buitenzorg agricultural school (founded by Scheffer in the 1860s), became the first president of Boedi Oetomo. The Sanskrit-derived name Boedi Oetomo was from the beginning translated by its creators as "Het schoone streven" in Dutch or "Beautiful Endeavor" in English. But as early as 1910 a commentator in the *Indische Gids* noted that an alternative translation of Boedi Oetomo was "knowledge is power" or "reason above all."[3] Boedi Oetomo's desire to lead native society by awakening it through Enlightenment thinking was hardly a secret. To be sure, this was to be a beautiful endeavor in that the steady and orderly hand of the Javanese elites would oversee it, working within the Dutch colonial system. It began as a professional organization for educated Javanese elites. Tjipto Mangoenkoesoemo, a graduate of the Batavia STOVIA, attempted in 1908 to turn Boedi Oetomo toward politics, but this idea was quickly rejected by other members. Although in the 1910s Boedi Oetomo began the transition to a more formal political institution, its politics were supportive of the status quo, certainly compared to the new parties and movements of the social revolutionaries.[4] It was a model of a modern institution meant to advance the interests of its members inside the colonial system. Still, it was the first, and for many years the only, institution that advocated expanding Enlightenment knowledge as a way to empower the colony's native inhabitants.

At the time of Boedi Oetomo's founding, there was broad agreement among the younger generation of native elites that a strong dose of Western knowledge and education was necessary if they were to meet the challenge of Dutch colonialism.[5] These young apostles of enlightenment, who had themselves received a Dutch education, believed traditional Indies cultures needed to be augmented with Western learning. Tjitpo Mangoenkoesoemo explained in 1911, "What the Javanese would really like, has moreover been demonstrated frequently and clearly. He feels only too well that from his original Eastern culture—no longer does one deny the Javanese his unique culture—something is lacking, and he cannot stand out, indeed cannot do justice to himself. That something is the development of the intellect, which precisely among various

Western peoples has unfolded in such an admirable way."[6] Education was expected to lead to a future of *kemajuan*, one of the watchwords of this period, meaning "progress and development." Progress via education was second nature to these apostles of enlightenment because most of them had personally advanced through a Western education. The editor of a new journal in 1916 devoted to kemajuan for the Javanese was most concrete when he got to educational policy: "We shall insist that education for our people be arranged in the best possible way so that our people will not be forced to travel to another country for their education."[7] The first debates inside Boedi Oetomo were about the content of an expanded and reworked colonial education.

Boedi Oetomo's first success, shortly after its founding, was convincing the colonial government to incorporate the Dutch language in the first-grade curriculum of the native Indies schools.[8] And it was debates about education that dominated the second congress of Boedi Oetomo in October of 1909. Dutch observers, and apparently Javanese audience members also, found this discussion much less satisfying than the exciting intellectual discussions about building toward modernity, which had been the center of the first Boedi Oetomo congress.[9] And indeed the second congress consisted of carefully scripted speeches about educational policy. Although the importance of mass education was raised, the focus was on proliferating leaders by, for example, expanding the number of young Javanese who studied in the Netherlands.[10] The wish that Tjipto expressed just before the congress started, that Boedi Oetomo would become a sun whose gift of warmth and light would end centuries of lethargy, encapsulated the thrust of this congress, namely, that Boedi Oetomo was to train the future apostles of enlightenment, who could then give others the gift of science without aid from the Dutch administration.[11] In 1912 E. F. E. Douwes Dekker, then a newspaper and magazine editor, also called on the government to expand higher education opportunities by creating a university in the colony.[12]

The apostles of enlightenment continued to link kemajuan to Western science. Of the few who had an opportunity to study in the Netherlands, the emphasis was on personal development, like "mastering the Western sciences" so as not to fall behind European colleagues, as the then student Ratoe Langie argued in 1916.[13] Soewardi Soerjaningrat added that students needed to immerse themselves in European culture broadly, both for their own development and for the broader struggle.[14] In 1918 F. X. Prawirotaroemo, also a student in the Netherlands, wrote, "Sound education is a powerful means for progress and prosperity. . . . For those who already work, may they continually expand their knowledge and science. It benefits themselves and their kin. Our beloved fatherland numbers 30,000,000 souls. How many Javanese doctors, lawyers, architects, physicists, chemists, etc., devote their energy to the uplift [*opbloei*] of our people?"[15] Students such as Soewardi and Prawirotaroemo were in the Netherlands not just for their own career advancement but also to be trained to become future apostles of enlightenment. As such, immersing themselves in Western culture was necessary if they were to gain access to Western science and technology. This was not in doubt, although Soewardi at the education congress in 1916 argued that once he and other students had received enough training the Dutch language would be replaced with either Javanese or Malay, even for Western subjects.[16]

Starting in 1911 and 1912, the apostles of enlightenment were challenged as civil society leaders by new political institutions, in particular Sarakat Islam and Indische Partij. The leaders of these organizations, known as *pergerakan* (the movement), sought to put colonial society in motion with the goal of moving the colonial inhabitants toward modernity through mass membership parties.[17] No full-fledged social revolutionary movement emerged, however, and many of the leaders of the pergerakan continued to be close to the Dutch colonial administration. The creation of the Volksraad in 1918 formalized an arrangement, by then under development by the Dutch administration for some years, under which the leaders of the officially recognized parties were treated as civil society leaders. That this was a Dutch colonial ploy was widely understood at the time. But it was sufficiently tempting to many native leaders, including those from the pergerakan, who saw a seat in the Volksraad as an excellent way to promote their own political agenda. Still, eight of the ten elected native members worked for the colonial government.[18] Most social revolutionaries stayed out of the Volksraad, not only because of their ideology but because they did not think that cooperating with the colonial government showed proper leadership of the native political movement. And even among those who participated in the Volksraad, the issue of leadership remained important, as revealed in a debate between two apostles of enlightenment, Soetatmo Soeriokoesoemo and Tjipto Mangoenkoesoemo, which emerged out of the elections for seats in the first Volksraad. Soetatmo and Tjipto agreed on much, including the superiority of Western science. And both agreed that progress, kemajuan, emerged out of education. Leadership styles distinguished them. Soetatmo argued for a Javanese movement in which sages of Javanese culture such as himself would serve as moral beacons, rearing Javanese society culturally through education. Tjipto cast leadership as the ability to lead the struggle against Dutch power. But even Tjipto, voicing the social revolutionary argument, did not deny the importance of Western science and education.[19] And he did not turn down a seat on the Volksraad when the governor general offered him one.

## Educational Leadership

What was educational leadership for these native apostles of enlightenment? In the first years of Boedi Oetomo, the answer had been about trying to influence the Dutch system of colonial education. But Boedi Oetomo's early success at effecting some small rule changes was not repeated. For example, the Dutch intellectual establishment in the colony, not just colonial officials but also most of the Dutch newspaper editors, were against a colonial university, citing financial, enrollment, and faculty recruiting issues. They instead supported expanded vocational schooling.[20] The apostles of enlightenment had more in mind than technical training. And as their ranks grew, while the Dutch system repelled serious change, more apostles of enlightenment envisioned leading through their own teaching. Noto Kworo, a young Boedi Oetomo member studying in the Netherlands, laid out his plans in 1918. He argued that the education (and regulation) of midwives be expanded under the care of Javanese doctors such as himself. Noto Kworo claimed that in the Indies "obstetric science as such does not exist. Its practice rests in the hands of healers [*doekoens*], who have not had a scientific education, base their knowledge on experience, acquired through their own pregnancies and births, or those of others."[21] After further expanding on what he saw as the inadequacies of Javanese midwives, Noto Kworo suggested Javanese medical doctors could educate midwives in the techniques of Western science and medicine. He further proposed that all midwives be registered with the state, that they receive a government salary, and that a midwife school be established. He concluded by suggesting that practicing Javanese doctors train local women in obstetrics and other medical practices, including the use of forceps.[22] Writing while still in medical training in the Netherlands, Noto Kworo envisioned his future professional career to include spreading the benefits of European science and medicine to Javanese society once he returned home.

Noto Kworo's notion of educational leadership, of bringing Western knowledge to the people of the colony, depended on a school setting. Throughout the 1910s, Boedi Oetomo's members believed the colonial government should take the initiative. As late as 1916, Soewardi, who a few years later founded the Taman Siswa schools without government aid, argued that the colonial government owed it to its inhabitants to provide this education.[23] And educational opportunities did greatly expand in the 1910s, most notably with more elementary schools staffed by native teachers. For Boedi Oetomo's members and other apostles of enlightenment, expanded primary education was not sufficient, as it did not lead to spreading science. During the first session of the Volksraad in 1918, Boedi Oetomo's members repeatedly raised the issue of the meager results of Western education. Wediodipoero Radjiman criticized the results of the agricultural school and the shallow penetration of Western medicine. Colonial education, he argued, did not lead to greater power for the native population:

> Does not the state have the responsibility to facilitate the orderly existence of society and to create the opportunities, through which society and individuals, who comprise it, to reach their highest development? . . . What is a person without intellect? Such a person is always the tool of someone else, and woe if

that person does not mean him well. So it is with the native population; the intellect of the masses is absent; the opportunities to attain that power are only sporadically available, and oftentimes presented so one-sidedly, that they are more confusing than formative. I mean that education is inadequate and does not fit with the native character, is not uplifting, but obscuring.[24]

Dr. Radjiman stepped back from calling this oppression (after mentioning that this might be appropriate), but he gave clear voice to his suspicion that the Dutch were intentionally tailoring education to prevent native uplift. The next day Boedi Oetomo member S. Sastrawidjono noted that the native intellectuals, in the struggle for the development of their nation, faced difficulties because of the lack of universities in the colony: "We natives have not had the opportunity, at least not as it should be, to develop these forces. For achieving this development, the best means is familiarity with the various sciences. But the opportunity to obtain this knowledge, i.e., through an institution of higher learning, regrettably does not yet exist."[25] Boedi Oetomo member Aboekasan Atmodirono added, "Now that the government has proven incapable of sufficiently providing for education and public health, it will be necessary to strongly support private initiatives."[26]

Indeed, intellectuals turned away from lobbying the state for educational change and began sponsoring private educational initiatives. Some new schools were built within the colonial curriculum and continued to receive support from the colonial state.[27] But after 1920 there were increasing numbers of schools, called "wild schools" by the Dutch, independent of both the state's curriculum and its funds and hence outside the influence of colonial policy. The most famous of these were the Taman Siswa schools founded by Soewardi in 1922, of which Sutatmo, a veteran Boedi Oetomo leader and Tjipto's interlocutor in 1918, went on to become the first president. In the Taman Siswa schools (in 1934 there were 172 branches), teachers were instituted as apostles of enlightenment under the schools' policy that teachers were to devote themselves to the people and their national aspirations.[28] Soedyono Djojopraitno, Taman Siswa's chief educational theorist in the early 1930s, compared the ideal teacher to the sun, "which shines in all directions and gives strength to all living things."[29] Although education would need to develop along traditional lines, he identified Western science as critical in the individual Indonesian's quest to acquire knowledge and wisdom. This education was distinguished from colonial schooling, which sought to create loyal colonial subjects.

For intellectuals in the 1920s, it slowly became apparent that the government would never really give them access to the power of science. Optimistic pronouncements about the progressive power of state-sponsored science, such as those by Noto Kworo about medical doctors training midwives with the state's blessing, disappeared. And with the pergerakan in motion, which in the early 1920s included strikes and other agitations against colonial power, many looked to the social revolutionaries to attain greater autonomy and power. As the apostles of enlightenment began to analyze their weak position, they devised measures to reassert civil society leadership. Creating social progress and development, kemajuan, now meant forging ahead without Dutch colonial support while remaining in touch with Western science. Boedi Oetomo member R. Sastrowidjono put the problem this way in a 1918 speech in the Volksraad:

> Within the political arena, Western democracy comes into contact with much that does not agree with the emotions and thoughts of the Javanese people. If this people had consciously realized its great past in the areas of the arts and sciences, we could patiently await the combination of Western democracy and Eastern culture. Now, however, I believe that the Javanese people are forced, under the pressure of the times, to acquire the Western sciences, so that they shall be able to take an honorable place in the family of civilized people. While their own culture still lives on unconsciously within the masses, means must be created to balance the Western and Eastern traditions. . . . Therefore Boedi Oetomo will propose that it is necessary to offer mass education on a broad scale, which will allow the people to be more capable of choosing the means necessary for advancing their living conditions. And the neglect of education, which means Indië is still filled with illiterates, is not the fault of the leaders of the national movement.[30]

Sastrowidjono went on to point to the need for increased education in both the Western and Eastern traditions. Even for members from Boedi Oetomo, this question about education, and how to expand it, had a renewed urgency as it became clear the government did not have the same political interests.

Intellectuals found out quickly that the Volksraad would not advance native power. Members' speeches were easily ignored by the Dutch colonial administration. As a consequence, by 1925 the Volksraad had ceased to be a center of nationalist politics.[31] Throughout the 1920s many apostles of enlightenment remained committed to kemajuan as they tried to fuse Enlightenment and education to the mass activist energy of the social revolutionaries. Kemajuan now meant self-education, self-help, and self-development. Indicative of this is Dr. Soetomo, who participated in the founding of Boedi Oetomo, mentored the founders of the nationalist Perhimpoenan Indonesia in Holland, and started the Surabaya Study Club in the 1920s. In the study club he emphasized the need for self-development of each individual Indonesian without outside help. Scholars have rightly pointed out that this meant a turn toward indigenous cultural strengths, although Western enlightenment—Soetomo was a teacher at the medical college in Surabaya, after all—by no means disappeared.[32] Malay, not Dutch, became the language of the apostles of enlightenment in the 1920s, following in the footsteps of the pergerakan, which had begun publishing journals in Malay a decade earlier. And increasingly the apostles of enlightenment were using the language and techniques of social revolutionaries, including publishing their own Malay-language publications.[33] The language, and threats, of social revolutionaries also appeared more frequently in calls for increased government education. One wrote, "In fact, it is reasonable and necessary for every government to enforce law and order, and also it has an obligation to organize and create development [*maju*] in land under colonization. And because it is the case that the Netherlands has already ruled our Indies for more than 300 years, it is proper and appropriate that we, children of the Indies, should be capable of reading and writing, and capable of ruling our own country, with the disciplines of Economics, Administration, Medicine, and others."[34] Still, the author quickly backed away from this radical claim and ended by asking the colonial government to provide more education. And in fact throughout the 1920s the apostles of enlightenment remained close to the Dutch administration.

## Official Scientific Nationalism

By 1910 the Netherlands East Indies state was adept at absorbing cultural innovations emerging from civil society. In the first decade of the ethical policy, innovations had largely come in the form of European challenges to the colonial status quo. This threat of Enlightenment idealism had mostly been neutralized by 1910, when room was made for these reformers to work within the colonial administration. As the early history of the Department of Agriculture shows, this transition was largely successful for the state, which then had access to colonial peasant society via technical experts. Moreover, Dutch associationist politics, an idea that originated with the orientalist C. Snouck Hurgronje, had led after 1900 to a policy of enlisting that potentially threatening group of native elites with a European education to help govern the colony.[35] This was coupled with expanding native educational opportunities under ethicist J. H. Abendanon.[36] Although the associationist ideologies became less and less important after World War I, this did not end the cooperation between professional native elites and the Dutch colonial system. Many native elites continued to be recruited into the colonial government through educational opportunities. And, if anything, in the 1920s more native elites gained a Western education and entered the colonial service than in the previous decade. What had changed, on both sides, was the political explanation. Native apostles of enlightenment continued, in many cases, as employees of the colonial state. But even the "conservative" Boedi Oetomo could see this was a professional opportunity only and was not likely to result in future political dividends. Nonetheless, the apostles of enlightenment were still years away from tangible accomplishments; for much of the 1920s they were out of the limelight, trying to build educational and publishing ventures outside of, or on the fringes of, the colonial administration. Instead, the political initiative among national leaders passed to those social revolutionaries who had been wary of cooperation with the government—and whose method of staging mass meetings and strikes politicized peasants and workers in the years between 1918 and 1926—and created an ideological movement built around mass-membership political parties.[37]

Overt, public, and unequivocal opposition to the Dutch colonial state was not the choice of most native elites, who saw their professional future in a continued Dutch colony. Nonetheless, after World War I the Dutch state's colonial ideology was incoherent, with the demise of the ethical policy and associationist policies. What replaced it by the late 1920s was a more naked statement of Dutch intentions to stay in the colony, most prominently expressed by the Vaderlandse Club, a political party founded in 1929, and throughout the 1930s supported by Dutch inhabitants opposed to increasing political autonomy for the Indies.[38] To be sure, the Vaderlandse Club was not synonymous with the colonial state, and the Netherlands East Indies continued to recruit local elites for administrative help. Still, after 1927 the state settled on a paternal relationship with native officials, who as functionaries had a Dutch education but worked inside the local cultural traditions.[39] The Dutch would tolerate and even encourage officials' local cultural and political authority under the ultimate and permanent protection of Dutch officialdom. The state's strategy of cultural relativism, which envisioned a native society that would remain forever outside of science, implemented policies according to local traditions, or *adat* customs.[40] This created what the historian Robert van Niel called a functional elite whose members sought Western diplomas and government careers even as the pressure to remake the colony along Western lines faded.[41] The functional elite was composed of technically trained professional officials whose value rested on their cultural relationship with local peoples while they could function within the Dutch administration. As pointed out earlier, these native elites wanted to use their Western education to change the native population and lift their countrymen into modernity. What we see is that the Dutch administration wanted to use them as a permanent interface between two cultural systems that would never change and never meet.

It was in the 1920s that the Dutch colonial state began to change the content and strategy of what Benedict Anderson has called official nationalism.[42] The Dutch official nationalism of the 1920s and 1930s was an ideology that envisioned continued Dutch control of the colony even as it created the illusion of a modern nation. It legitimated colonial rule and also created a mechanism for bringing native professionals into the government. Expanding educational opportunities for natives was critical to the functioning of official nationalism, as this outfitted the future functional elite with a professional education, which would promote official nationalism.[43] Moreover, elementary education in native languages would help bring official nationalism to the masses. H. Colijn, whose 1928 book *Koloniale vraagstukken van heden en morgen* (Colonial Problems of Today and Tomorrow) summarized the conventional wisdom of colonial politics in the 1920s, saw a special role for education even as he advocated keeping the Indies population in their traditional institutions.[44] And in fact schooling and schools with state support expanded in the 1920s and 1930s.[45] Diplomas from Dutch-language schools remained necessary for the best state jobs, and after 1921 it actually got easier for native students to stream from a vernacular school into a Dutch secondary school by means of a bridge school, that is, a transitional school between vernacular-language schools and Dutch-language schools.[46] The Dutch investment in Dutch-language schooling after 1920 led to a greater number of functional elites trained to be links between Dutch officialdom and native society.

An important component of educating more functional elites was the creation of institutions of higher education in the colony, which came to provide professional training to elite Indonesians. In 1920 a consortium of trade and agricultural entrepreneurs formed the Faculty of Technical Sciences in Bandung, although initially it was developed as a trade school with only one degree curriculum, civil engineering. The government took over the school in 1924, and henceforth graduates received a certificate equivalent to one from the engineering school in Delft. It awarded its first doctorate in 1930.[47] In the 1920s only a handful of non-Europeans attended, but the future nationalist Sukarno earned his engineering diploma in 1926.[48] In 1924 the government officially opened a law school in Batavia, whose graduates could and did enter the bar and were able to practice law in the Indies. And in 1927 the government reorganized the STOVIA medical school in Batavia into a Faculty of Medicine with a curriculum roughly equivalent to those of its Dutch counterparts. All these schools were under jurisdiction of the Department of Education and Religious affairs.[49] These institutes of higher learning, encompassing engineering, law, and medicine, were meant to train the elite ranks of professional experts in the colonial bureaucracy without the cost and difficulty of arranging for study in Europe. Most students were Europeans, and all the professors were, but a small number of elite native students studied at these schools in the 1920s and 1930s.

After World War I, colonial schools, from the primary level through higher education, were an important site for crafting official nationalism, and these schools began to associate themselves with national symbols expressed in Malay or Indonesian, though of course on Dutch terms. The evolving agricultural schools are an example of this. Agricultural education, begun tentatively by Treub, was expanded after Lovink became the department director in 1910. The department turned its collection of irregular agricultural courses into an agricultural high school in Sukabumi. The department added another agricultural high school in 1918 in Malang. All these schools provided trained personnel for both the government and private sides of colonial agribusiness, although native graduates were streamed toward positions in the department's own extension service.[50] Other agricultural schools were founded on private initiatives but with government subsidies, and by 1915 there were eight agricultural high

Students studying in the laboratory at the Agricultural College in Buitenzorg, ca. 1910 (Royal Tropical Institute/Collectie Tropenmuseum, Amsterdam, coll. no. 60041684)

schools on Java. The curriculum included basic biology classes augmented with practical education about cultivation techniques and technologies.[51]

At the same time the Agricultural College, established in 1903, was relaunched. It remained open to both Europeans and non-Europeans with a high school diploma. J. C. Koningsberger, the director of the school under Treub, was replaced by W. G. Boorsma. Under Lovink, who had overseen the expansion of agricultural education in the Netherlands in the early 1900s, the Agricultural College was modeled on the Dutch agricultural colleges in Wageningen and Deventer. Article 1 of the schools regulations announced that "with all the education the training for practice is at the forefront."[52] The department made more resources available to the Agricultural College, including appointing permanent teachers for the basic curriculum, and by 1916 a new building and dedicated land and fields for the practicums had been added.[53] Science remained front and center; physics was a required course during the first two years, and chemistry and biology were required all three years. The remaining courses were in other agricultural sciences, including about a quarter of the course work as practicum.[54]

The school never caught on with the Europeans, and during the 1915–16 school year out of sixty-eight students enrolled there was not a single European.[55] In these early years almost all the students were destined to join the agricultural extension service after graduation, where they were required to serve for five years. As a result the agricultural school became associated with preparing educated natives for colonial officialdom. In 1927 an article touting the Agricultural College appeared in the Malay-language magazine *Pandji Poestaka*, published by Balai Poestaka, the government publishing house. The article was addressed to prospective students and their parents, sketching the opportunities and advantages of the school, and included eight large pictures of students studiously at work in the classrooms, laboratories, and experimental fields. The author, listed only as the Department of Agriculture, Industry and Trade, outlined the career in the agricultural extension service that awaited the graduate and then described the social benefits:

> [T]he work of the Adjunct-Agricultural consultant is not only personally ful-
> filling, but very beneficial for the prosperity of the country and the health of its
> inhabitants; moreover, you will see with your own eyes the true living condi-
> tions of the people, and will learn the needs and desires of the people; and you
> will learn also if the Government's actions are fulfilling those needs and desires
> adequately.[56]

Here was an opportunity, the article argued, to do the work of developing the nation and its people on the side of the colonial government. This language was, to be sure, neither the language of Boedi Oetomo's Enlightenment nor the social radicalism of the pergerakan. But its official nationalism was de-signed to compete with those models.

The graduates of the agricultural schools functioned as outreach represen-tatives to the native economy. They were the ones who brought new agricul-tural technologies to peasant rice cultivators. This was not a relationship built on science. Although science remained important in the schools, it was only important as professional preparation for the functional elite. The relationship between state and peasant centered on experimental fields, where colonial offi-cials (many of them Javanese) worked with local cultivators to develop better rice cultivation techniques.[57] Scientific training or knowledge was not passed down to the farmers. By creating a career for native scientists in which they were to be placed in between the people and the colonial government, the Netherlands East Indies state built its own safe system of national science.[58] And there was room for advancement in agricultural science. Research jobs were usually reserved for Europeans. Only a handful of Indonesians attained senior positions within the agricultural bureaucracy, including Wisaksono Wirjodihardjo, a graduate of the Buitenzorg Agricultural College who in 1940 became a senior soil engineer at the government Soil Research Institute and after independence became a leading Indonesian scientist. A small number of Indonesian students attended the agricultural degree programs at Wageningen University in the 1920s and 1930s, and after graduation most of them returned to the Indies to work in senior administrative and research positions in Buitenzorg.[59]

Native elites were to be co-opted by a stable and fulfilling career but also with the promise of helping to further the development and prosperity of the people from within the boundaries of the colonial government. Nationalist aspirations could be pursued as well but only to a certain extent.

## Cosmopolitan Nationalism

Starting in the mid-1920s, Sukarno mixed the social radicalism of mass membership parties with the Enlightenment idealism of study clubs and defined the struggle against the Dutch state in national terms. He argued that Dutch colonialism, not capitalism or Christianity, was the enemy, and it would need to be faced with the unity of national opposition. After 1926, Sukarno, through magazines, newspapers, the Indonesian Nationalist Party (PNI), and an alliance of Indonesian political parties, began to unite the nationalist movement under the common goal of striving for *merdeka*, "political independence," from Dutch occupation. The power struggle was paramount.[60] Sukarno's nationalism also redefined native leadership, dividing native elites into noncooperators (such as himself), who fought for merdeka, and cooperators, the functional elite, who continued to advance colonial occupation.[61] Sukarno's politics of cooperation and noncooperation provides a key insight into the working of the Dutch state, which had for decades administered the colony with the help of native elites. If, as Sukarno and others advocated, this cooperation stopped and became the basis of an anticolonial movement, the colonial state would falter. These ideas spread to students, culminating in a youth congress in Batavia during October of 1928, which produced the famous youth pledge, stressing Indonesia's common and united history and destiny.[62] The harsh Dutch response of jailing Sukarno and his followers suggests that they feared Sukarno and this new nationalism and confirms much about Sukarno's political analysis. But in the short run Sukarno did not create a mass movement; it was not until the Japanese occupation in 1942 (and the collapse of the Netherlands East Indies) that Indonesian nationalism became a mass movement opposed to European colonialism.[63] Moreover, Sukarno's noncooperation politics did not even unite all the native elites. While hundreds of political leaders, including eventually Sukarno, were interned, the core of the functional elite continued to work with and for the Dutch colonial state.[64]

For those among the functional elite who planned for a future without Dutch power, the political options were limited if they wished to stay out of jail.[65] Like Sukarno and his followers, the educated native elite framed the discussion around the Indonesian nation. Throughout the 1930s M. H. Thamrin, a leader of the functional elite in the Volksraad, talked publicly about bringing Indonesia to independence as quickly as possible.[66] But even as the native representation grew in the Volkraad, their political results were meager, culminating in the Dutch government's rejection of the Soetardjo petition of 1936, which had only called for a forum to discuss the possibility of autonomy for the Indies.[67] Thamrin, Soetardjo, and others were well aware of Indonesians' weak political position but looked for ways to strengthen it before opposing Dutch power. They saw the weakness as stemming from a lack of numbers. Thus, they searched for a means to nurture and uplift enough Indonesians to change the European view of what an Indonesian was. With an enlarged population of enlightened Indonesians whose numbers would command respect as a modern people, the apostles of enlightenment believed they could eventually dictate political terms. In 1927 Dr. Soetomo's journal *Soeloeh Indonesia* printed a letter arguing that in the future "the group of so-called intellectual leaders shall get stronger and more reliable, and paired with that will be their increasing influence on the people."[68] Cultural uplift would make political opposition to the Dutch viable. Throughout the 1930s the apostles of enlightenment continued to pursue their educational missions through teaching and publishing, all of it promoting progress, kemajuan.

By the 1930s it was common currency that education would prepare Indonesia to take its place alongside other nations of the world. While there was broad consensus about the need to spread a popular Enlightenment, and that it would be the educated native elite who would lead this endeavor, there was uncertainty about what kind of enlightenment was best. No single foreign ideology was a perfect fit, given the unique Indonesian traditions and history, and a wide variety of blends of East and West emerged. The pergerakan leader Hadji Mohammad Misbach argued that communism would guide Muslims on the path to independence, Soewardi advocated shaping a sacred family around Javanese culture with a dose of theosophy, and Soetomo aimed to raise a self-disciplined cadre of socially responsible intellectuals who sought practical means for addressing national issues.[69] And there was considerable experimentation, seeking ways to connect with the Indonesian populace. John Ingleson has recently shown how the Surabaya chapter of the Indonesian Study Club continued to try to organize an urban labor movement after its political defeat in 1926. Instead of directly challenging the power of Dutch capital and the colonial administration, its members sought to improve the social and economic conditions of Surabayan workers. In particular this entailed providing adult— that is, practical—education, as well as organizing a mutual aid society that supported labor union members.[70]

An important development in the 1930s was the invention of a new kind of journal, the Indonesian-language cultural magazine.[71] In early 1934 M. Amir and G. S. S. J. Ratoe Langie began publishing the weekly *Peninjauan* (Observation).[72] Their observations were self-consciously cosmopolitan, with news from all over the world, including in the first year a special series about the Philippines and India, a sports page, a women's section, and weekly updates from the world of medicine. Although its authors and editors wrote about politics and political culture, they were cautious and circumspect about their own political advocacy, that is, conservative with regard to local politics. The goal for their readers was to learn from the outside world. After learning of the French chemist Marie Curie's death, they published an obituary, highlighting not only her scientific accomplishments but the importance of her pathbreaking appointment as the first female professor at the Sorbonne.[73] They intentionally highlighted Madame Curie's transformation of the possible roles for the female underclass of Europe through her own personal integrity, brilliance, and pursuit of scientific knowledge. One can hear the inferred message: if she can do it, so can we. Certainly *Peninjauan* celebrated the West, not only with admiration for Madame Curie and photos of Jean Harlow, but more generally Western culture, stating, "The more western science travels to the East, the better. The West, even though it is the side of the ferocious imperialist — is still the storehouse of knowledge for the world, also for the people of the East. Japan has not examined the problem of internalizing the world of east and west, like our people have at the significant 'culture congresses'; rather Japan has swum across the sea of western knowledge!"[74] The writer argued that the old ways of holding endless meeting in search of the perfect synthesis of East and West had been full of enthusiasm but had not led to real progress. Progress, he argued, should be targeted, and "if the enslaved wants progress, wants honor, he must think about what kind of progress and honor. Don't just cheer hip-hip hurray like in the year 1910. That was an earlier generation's way, ancient history."[75] He further dismissed those who dreamed of returning to a golden age or those who in the last twenty-five years had been advocating a return to traditional culture. Instead, hard individual work would create modern individuals, for "only a strong spirit, only 'personality,' can deliver new results and be useful for the development of our people."[76] This cosmopolitan journal took as its task producing that new spirit by trying to pull the mass of Indonesians across the sea to Western culture.[77]

The short-lived publication remained optimistic that it was staking out a new, viable path toward development. Less than a year into publishing their weekly observations, the editors celebrated gaining their three-thousandth subscriber, writing, "We all at *Penindjauan* can now continue on with the task of raising the magazine to the best it can be, a true friend to the Indonesian youth and elders who wish to carry through on the road to progress [*kemajuan*]."[78] Still, notwithstanding the innovations — writing in Indonesian, printing photos of women in bathing suits, and an orientation toward self-development — the leadership role the functional elite imagined for themselves remained intact. The journal was still the mouthpiece of those cooperative elites who were educated by, and invested in, the Dutch system. This was clear to all readers; the third issue included a report, with printed speeches encouraging the state to expand higher education, of a recent joint meeting organized by Ratoe Langie of the Association and Indonesian Academics and native members of the Volksraad.[79]

Another journal, *Kemadjoean Ra'jat*, "People's Progress," was pitched as a self-help magazine; its Surabaya-based editor, R. M. Hardjokoesoemo, announced under a logo of a rising sun: "Improve your situation, position, and life!"[80] Beginning in January of 1937, it ran for at least two years, and all articles were in Indonesian. It switched from once to twice monthly in early 1938, at which point it reported having ten thousand subscribers.[81] Its pages were filled with short articles of practical information and advice directed toward helping individuals achieve success. There is no direct discussion of religion or colonial politics, and even ideology seems missing, although an article celebrating the wedding of Princess Juliana and Prince Bernhard of the Netherlands suggests that it was addressed to the functional elite. In fact, though, as the article pointed out, this, too, was meant to inform. Urban Indonesians had all seen the flags and other decorations announcing the wedding. The article answers a question not addressed in official Dutch celebrations: who is this man entrusted to become the husband of the future queen of the Netherlands (and its colonies)?[82] News from around the world — Adolf Hitler's speech about lands lost during the Great War, the illness and recovery of Chiang Kai-shek and the pope, and a new lower price for sending postcards in the colony — was delivered in one-sentence summaries.[83] More than half of the publication was devoted to teaching its readers about global culture, in particular explaining the rules of Western science and politics.

Most of the articles in *Kemadjoean Ra'jat* had an explicit pedagogical tone; that this was intentional is suggested by an article in the first issue about the importance of children's education.[84] During the first year, about one-third of the magazine was devoted to science lessons. These articles were not really theoretical or practical but grandly taxonomic, enumerating all the bits of worldly matter and in the process introducing readers to the scientific jargon needed to describe them. It reads almost as if the editors were summarizing and translating sections of Dutch high school science textbooks. Its first biology lesson distinguished between live and not-live matter and then taught the Dutch terms to describe minerals.[85] Shortly thereafter followed the first botany lesson. Subsequent issues went into greater depth; by the end of the year, they were up to vertebrates, pointing out, of course, the presence of backbones but also explaining the ease with which vertebrates moved.[86] At first the emphasis was on building a basic foundation in physics, biology, and the earth sciences, but this was only the beginning. At year's end the editor explained his ambitions to cover the disciplines of "Law, English, Economics, Business Administration, Psychology, Photography, Radio, Electricity, Music, Public Administration, Cosmography, Agriculture, Sociology, Occult sciences."[87] And in the second year there was a considerable expansion of topics covered, moving beyond science's building blocks to include an introduction to law and society, which began with an overview of civil law (and was notable for historicizing in general the reason why all societies create legal systems).[88]

Analysis of colonial politics was out-of-bounds, but politics was not completely absent. The editors introduced readers to global (mostly Western) trends in politics, and in such discussions there were hints of local concerns. Feminism, for example, was introduced by referencing the achievements of Kartini, the daughter of a Javanese aristocrat whose Dutch letters had been published in 1911 as *Door duisternis to licht: gedachten over en voor het Javaanse volk* (From darkness to light: Thoughts about and on behalf of the Javanese people).[89] But then the author argued that in Europe some were casting the issue in terms of creating legal equality for women. The author cautioned that earlier generations across the world had made mistakes by treating women as slaves useful only for birthing the next generation. This had shattered any possible social progress. Social progress, the article argued, was only possible if women had equal access to work and the law. *Kemajuan* meant freedom (*kemerdekaan*) for women.[90]

Each issue began with an uplifting sermon about self-improvement. For example the issue of May 1, 1938, contained the statement "Broad education and widespread knowledge is actually more valuable than a mountain of riches."[91] This and much other similarly pitched advice would, the argument went, allow readers to attain their professional aspirations. Most surprising is the lack of cultural specificity; there are few anecdotes or examples to drive the point home. Most of the advice remains at the level of exhortation. And when there are specific examples or role models, they tend to be Western, including passing references to Napoleon. This is most clearly illustrated by a long-running series about Benjamin Franklin's life and accomplishments.[92] Although there was a self-imposed prohibition against controversial topics, most articles had no specificity to Indonesian cultures, colonial or otherwise. This may partly have been a function of what the editors and authors were reading, as the models for writing in the self-help genre would have come from outside the colony. The journal's articles suggest that achievement, happiness, and success are not tied to any locality but come from individuals striving toward excellence and enlightenment.

How well did *Kemadjoean Ra'jat* succeed in its goal of social progress? It enjoyed reader support, at least for two years. And it backs up two recent arguments about Indonesian nationalism. James Siegel has argued that a "fetish of modernity," which opened up communication separate from colonial hierarchies, led individuals to find their identity as Indonesians in the early decades of the twentieth century.[93] This was expressly the point of R. M. Hardjokoesoemo's journal, and it apparently found Indonesian readers who shared this "fetish of modernity" as a means of finding Indonesians a place in the world. Moreover, as Rudolf Mrázek has recently argued, technology brought a sense of power to Indonesian nationalism in the sense of its suggestion that technology had the potential to break the straight lines of colonial orthodoxy.[94] *Kemadjoean Ra'jat*, with its celebration of science and technology, for example, a discussion about the possibilities of interacting with the world created by expanded airplane service, is testament to this confidence that embracing the modern will result in progress.[95]

Creating a scientific and technological nation, a "fetish of modernity," as a means of promoting progress had broad appeal, at least among those able to subscribe and read an Indonesian newspaper or magazine. But what of the political results? Did the fetish of modernity create a national, modern community? Did science and technology give nationalists the power to punch their way out of colonial subjugation? Were the pictures, lectures, and lessons of *Kemadjoean Ra'jat* ever more than a dream or fantasy? Given that *Kemadjoean Ra'jat* has been forgotten by historians and others, and that Indonesian independence was won on the basis of revolution and war with the Dutch, the answer seems to be no. Still, even if *Kemadjoean Ra'jat* in particular, and the fetish of modernity more generally, did not shape Indonesian political reality, as a dream it continued to exert influence on Indonesian intellectuals, even after Indonesians won independence from the Dutch in 1949. And an examination of the responses to the fetish of modernity illuminates the continued weakness of the apostles of enlightenment, who failed to create a large number of cosmopolitan nationalists.

Easiest to gauge is the state's response to scientific nationalism; the colonial and postcolonial state's largely successful co-optation of nationalist scientists is examined in the next two chapters. More difficult to measure is the public's response to the popular Enlightenment of *Kemadjoean Ra'jat* and like-minded attempts to spread it. The short life of *Kemadjoean Ra'jat*, and most publications meant to promote cultural modernity, is telling. The model of building bridges outside of politics never became popular enough to sustain itself. Certainly *Kemadjoean Ra'jat* wished for popularity; at the end of its first year it suggested that readers send in postcards informing the editors of the topics they would like to see addressed. At the same time it sought new subscribers by using existing readers as recruiters; it offered gifts for those who found five new subscribers and further explained that current readers would be acting as a useful and important torch to mankind by spreading the wisdom of *Kemadjoean Ra'jat* to their friends.[96] The editors had a monetary incentive to expand the subscription base, but there was also the desire to create a national community, creating progress for the people of Indonesia. The results of their plea is unknown, as there were no further reports of subscribers and the magazine was not in the habit of printing letters or postcards it received. Judging by the index of articles from 1938, there was a considerable expansion of material compared to the year earlier, not just lectures about chemistry, Benjamin Franklin, and education but also more about technology, world cultures, and business success.[97] In 1938, still, the world of *Kemadjoean Ra'jat*, even for those Indonesians who thought it a beautiful torch of enlightenment, would not have found many ways to connect the articles to their lives even if the magazine exhorted each individual reader to do just that. Without a means of connecting the fetish of modernity to the societies of the Netherlands East Indies, the magazine's political impact was small.

## Conclusion

After 1935 for nationalists to remain out of the Boven Digoel internment camp their speech could not betray anything nearing political criticism. One topic still open for discussion was education. Many nationalists continued to expect that increased education was necessary for national progress. This was true as late as 1938, when a nationalist compared higher education in the Dutch colony with higher education in British India and found the former wanting, especially since only one unnamed Indonesian native had ever been appointed as a professor in the colony and he had since then quit. Although the article included a cautious criticism of the state—"Establishing university education was begun far too late in this land"—it concluded that aspiring native professors needed to produce more, to show they were as qualified to be scientific researchers as anyone else. The writer exhorted his people to try harder to become qualified as university professors. Then, if the Dutch did not insist upon a "race-criterion," their opportunities for advancing native higher education would become reality.[98]

There was no dissent among nationalists about the need for a multidisciplinary university, where learning would be celebrated rather than channeled toward one useful and functional profession. Ratoe Langie, the previous member of the Volksraad and veteran magazine publisher, wrote in his widely read journal *Nationale Commentaren* that there was a great need for a national university, which would allow Indonesia to grow into a nation of the world. He pointed to the example of Japan, the only Asian country that had devoted itself to commanding Western knowledge and was hence now one of the Great Powers.[99] A week later, Ratoe Langie elaborated:

> Every nation, which wants to be counted in the international appreciation, must also have its disciples in the scientific world, to keep its national standing high, in all areas in which the human spirit and the human intellect can be developed. . . . And in this task, for the upcoming Indonesian intelligentsia, the university is simply an indispensable institute. . . . Thusly put, it will build the bridge between our precious scientific workers and the national movement itself. The question of the Indonesian university is the problem of the intellectual emancipation of the Indonesian nation; it is the problem of the organization of the Indonesian intelligentsia. . . . And finally, this is the heart of the problem of nationalism. In modern world relations it is the intellect, and its enlightenment [*doorstraaling*] of social development and social organization, that settles it.[100]

The naming of an Indonesian nation and society has made the reasons for scientific education sharper, but at its core Ratoe Langie's plea is not substantially different from the educational reforms discussed at the first Boedi Oetomo conference in 1908. The central problem remains extending leadership from the Western educated to the Indonesian people. In 1939 the national bridge connecting diverse local peoples and communities through the intelligentsia still did not exist. But there remained the potential for science to create that bridge if only the opportunities were afforded. Indonesian independence would offer far more.

Nationalists of all stripes were disappointed and frustrated that their ranks remained small and powerless at the end of the 1930s. For Sukarno and other social revolutionaries, this only reinforced their belief that mass opposition would end Dutch colonial power. The apostles of enlightenment envisioned a different solution, one in which an expansion of the ranks of educated Indonesians would create a credible bulwark to Dutch power. Thus, throughout the 1920s and 1930s they looked for new ways to shape and enlarge a society of enlightened nationalists. In Ratoe Langie's formulation, the intellect would determine the relationship between colonizer and colonized, and if the Indonesian side had enough intellectuals it could settle the problem of colonialism in its favor. This priority, of expanding the ranks of cosmopolitan and enlightened nationalists, would continue after independence.

# 6

# Technocratic Dreams

Since the middle of the nineteenth century, colonial scholars, including scientists such as Junghuhn and Treub, envisioned a politics in which technical experts, guided by science, would set administrative policies, and thus enact enlightened leadership. With the spirit of frontiersmen, these men held the illusion that they could rise as high as their highest ambitions. However, Junghuhn's naturalism, Treub's scientific agriculture, and other technocratic ambitions were never fulfilled. As I have argued, this failure stemmed from the innate economic pragmatism of the colonial state, and its clever co-opting of the apostles of enlightenment for administrative tasks. This is a reason why the Indies remained, as the historical sociologist J. A. A. van Doorn has pointed out, an incomplete technocracy.[1] This did not stop idealistic envisioning, though, which multiplied during the first three decades of the twentieth century. Social, political, and cultural reformers proposed innumerable improvement schemes in which reason and science would generate prosperity without recourse to the colonial politics of divide and rule. This was, for example, the case with nationalist institutions such as Boedi Oetomo and journals such as *Kemadjoean Ra'jat*, whose leaders saw in Western science the possibility for prosperity and progress. Similarly, European reformers such as H. F. Tillema dreamed in his six-volume *Kromoblanda* of an engineered Indies with clean roads and impeccable hygiene, able to unite native and European society.[2] This reformist spirit remained part of the colonial society, right through the worldwide depression of the 1930s.

Reform after 1930 meant less idealism and more central control, especially of the economy. Central economic planning, which only since World War I had been in the hands of local colonial officials, became a critical function of the colonial state. When the world stopped buying the colony's export crops, officials took measures to restrict their output and protect markets and manufacturers. The government established import quotas and licensing schemes, and increased regulation of exports. The governor general's staff was responsible for these policies but over time was increasingly aided by economists at the Department of Agriculture, Industry and Trade. By the end of 1933 the Department of Agriculture, Industry and Trade was reorganized as the Department of Economic Affairs, and in the following year the office was relocated to Batavia.[3] The economist E. P. Wellenstein replaced the biologist C. J. Bernard as director at the time of the change.[4] The following year, after Wellenstein's death, the lawyer G. H. C. Hart became the director. The new department was responsible for setting and implementing crisis measures.[5] After 1933 its power increased as it began to implement new protectionist trade policies.[6]

Even before the onset of the depression, the Department of Agriculture, Industry and Trade had been revisiting economic policies meant to advance the native economy. As Suzanne Moon has recently shown, challenges to the economic theories of the ethical policy of the 1910s and 1920s led to a new set of neo-ethical policies meant to encourage native agriculture more effectively. These policies developed as a response to J. H. Boeke's dual economy thesis, which posited a separation between the Javanese peasant economy and the world market. Boeke argued that Javanese peasants did not have an individualistic economic outlook and hence did not respond to the kinds of economic incentives or policies created by the ethical policy. This critique of the ethical policy pushed officials in the Department of Agriculture, Industry and Trade to devise new strategies for focusing development programs. As Moon shows, this resulted in officials at the Department of Agriculture, Industry and Trade concentrating education and agricultural extension programs on peasants who had regular contact with Europeans and hence straddled the dual economies.[7] Boeke, upon ascending to the new chair of tropical-colonial economics at Leiden University in 1930, set forth in his inaugural lecture the claim that the dualistic nature of Asian colonial economies was a permanent feature of the European colonies and that a single economy, built around capitalism, was unattainable. An implication of Boeke's argument was that teaching Western agriculture techniques to native peasants, whom he saw as immobile masses outside the creative influence of capitalism, was futile. This speech put the extension officials in the Indies on the defensive, and this led to a new urgency in reshaping government policy toward the indigenous economy. Wellenstein, later the director of the Department of Economic Affairs, responded to Boeke's lecture in 1930 in the Indies journal *Koloniale Studiën* (Colonial Studies), pointing to the danger of reifying the difference between East and West. Wellenstein argued that for the Dutch to right the exploitative wrongs of the past, "development will become even more difficult if attention is centred on antitheses that do not actually exist and the roots of genuine antitheses are not clearly discerned."[8] Boeke's arguments helped spur Wellenstein and his successors at the Department of Economic Affairs to find improved means for the colonial government to create economic progress. One long-term result of the changes at the department was that professional scientists and other experts were given increased prominence as agents of economic change.

In the years between 1933 and 1942, the responsibilities beyond agriculture assigned to the Department of Economic Affairs grew steadily. This also created opportunities for the department to centralize political power. The period began, though, with the field officials of the Department of Civil Administration returning to prominence, wielding near absolute power in their districts. But as the responsibilities of the Department of Economic Affairs grew, it emerged as an alternative lever of power. The colonial official Hubertus J. van Mook, who worked at the department after 1934, believed that he could shift power away from field officials of the civil administration to centralized government institutes such as the Department of Economic Affairs. He envisioned idealistically that the department could become the basis for expert rule, where specialists in social policy, economics, and science had direct influence upon the colonial executive. This arrangement, he believed, would create an enlightened colonial regime. By the late 1930s he had begun to implement his plans by reestablishing the Buitenzorg Botanical Gardens as the central biological institute in the colony. His larger plans were upset by the war, but after 1945, as the chief executive of the colony, he continued the centralization of science in Buitenzorg and Batavia. While the Japanese invasion in 1942 destroyed the hegemony of the Dutch civil administrator, never to be rebuilt, the scientific institutions of the Department of Economic Affairs remained viable and intact throughout the 1940s. And because of their central locations in Batavia and Buitenzorg, the department's components were quickly rebuilt after the Dutch return in 1945. Van Mook's ideal of enlightened rule through technocrats was unfulfilled, but the institutional apparatus of the former Department of Economic Affairs smoothly would become part of the administration of the Republic of Indonesia. And as a result, science, which had been in a very weak position in the early 1930s, would become institutionalized as a technical component of the new Indonesian state.

## Science Imploded

The Depression challenged the relationship between science and the state that had emerged after Treub. The colonial budget, which drew much of its revenue from the now dwindling export economy, contracted. The bulk of savings were to come from personnel reductions.[9] The Department of Civil Administration escaped the downsizing because officials in Batavia and Buitenzorg saw its administration of the colonial people as too important to undermine through cuts.[10] Hardest hit were those state institutions and personnel meant to develop native society economically.[11] They were considered expensive by the central administration and supplemental to the basic power structure in the colony. This challenged the existence of all the government agencies created or expanded under the sway of the ethical policy. And, in particular, scientists came under great pressure to justify their existence and account for their expenditures.

At nearly the same time a conservative government in the Netherlands appointed B. C. de Jonge as governor general in 1931. After his arrival in Buitenzorg, a new pattern of administrative reform characterized the Indies. The decentralization of power, which had begun in the previous decade and had meant to pass along great political autonomy to the colonies' inhabitants, was continued but to a different purpose. Instead of devolving political power to the colony's inhabitants, the policies were designed to augment the administrative power of the Dutch officials in the Department of Civil Administration. This was particularly effective outside of Java, where advisory councils, originally meant to empower locals in political decision making, were staffed by senior-level Dutch administrators whose loyalty to the colonial state was strong. Within their district, these officials of civil administration had virtually unchecked power. That this was meant to create long-term positions for Dutch officials was by the mid-1930s an open secret; it was governor general De Jonge who in 1936 declared in an interview with a Sumatran newspaper, "I believe that since we have worked in the Indies for some 300 years, another 300 will have to be added before they may perhaps be ready for some kind of autonomy."[12] After 1933 neither Dutch nor Indonesian nationalist criticism of the Department of Civil Administration's hegemony was tolerated.[13] Modernization, political, social, or otherwise, was on the back burner.

As late as 1929, colonial science was riding high. In 1929 the budget for natural sciences run by the Department of Agriculture, Industry and Trade (not including the General Experiment Station) was f.541,000; this included about f.100,000 to pay for the Pacific Science Congress, which with much fanfare was brought to the colony that year.[14] Since Treub had begun building his scientific empire thirty years earlier, the number of scientists had greatly expanded, and in 1930 there were 700 in Buitenzorg, most of them working in the applied science sector. Under the care of W. M. Docters van Leeuwen, director of the Buitenzorg Botanical Gardens from 1918 to 1932, research science, too, gained ground, started by a new generation of young, university-trained biologists who came to Buitenzorg from the Netherlands in the 1910s and 1920s. Moreover, pure science gained an institutional basis, as it was now held up by colonial administrators as evidence that the Dutch were doing a decent job bringing modern civilization to the colony.[15] In the 1920s, as the practice and discipline of botany and ecology in Buitenzorg were shaped to correspond with European models, for the first time it was possible to jump from a job in Buitenzorg to a European or North American career—H. J. Lam left Buitenzorg in 1933 to become the head of the Rijksherbarium in Leiden, and F. W.

Botanist H. J. Lam at his desk in the Buitenzorg herbarium, ca. 1930 (Courtesy of the National Herbarium of the Netherlands, Leiden University branch)

Went moved to the California Institute of Technology in 1932 on the strength of his ecological research. And bringing the 1929 Pacific Science Congress to the colony, including 142 foreign scientists, suggested that science in the Netherlands East Indies was entering a golden age, one in which Indies science was farther advanced than in other dependent regions and was approaching the standards of European science.

Nonetheless, the weakness of colonial science's position was revealed by the slumping economic times. Being so reliant on the state for patronage meant colonial scientists were helpless as funding was cut. The Natuurwetenschappelijke Raad, the official Scientific Council in the colony since 1928, was ignored by both Governor General De Jonge and C. J. Bernard, head of the Department of Agriculture, Industry and Trade.[16] And, indeed, colonial scientists had trouble articulating solutions. At the lowest point of the slump in 1932, the Scientific Council submitted to the colonial state the results of a survey of scientists' proposals to save money. The surveys had come back foretelling doom if already depleted resources were cut further. K. W. Dammerman, then head of the botanical gardens and president of the Scientific Council, warned the government of breaking the old arrangements between science and the state: "We must hold to the conviction that science, even if the direct application or the economic yield is not initially obvious, is not just desirable but necessary for redevelopment and continued economic progress."[17] Neither Dammerman nor any of the other scientists wanted to reinvent colonial science as their bosses wanted them to. And later writings suggest that even those scientists that did want to reinvent science were hardly in a position to do so. Dammerman, tellingly, used the crisis to try to improve the position of the botanical gardens when he suggested that all the scientific research institutes be concentrated under him. And, probably unwisely, he whined about the heavy administrative duties suffered by institute directors, which, he claimed, kept them from exercising proper scientific leadership.[18] These suggestions, reasonable as they must have seemed to Dammerman and his colleagues, largely argued the relevance of science based on an earlier model.

In 1932 the botanical gardens, with its research focus, came under particular scrutiny.[19] A letter from early 1932 to the newspaper *Bataviaasch Nieuwsblad*, apparently written by an official within the department who had previously been slighted by the gardens, argued that not the botanical gardens but other scientific institutes had produced innovations in agriculture.[20] More serious than this anonymous attack was that this view was shared by other officials at the Department of Agriculture, Industry and Trade and the Department of Civil Administration, who doubted the gardens' value.[21] Although officials decided not to scrap the gardens altogether, Dammerman grew concerned, and he enlisted outside help, alerting foreign biologists with ties to the gardens of the crisis. Biologists in the United States and Europe responded positively to his plea for financial assistance.[22] As the old botanical gardens was disappearing, what would replace it was not known.

In a series of articles published during the mid-1930s, the future of the gardens became an opportunity to debate what the relationship between science and the state should be. The botanist V. J. Koningsberger—former director of the Experiment Station for Sugar in Pasuruan, newly appointed as a professor at Utrecht University, and son of J. C. Koningsberger, former director of the Buitenzorg Botanical Gardens—opened the debate in 1934 in the widely read journal *Koloniale Studiën*. He claimed that the destructive power of the Depression had already finished off a badly wounded botanical gardens. He blamed this on its leaders, especially the kind of scientist who "held that Science was that which under no circumstance could have practical meaning for mankind," although he did not mention either Dammerman or Docters van Leeuwen by name.[23] He suggested that only a scientific visionary, someone like Melchior Treub, with a powerful and innovative scientific vision, could again put the botanical gardens on "a scientific basis, upon which we can build."[24] In fact Koningsberger's suggestion—turn back the clock to the golden age of Treub—was not so different from Dammerman's (who responded in two subsequent issues of *Koloniale Studiën*): continue the kind of pure, research science of the 1920s, perhaps by moving the gardens to the Department of Education.[25] Neither was able to lay out a new identity for the gardens though. This was noticed,

Kees van Steenis in Buitenzorg, ca. 1935 (Courtesy of the National
Herbarium of the Netherlands, Leiden University branch)

and other nonscientists joined the fray with different proposals suggesting new
ways to combine "applied" and "pure" science at the gardens.[26] This distinction
between applied and pure knowledge, shared by all the authors seeking a solution, echoed earlier conflicts but was meaningless to officials at the Department of Economic Affairs in the mid-1930s, who had no need for either.[27]

Scientists, particularly those at the botanical gardens, were not in a good
position to reinvent science. For most of the 1930s the botanical gardens drifted.
The character of Buitenzorg changed drastically when in 1934 the Department
of Agriculture, Industry and Trade departed for new offices in Batavia. One
observer wrote, "Buitenzorg looks like a ghost town, littered everywhere with
empty houses."[28] In the same year, a Goodyear tire factory moved in and with
it sixty American managers. The mayor planned to turn Buitenzorg into a factory city and hoped to lure Firestone as well.[29] Dammerman tried to reignite
interest in the botanical gardens by opening a restaurant in it.[30] His scientific
administration was lackluster, and he bungled the appointments, which had

been especially arranged in the Netherlands, of two systematists.[31] Individual scientists kept a lower profile and went their own way; the junior systematic botanist, Kees van Steenis, was left to his own devices and began work on general plant surveys of the Indies.[32] Science and the colonial government decoupled after decades of close cooperation.

## Reinventing Colonial Science

Despite its collapse the botanical gardens retained flagship potential. Even in the late 1930s it was one of the few scientific institutions known outside the Dutch empire. And in the mid-1930s it was this potential that drew a rising star in the colonial administration, Hubertus J. van Mook, to the problems of the gardens. It would be Van Mook who, using the botanical gardens as a template, would strengthen the relationship between science and the government. Throughout the 1930s scientists were largely unwilling participants, chafing under increased government oversight and a changed charter, but eventually they came around, seeing the renewed authority and promise that government science afforded them in Van Mook's schemes of centralized science. This process would not be completed until the late 1940s, when under the leadership of Van Mook, one part of his dream of technocracy, centralized state-sponsored science, became a reality.

Van Mook had risen to prominence in the early 1930s as a leader of colonial reformers, who published in their journal *De Stuw* (The Push) and sought to revive ethical policy ideals, including greater colonial independence.[33] In addition to contributing to the reformist mouthpiece *De Stuw*, Van Mook was a regular contributor to *Koloniale Studiën* and the editor of *Koloniale Tijdschrift* (Colonial Journal), writing frequently about government reforms. In a 1932 brochure he proposed turning what he saw as ineffective scattering of unmanageable autocrats ruling the colonial bureaucracy into an entity that could exert leadership for the whole of colonial society. He suggested the creation of an executive council, which would coordinate colonial policy in tandem with the governor general and in consultation with the Volksraad. Centralizing governing policy could be affected by creating a "concrete, routinely updated work-plan for every governing period." At the same time department directors would be given more responsibility, becoming administrative statesmen, not just administrative heads, who advised the Volksraad and governor general directly.[34] Thus the chain of accountability would run directly from each department to the governor general. He concluded that "an Indies government, which is able to enact leadership, is a necessary prerequisite for the rapid development and emancipation of these lands."[35] Although the rhetoric of development and autonomy was that of the neo-ethicists and overlapped with Indonesian nationalist ideals, his reforms proposed strengthening the power of state officials in Batavia and Buitenzorg and was a far cry from democratic principles. This may have been one reason he caught the eye of Governor General De Jonge.

Van Mook's first reform program concerned colonial science, biology in particular, which he began to examine after he moved into the central administration. In 1934 Van Mook had been suggested as the dean of the law college in Batavia, but Governor General De Jonge instead brought Van Mook into the Department of Economic Affairs.[36] Van Mook initially had an ad hoc position there, and he was given some latitude to revise the workings of the department. It was out of his power to reinvent the colonial state, as he probably wished to do. But given the consensus that colonial biology was broken, this area was probably an obvious place to start. He followed the program outlined in his 1932 brochure, and aimed to create a more unified biology that was integrated with other parts of the colonial administration. He began by finding allies in the Netherlands. While Van Mook was in the Netherlands in the fall of 1936, he sought the guidance of leading botanists there, asking them in particular about the botanical gardens.[37] All the respondents emphasized different problems, but they were unanimous in recommending that Dammerman be replaced. Not only would a new leader right the ship, but he would be in a position to establish the gardens as the administrative and scientific center of colonial biology.[38]

Van Mook became outright head of the Department of Economic Affairs in August of 1937, and, although science was hardly his only responsibility, he quickly carried through on reinventing colonial biology. He had wide latitude to do so, as overnight he had become one of the most powerful bureaucrats in the colony.[39] In 1938 he chose his old college friend L. G. M. Baas Becking to clean house at the botanical gardens.[40] Baas Becking arrived in 1939 as a special official in the Department of Economic Affairs directly responsible to Van Mook. When Dammerman retired in February of 1939, Baas Becking was appointed temporary director of the botanical gardens.[41] He wasted no time submitting advice about the duty of the gardens, the coordination of research between the gardens and the various experiment stations, and the exchange of scientific personnel.[42] Administrative follow-through was not Baas Becking's strong suit though.[43] And he did not like dealing with the officials at the Department of Economic Affairs, lecturing them instead on the importance of biology.[44] Nonetheless, Van Mook carried through with the reorganization plans, dragging Baas Becking and everyone else along.[45]

As is clear from a long memo to the governor general, Van Mook was convinced that science could become again a critical component of colonial administration. If the botanical gardens were usefully integrated into the state structure, biology could not only flourish but would serve the colonial administration as well by enlightening colonial policy. The key was coordinated cooperation. Here Van Mook cut through the "applied" versus "pure" dichotomy that had defined the debate about colonial science since the 1910s. Van

Mook's solution was to raise the botanical gardens to the status of a coordinating body where it could pursue a true colonywide scientific policy. He made much of the utilitarian benefits. But he also argued:

> In the Netherlands Indies, which perhaps contains the richest tropical-biological diversity in the world, the Government must also lead in the explorations of those riches. This responsibility goes especially for that research which can only be done in tropical regions. The Botanical Gardens is the central institution in the practice of biological natural science, intrinsically and because of its cultural meaning. . . . That at this moment precisely in the Netherlands-Indies the cultural responsibility weighs heavily, is clear, when one remembers how much study of the object—the Indies' nature—must still be done, and how far in general the subject—the people—is still removed from the modern natural sciences.[46]

Van Mook proposed a new leadership role for the Buitenzorg floracrats, who would lead colonial society in its interactions with nature. He remained vague at this time about his long-term plans beyond a discussion of the likely continuing economic benefits, but he advised the governor general that the Buitenzorg floracrats were to become the leaders of the scientific administration in the colony. This meant putting more resources into the botanical gardens, reversing years of cutbacks.[47] This was good news for the staff at the botanical gardens.[48]

Rescuing the botanical gardens meant, for Van Mook, publicly reauthorizing colonial biology. Because of the Buitenzorg Botanical Gardens' past achievements, he believed that institution could return to its earlier glory, and in the future would serve as the central government institute responsible for coordinating all colonial biology. Hence his wish that the Volksraad discuss and approve the reorganization instead of just having the governor general sign off on the measures. Van Mook's proposal pointed especially to the cultural importance of the botanical gardens, in particular its capacity to lead the colony scientifically: "[T]he spread of general scientific knowledge and the passing along of this knowledge and enthusiasm for it to the next generation is of fundamental importance to society. . . . [T]he results of scientific and technological research shall belong to the most powerful means for reaching and maintaining the dynamic balance between people, earth, and living nature, which is necessary for the further development of prosperity in these lands."[49] After the proposal came back from the Volksraad committee, members made the point that few details had been included and many of the projections about the centrality of the botanical gardens had not been proven.[50] Further opposition to Van Mook's proposal showed that most colonial elites still thought of colonial science in its old paradigm, in particular the desire to keep pure and applied science separate.[51] But Van Mook was not (could not?) be denied his supplemental budget request, and it passed unanimously by voice vote.[52] This model of centralizing scientific authority in the premier scientific institutions would be the template for all postwar government science, revived by Van Mook in late 1945 and inherited by the Indonesian Republic in 1950.

## Japanese Colonial Science

During the 1940s, World War II and the Indonesian revolution disrupted science in Indonesia. This is not surprising. What requires explanation is that this disruption was limited. Throughout the 1940s, biologists and other scientists reported to work on most days, many of them pursuing research they had begun before 1941. This was particularly true in Buitenzorg. This counterintuitive arrangement was possible because all state authorities—not just the Dutch colonial state but the Japanese imperial administration and the Indonesian Republic as well—were committed to keeping government science alive in the 1940s, even at a time when it had no direct economic or military benefit. This extended beyond preventing the physical destruction of science's buildings and institutions. Government authorities were all planning for the postwar period when, it was imagined, science would retake its position as the principle arm of government's control of nature. Scientists, hardly surprisingly, welcomed these arrangements and worked with whatever state authority held sway.

Colonial Southeast Asia, including the Netherlands East Indies, fell rapidly after the Japanese invasion began on December 8, 1941. The Japanese empire, once its military forces began to arrive on Java in February of 1942, moved quickly to restaff much of the colonial bureaucracy, including the scientific institutions of Java. Dutch officials stayed in their jobs for a few months, but in areas such as the civil administration inevitable loyalty conflicts made it impossible for Dutch bureaucrats to continue. They were replaced with Indonesian and Japanese officials. Large-scale internment of Dutch inhabitants began in July.[53] But as the Japanese considered the Dutch inhabitants to be imperial subjects, they could in certain circumstances continue to work under Japanese authority.

Research science had been intertwined with Japanese imperial practice as far back as the South Manchurian Railroad Company in the early twentieth century.[54] By the 1930s there were numerous research institutes in the Japanese colonies examining the natural and social world and meant to aid the colonial administrations. Moreover, scientists became important imperial officials who did not work under the direct authority of the Japanese colonial administrations. For example, in 1935 the Continental Science Institute in Harbin, Manchuria, opened with a director whose position was equivalent to that of a minister. Its first director justified Japanese colonial rule in Manchuria by emphasizing the importance of developing the colonies scientifically. By incorporating this civilizing mission ideology, Japanese colonial scientists had greater independence than their colleagues in Japan. Moreover, their political positions were not narrowly defined by their economic usefulness.[55]

On Java the Japanese continued to run a number of the Dutch scientific institutions, especially those with "pure science" attributes, including most of the institutes attached to the Buitenzorg Botanical Gardens. Interestingly, the Japanese neglected the export-crop experiment stations, most of which were idle during the war. To be sure, this was a result of Japanese imperial policy, which ignored the export-crop agricultural sector. Nonetheless, institutions of biology and astronomy had no economic value, yet the Japanese authorities diligently supported them. The optical observatory at Lembang near Bandung continued to observe double stars in the night sky under the direction of the longtime director there, Joan Voûte, working under a Japanese boss.[56] Likely this support was made available because astronomy and biology fit into the larger Japanese political ideology of Greater Asia, which sought to create a Japanese Asian modernity opposed to, and better than, Western modernity.[57] Investing in scientific institutions such as the ones in Buitenzorg demonstrated that the Japanese were serious about developing this new civilization, a modernity built around Japanese leadership.

In Buitenzorg, shortly after the Dutch surrender, the Japanese engineer Narusawa, who had worked in Buitenzorg before the war, was given military oversight over the scientific institutes now considered property of the emperor.[58] Those Dutch scientists who had not been conscripted to protect the colony reported to work almost as normal. D. F. van Slooten, the acting director of the gardens since 1940, stayed on and was even allowed to appoint scientists, complete with salary.[59] The Japanese had hoped to staff the various institutes in Buitenzorg with either Indonesians or Japanese, but as the Dutch had trained very few Indonesian scientists and a boatload of Japanese scientists apparently was torpedoed on the way to Southeast Asia, the Japanese quickly turned to the Dutch scientists. About one hundred were brought from internment camps to the various biological institutes of Buitenzorg in August of 1942, and, although they lived in a separate camp away from their families and under constant guard, they were also paid a salary.[60] There is no evidence that the Dutch biologists who went to work for the Japanese resisted these appointments, although they did not have much choice and had no freedom of movement. In March of 1943 a new batch of Japanese biologists came to Buitenzorg, including Nakai, the new director of the gardens; R. Kanehira, head of the herbarium; the botanist S. Hatushima; and a few months later J. Ohwi. Nakai and Kanehira were generals, and certainly initially they had large amounts of autonomy to arrange matters to their own choosing without interference from the Japanese military police, the Kempeitai. Dutch scientists continued work in Buitenzorg through early 1945, after which they were returned to the internment camps.

What did the Japanese want from the botanical gardens? It was not about either applied or pure research. It appears the Japanese biologists wished to make known the new riches of the Japanese empire using the resources of the recently acquired Buitenzorg Botanical Gardens, now an imperial institution. Kees van Steenis reported that he and the other Dutch botanists worked six days a week in the herbarium. Van Steenis wrote after the war that he had not collaborated with the Japanese. His work "was of course completely for the benefit of the institute, and had nothing to do with the war efforts of the Japanese."[61] There is no indication that his work, or that of his Dutch colleagues, aided the Japanese war effort in any significant way. Still, at least some of the research done by Van Steenis and other Dutch scholars was done on Japanese terms and was ultimately to be published in Japan as reflecting the work of a Japanese imperial institution. The ease with which the two sides settled into this peculiar professional arrangement was because scientific research was the common agenda. The Dutch scientists worked under other scientists, some of whom they knew. There was a collegial atmosphere at the herbarium, with English the common language.[62] Kanehira, the head of the herbarium, had been to Buitenzorg before the war, at which time he had worked with Van Steenis classifying New Guinea plants. At the time of Kanehira's death after the war, the botanist H. C. D. de Wit wrote, "Kanehira protected the Herbarium and Library during the years of the Japanese occupation, with all that was in his power. We have no doubt, that he was repeatedly on not too good terms with the Military Police on account of his efforts on behalf of European scientists and the preservation of the scientific treasures put to his charge."[63] Even with these shared scientific values, however, there were official Japanese scientific projects, and even if they did not directly benefit the Japanese war and occupation, the Dutch scientists worked some of the time on assignments for Japanese biology and natural history. Kanehira was interested in writing general taxonomic pieces about the flora of the Indies, as well as a history of botanical research there. All of this was the basis of general texts about Malay flora and its history, and Kanehira was preparing a three-volume work about the botany of Indonesia.[64] Much of this work involved library research and the writing of summary essays, which is part of what the Dutch scientific staff was ordered to do, while the Japanese botanists continued their systematic work, including preparation of a taxonomy of the tropical grasses of the region.[65] Kanehira asked for a series of short biographies about plant explorers of the archipelago; Van Steenis wrote them in Dutch, and De Wit translated them into English, which one of the Japanese biologists then translated into Japanese. They were destined to become short newspaper articles in Japan.[66] Kanehira and Nakai were apparently hands-off administrators, allowing the Dutch scientists to work independently. Moreover, Kanehira and Nakai's ambition of writing a general-level flora of the Indies was in line with Van Steenis's research. And apparently there was mutual respect, up to a point, as Kanehira promised the herbarium botanists full-time jobs after the war ended.[67] A month after the Pacific War ended, Van Steenis wrote his friend Verdoorn in the Netherlands, reporting on himself, his wife, and son, "We 3 OK owing to botany [lost?] 1 year work but worked harder than in any other period."[68] The material he completed during the war would form the basis for the Flora Malesiana, which Van Steenis began publishing in 1948.

World War II ended suddenly in Indonesia, with no fighting leading up to the Japanese unconditional surrender on August 15, 1945. Sukarno and Mohammad Hatta declared Indonesian independence two days later, and with that the Indonesian revolutionaries broke into the power vacuum left by Japanese defeat. Tension flared further after British troops landed in the capital in October of 1945. Young revolutionaries moved their agitation to the cities and towns surrounding the capital, and Buitenzorg was tense in October and November.[69] Because Buitenzorg was on the main road between Jakarta and Bandung, British troops from the Twenty-third Indian Division occupied the part of Buitenzorg nearest the main road. In December they drove the republican military units out of town and extended their control to include the governor general's palace and the botanical gardens, although some parts of Buitenzorg remained under the Indonesian Republic's administration.[70] During the next few months the British military administration managed some of the scientific institutions in Buitenzorg.[71] They showed little interest in science. The British brigadier John Mellsop ignored Van Steenis and followed his own council in dealing with the scientific institutes, at one point even considering handing over a chemical laboratory to the Indonesians.[72]

## Caught between the British and the Nationalists

Many in Indonesia felt that after the Japanese capitulation a new political future was dawning. In the wake of World War II, there was widespread hope that a new, more developed, more democratic, and more prosperous Indonesia could be shaped, altering the political and social dynamics of the past. As the Dutch returned to reclaim their colony, some colonial leaders saw it as a chance to revise the colonial relationship, creating a colonial system in which the Dutch shared power with Indonesian national leaders. Of course these visions of the future were contested, in particular by those Dutch who wanted to set the clock back to 1941, as well as by Indonesian nationalists and revolutionaries. In the middle of this stood Van Mook, who headed the Indies government in exile in Australia from 1941 to 1945, and who, as lieutenant governor general was the chief Dutch executive from 1945 to 1948.[73] He wished to safeguard Dutch prerogatives, economic as well as political, but believed a power-sharing arrangement with Indonesian nationalists was desirable. But there were many Dutch, both in the Netherlands and Indonesia, who resented Van Mook's strategy of negotiating with the nationalists. Indonesian nationalist leaders from across the archipelago had different dreams and plans for Indonesia. And among the national revolutionaries there remained considerable differences in strategy regarding fighting and negotiating with the Dutch based on their ideological and geographical interests. Still, starting with the Linggajati talks in November of 1946, and lasting into 1948, the chief protagonists in the conflict were Van Mook, who envisioned continued Dutch participation in a federal Indonesia, and the Indonesian Republican leaders, in particular Sutan Sjahrir, who desired autonomy and sovereignty under Republican leadership.[74]

In the contest between Dutch and Indonesian power, scientists were bit players. Nonetheless, in the conflict between the Dutch and the Indonesians, science and, more broadly, Western culture, were hotly contested. This began shortly after the British reconquest in late 1945, as Indonesian professionals, both inside and outside the British zone, were torn between their loyalties to the nation, modernity, and order over chaos. Indonesian administrators and officials took shelter in the institutions where they had worked during the Japanese and the earlier Dutch period, finding ways to reinvent those institutions' hierarchies as nationalist yet not as an explicit challenge to Western culture. Some went to the Netherlands on scholarships; others founded high schools for the Indonesian students. At times they risked themselves for the revolutionaries but not if it meant stopping their own work.[75] This was true in Buitenzorg as well, where the botanical gardens was split between British and Republican territory. The Dutch botanist Van Steenis, who lived inside the British camp at the botanical gardens proper, regularly visited the herbarium in 1946, which was across the street and in Republican territory. At the herbarium Toha and Margono (the first the leader of the *pemuda* [youth] faction in the herbarium, the latter the eldest staff member) apparently agreed with Van Steenis that the herbarium had an international value and strived to maintain it as a scientific institute until peacetime. The Republic paid the herbarium staff a salary and provided some rice. Despite hardships, they had kept the collections intact and only sold off small valuables to supplement their salaries.[76] Initially this was possible because a stalemate between the British and the Republic in West Java kept Buitenzorg calm. But the informal cooperation between Van Steenis, who was living under British protection, and Toha and Margono in Republican territory became strained as the British steadily increased their troop numbers in West Java in the early months of 1946, leading eventually to the British capture of Bandung in late March of 1946. After that, institutions in Republican territory such as the herbarium could no longer maintain both a Western identity and a commitment to the revolution.[77] Van Steenis was no longer able to work in the herbarium. Science still bound him to the Indonesians there, who agreed that he could take possession of his personal library, which was still in the building. Toha and Margono felt they could not hand over such assets to an enemy scientist. Nonetheless, they agreed to pack his belongings, and on the appointed day they did not go to work, allowing the British to stage an attack on the herbarium. During the temporary occupation Van Steenis's library was moved across the street to Allied territory, after which the British retreated to the gardens, relinquishing the herbarium to the nationalists.[78] The herbarium remained in Republican territory until the first so-called police action of 1947.

In 1945 and 1946, Dutch scientists had more freedom of movement than nationalist scientists and far greater opportunities. Their individual interests were usually best served by seizing more administrative power. Baas Becking, who returned to Java shortly after the British reconquest, suggested a scientific hierarchy in which research scientists like himself would orchestrate progress through their coordination of science's agenda, including the work of applied scientists.[79] Veterans of the prewar experiment stations had different plans. They sought to maintain some of the independence of the experiment stations, although there was a consensus that the colonial state should have a stronger role than before 1942. In early 1946 H. M. J. Hart, the director of the newly created Coordinating-Commission for Scientific Affairs, prepared a memo about the meaning of colonial science, which laid out plans for rebuilding the physical and intellectual infrastructure of science. He suggested a larger social role for scientists, with greater effort devoted to applying scientific advances to society: "The standard of living is a direct result of the degree to which science is able to exploit and develop the sources of economic life in a land."[80] And in fact scientists were given greater access to the levers of state power, especially when in late 1945 Pieter Honig, who was before the war director of the Pasuruan sugar experiment station, was appointed director of economic affairs. He and V. J. Koningsberger worked out a scheme in which private (and semiprivate) experiment stations would coordinate their work with other scientific institutes under limited government supervision, hence increasing state control over science.[81]

## Van Mook's Technocracy

Science had a great ally in H. J. van Mook, the Netherlands' chief civilian colonial official in the Indies. Shortly after arriving on Java in 1945, he placed Baas Becking in charge of all scientific matters, including those at the experiment stations.[82] Shortly thereafter Van Mook and Baas Becking agreed to follow the overall logic of the botanical gardens' reorganization plans in 1939 and 1940, with the Buitenzorg Botanical Gardens the official center of state-sponsored science and with other biological institutes subordinate to it. They thus labored to further expand the government's control of the colony's biological sciences.[83] Van Mook had spent the war in Australia, the United States, and Great Britain, and after returning to Indonesia in October of 1945 he came to believe that the Dutch would need to cede ground to the Republic even as he planned a federal system of power sharing.[84] He envisioned a new Indonesia, more multicultural and decentralized but ultimately a reconstructed Indies with Dutch interest and power ensured.[85] Science would be one of the institutions in which Dutch agency and power would remain. He initially wielded this power through Baas Becking and a science coordinating commission. During a meeting of the Coordinating-Commission for Scientific Affairs in March 1946, Van Mook explained that he believed "a country rests upon its scientific capacity." Because of the importance of scientific recovery, he swept aside concerns by the sugar manufacturers that the government under Baas Becking was taking over the experiment stations without proper consultation. Van Mook argued that immediate action was necessary in order to increase scientific capacity so that it could meet the social, economic, and political needs of the country as it existed now.[86] His scientific commission was sufficiently important to his vision for a new Indonesia that he made his ambitions for state-sponsored science public at the Malino conference on Sulawesi, where in mid-1946 he unveiled his plan for a federal Indonesia, promising that the commission would take the lead in shaping the new federation's scientific policy.[87]

Van Mook's reputation did not survive the breakdown of peace in 1947, certainly not among the Indonesian Republicans, who came to see him as a double-faced colonial official, projecting sweetness while secretly expanding Dutch power, nor among the Dutch, especially those in Europe who scapegoated him for losing the empire. Most scholarship about Van Mook has taken up the question of his intentions and importance in the context of the struggle between the Dutch and the Republic. Given his ultimate political failure — not only did his federal Indonesia fail to come about but he was also replaced in late 1948 — it is understandable that his legacy in shaping the administration has been downplayed. Nonetheless, in the three years that he managed the Dutch colonial state he pushed through a reorganization of the scientific administration, and many of these interventions were tacitly overtaken by his successors, both subsequent Dutch administrations in 1948 and 1949 and ultimately the Republic of Indonesia.

With his scientific reorganization Van Mook set out to create a means for centralizing scientific decision making under the governing executive, which in his theory would mean improved implementation of scientific knowledge. The scientists at the Department of Economic Affairs accepted this; certainly they noticed it placed them in a powerful position, and in late 1946 the Coordination Commission sent Van Mook its proposal for a Netherlands-Indies Scientific Service (Nederlandsch-Indischen Natuurwetenschappelijken Dienst). This was to be a powerful center of state-sponsored science, whose director would answer directly to the head of government and have the responsibility of orchestrating science for the whole country regardless of who owned or ran those institutes. This would minimize bureaucratic wrangling and allow the government, aided by scientists, to give direction to the course of Indies science. Officials from the service would be given wide latitude to enforce compliance.[88] From the commission's point of view, the Scientific Service would ensure that Dutch scientists would keep their institutions, and their control of science, even under Van Mook's federal state. This was made explicit in a paragraph of their proposal where the decline of Philippine science was linked to the hand-over of scientific research and leadership to Filipinos, who, they wrote, were of a mediocre caliber, even those educated in the United States.[89] But by then, Van Mook's influence and power were already circumscribed. After viewing the proposal for the Scientific Service, other parts of the colonial bureaucracy smelled trouble, and lawyers in the Department of Justice after reading the proposal decided that legally the service had to be either its own department or belong to another department.[90] This amendment negated the whole proposal, as the original service had called for a science czar, not just another department. Nonetheless, Pieter Honig, since mid-1947 the head of the Coordination Commission, plodded along. And on May 1, 1948, Honig became the new director of the Organization for Scientific Research (OSR), whose purpose was still the coordination of science.[91] Its real power was nil.[92] Still, it existed as a shell, and the government spent a considerable f.130,000 in its first year to construct a new building.[93] Even without real power, though, its prominence in the bureaucracy gave it wide theoretical power, and it became a blueprint for organizing science in Indonesia after 1950.

## Dutch Colonialism Adrift

For most of 1946 Van Mook and the Dutch held fast to a diplomatic solution in Indonesia. From their perspective there was reason to expect a peaceful de-colonization in which the Dutch would continue in important capacities afterward. What the Dutch saw in the new Republic were cosmopolitan nationalist leaders pursuing talks with them. A cease-fire in late 1946 brought a lull in hostilities. In the Linggajati Agreement of November 12, 1946, the Dutch recognized Republican authority over Java, Madura, and Sumatra in a future United States of Indonesia. At the same time, Van Mook's federal Indonesia was gaining momentum with preparations for forming federal states in Borneo and eastern Indonesia. The two sides seemed so close to an agreement that the British troops left Java at the end of 1946.

All this progress masked a great distance between the two sides. At Linggajati, the negotiators had not addressed some of the hard questions about a future federal army combining Dutch and Republican forces or the continued existence of Dutch enclaves on Java and Sumatra. Worse, the Linggajati Agreement, signed by the Dutch government on March 25, 1947, had wording changes from the earlier agreement—changes not approved by representatives of the Republic—increasing the Dutch role in the future union. Pemuda military pressure revived in early 1947. At the same time, the Dutch troop presence was growing steadily, and the generals planned for war. The Dutch government now saw that the military occupation was too expensive without revenue from key economic resources on Java and Sumatra. By May of 1947, within the Dutch government leadership, the tide had shifted toward a military solution.[94]

That Dutch public opinion pushed them farther in that direction is shown by a letter written in June of 1947 by former colonial scientists in the Netherlands, asking the Dutch government to do something about the Republic. V. J. Koningsberger and sixty-one others (including Docters van Leeuwen, Koningsberger Sr., Pulle, Lam, Backer, and Dammerman) prodded the Dutch government to take strong action to avert disaster in the scientific institutions under Republican control. They argued that these scientific institutions had kept nature in balance with the population explosion of the last few decades and if this balance were disrupted, for example, by changing agricultural practices, it would cause irreparable damage to the population. Since the Republic refused to return the scientific institutes, it was up to the Dutch to retake them.[95]

The Dutch scientists' letter reached Van Mook after he had decided to pursue a military solution. Still, the arguments expressed by the Dutch scientists were shared by Van Mook. As the historian H. W. van den Doel has recently argued, Van Mook saw the Dutch colonial mission as one to protect, civilize, and uplift the people of Indonesia. In mid-1947, unable to exert influence inside the Republic any other way, this meant military intervention, even if Van Mook knew better than most Dutch the likely hardening of Indonesians' negative feelings against the Dutch that would result.[96] Starting on the evening of July 20, 1947, Dutch troops began an invasion of Republican territory on Java and Sumatra, formally breaching the Linggajati Agreement. The Dutch referred to this military operation as a police action, meaning to suggest that it was an internal colonial matter. They captured areas of Republican territory in West and East Java, in particular its economically important areas, as well as the plantation areas in Sumatra, before being halted by international pressure. Still, as Van Mook continued to speak and plan for a federal Indonesia, at its root the conflict was now a contest for power without much room for power sharing. A new compact, for a new federation, was signed between the two sides, the Renville Agreement. This, too, failed when Dutch troops invaded Republican territories again in late 1948 in the so-called second police action (Van Mook had been replaced by L. J. M. Beel in early November of 1948). But the Dutch were unable to establish long-term military or political control of the colony, and international pressure on the Dutch government forced talks that eventually would lead to the transfer of sovereignty in 1949.

After mid-1947, Dutch inhabitants returned to their homes and workplaces in the towns and cities of Indonesia that were under Dutch control after the so-called first police action. Theoretically things were normal. Government officials, though, had little sense of how to proceed. Confusingly, they were supposed to work on creating institutions meant to be part of a federal Indonesia, but they usually worked inside the institutions of the Netherlands East Indies, and they knew only the systems of the colonial state. Old things got new names, and those new federal institutions that did come into existence were much like the OSR, empty shells without clear jurisdictions or responsibilities.

The Buitenzorg Botanical Gardens epitomizes this trend. In mid-1947 Baas Becking again took up his job as the garden's director, and he tried to take up where he had left off in 1940. As he wrote at year's end, the "best news" of 1948 came when a telegram arrived from the Netherlands announcing that the botanical gardens could now identify itself as a "royal" institution.[97] In August 1948 Baas Becking published an English-language pamphlet, complete with attractive woodcut illustrations of the gardens, to demonstrate to the international scientific community that the Buitenzorg Botanical Gardens was back. Baas Becking claimed in this pamphlet that the gardens was returning quickly to its prewar condition with even better funding than in the 1930s. He boasted that the 1948 budget for the botanical gardens could fund thirty-seven permanent academic positions with nineteen already filled.[98] Yet at the same time his subordinates knew that Baas Becking was making no serious effort to recruit new scientific personnel and was hindering the one active research project, the Flora Malesiana, probably because he did not believe in the value of plant systematics.[99] On the other hand he had no rival research program. He quit in late 1948, leaving for the South Pacific after conflicts developed with Honig, as well as with Wisaksono Wirjodihardjo, who had become director of the Department of Agriculture and Fisheries, under which the Botanical Gardens had been placed.[100] Wisaksono was a soil chemist who had graduated from the agricultural high school in Buitenzorg and had since 1920 worked at the soil institute in Buitenzorg, where he was one of the very few Indonesians who before 1942 had held a scientific position in Buitenzorg.[101] Van Slooten took over as director of the botanical gardens, but all he could do was look backward to the ideas of Van Mook and Baas Becking.[102] Dutch colonial science, like Dutch colonial life, was completely adrift.

### Nationalist Science

The transfer of sovereignty to the Republic of the United States of Indonesia on December 27, 1949, was greeted with excitement by Indonesian intellectuals. A buoyant spirit was prevalent, not just among the victorious Republican leaders but among the Indonesian intelligentsia. Pramoedya Ananta Toer remembered how a great pressure lifted when the sovereignty of Indonesia became a reality, creating the freedom to circulate his ideas broadly. He felt that as a young nationalist writer he had a great future.[103] The Republic quickly began to take over the institutions of the colonial state, making nationalism the prescribed ideology. In February of 1950, the new state issued a nationalization decree for science, requiring it to reflect the national tendencies of Indonesia. This new decree entailed especially personnel changes, and over the next few years Dutch personnel were replaced with Indonesians in all the scientific institutions. Dutch administrators were replaced immediately, while Dutch researchers and educators stayed on. In early 1950 at the botanical gardens, Koesnoto

Setyodiwiryo replaced Van Slooten, although the various units of the gardens were still directed by and staffed with Dutch scientists. The political change, in science also, reflected the Dutch defeat. In 1949 the OSR governing board had two Indonesian and twelve Dutch members; in 1953 it had six Indonesian and two Dutch members. Just being Indonesian was not enough, though, as the decree had also prescribed that science would serve the nation.[104] The Dutch scientists grumbled, predictably. But many intellectuals believed, like Pramoedya, that they now had the power and opportunity to shape the political future of the country. This meant instilling the spirit of nationalism in the halls of old colonial institutions.

Nationalism fueled the institutions of science in the early Republic of Indonesia. It would be up to the very few Indonesian scientists—in the early 1950s they were either, like Koesnoto, veterans of the Dutch colonial agricultural institutes or doctors trained in the colonial Dutch medical schools—to create Indonesian science out of the Dutch scientific institutions. But given the limited resources and their own background as colonial officials, shaping a new Indonesian science did not constitute a radical break. Moreover, the Indonesian scientists concentrated on building professional institutes with government funding, much like their colonial forebears had. As I will show, the science of the early Republic drew on colonial models, including the technocratic administration partially established by Van Mook. Nonetheless, Indonesian science, useful to the people and supported by the government, was to be defined in national terms. This did change the character of Indonesian science in more ways than sending Dutch scientists packing. It also explains the transformation of a peripheral research dream about the archipelago's flora into Indonesian biology's first official project. In the early 1950s Kees van Steenis's flora project, the Flora Malesiana, went from a one-man obsession to an authorized scientific project paid for by the Republic of Indonesia. The Flora Malesiana was to be a compilation of all the data ever collected on the plants of the archipelago, synthesized, analyzed, and organized in one major publication. Van Steenis had been laying the groundwork for the Flora Malesiana since the 1930s, and after he left Buitenzorg in 1946 he had lined up support from some tropical plant systematists in Australia and the United States by very aggressively advertising the project internationally, suggesting that it had government support.[105] In fact through 1948 he had been paid only on an ad hoc basis and was spending money not yet allocated by Van Mook's government.[106] To Dutch biologists still in Buitenzorg, it seemed logical that the Flora Malesiana would be absorbed by the botanical gardens as one of their own projects.[107] Part of this was a turf war, which Van Steenis was losing.[108] But before this time his Flora Malesiana had not served much of a social or political purpose. Van Steenis seemed to be aware of this, and in a piece written late in 1948 he stated a social purpose for the Flora Malesiana for the first time: "None of us can predict the industrial future of some neglected plant species, but we should be prepared for any coming rush on the botanical wealth of this vast Archipelago."[109] Nonetheless, this kind of argument about inventorying the known plants of the islands was hardly a strong claim for supporting the Flora Malesiana over the botanical gardens, which had been making the plant species of the archipelago known for a century. By mid-1949 Van Steenis conceded that the Flora Malesiana was in serious trouble and might only be saved if he were appointed director of the gardens.[110]

So it is all the more surprising that in early 1950, the Flora Malesiana received funding independent of the botanical gardens. One champion of the project was Wisaksono Wirjodihardjo, who had become the secretary of state for the Department of Agriculture and Fisheries, under which the botanical gardens was placed.[111] When Wisaksono and his Dutch subaltern adviser (known as his *pembantu*), J. van der Ploeg, visited the Netherlands in mid-1949, they met with Van Steenis and suggested creating a foundation (*stichting*) to oversee the Flora Malesiana. Van Steenis immediately consented and hired a lawyer to help write its statutes.[112] He returned to Indonesia a month before the transfer of sovereignty and, armed with the statutes, shopped the Flora Malesiana around to the various new Republic of Indonesia offices in Batavia, now renamed Jakarta. Of particular importance were Hermen Kartowisastro, the director general of agriculture and fisheries, and Djuanda Kartawidjaja, a graduate of the Faculty of Technical Science in Bandung, who as the minister of prosperity in the Hatta cabinet (December 1949–August 1950) had jurisdiction over agriculture. From all these meetings Van Steenis learned that the Indonesian officials in Jakarta were not against science but against Dutch leadership of science. The institutions of Indonesian science were to be administered by Indonesian scientists even if they enlisted Dutch or foreign help. The Indonesians would never approve of a Flora Malesiana Foundation unless Koesnoto, who as the head of the botanical gardens was the chief academic biologist in the country, controlled it.[113] This did not bother Van Steenis, who was starting to believe that under the colonial government the Flora Malesiana would never have come about.[114] The Indonesian scientists and officials liked the idea of Indonesia sponsoring an international research project controlled by Indonesian scientists and ultimately about the plants of Indonesia.[115] The Flora Malesiana doubtless had another appealing aspect for Republic leaders such as Djuanda and other officials in the early Republic.[116] For one hundred years a series of Dutch experts had collected and cataloged the natural organisms of the archipelago, sometimes agreeing on the taxonomies, sometimes vociferously not. The documentation of these resources was not, in fact, simple to access. The botanists had become guides not only to the natural realm around them but to the system of paper that explained it. They argued repeatedly that only they, the Dutch scientists trained and inculcated in the system of documents, could properly access knowledge and must therefore stay at the gardens. They argued that no Indonesian could replace them. Doubtless some officials could see that the Flora Malesiana, with its arcane knowledge, was a way to sidestep the old boys club, and place it, easily accessible, into the hands of Indonesian scientists. Dutch involvement, such as that of Van Steenis, was fine as long as it acknowledged Indonesian control.

Van Steenis, who by his own admission had not been interested in colonial politics, was astute (still not articulate though) about the Indonesian political position on science: "They know that the whole world is dependent upon each other, and that precisely their independence enhances this dependence, but they do not want to see it said."[117] When the Flora Malesiana Foundation finally came into existence in October of 1950, after much wrangling about the wording, its charter said that the research would "be made possible by a Government grant to the Foundation, [and] the Foundation shall be an Indonesian organization."[118] With this the first Indonesian science project was launched.

The Indonesian scientists would soon take up where the Dutch left off in scientific institutions everywhere. They did not yet, however, hold sway. At the center was the OSR. It had been printing a regular journal since 1949, and this was continued. In 1950 and 1951 it served the purpose of printing parting shots by Dutch scientists who were leaving, or had already left, the country. For example, a Dutch professor who had taught at the former medical college in Surabaya, and since 1951 had been on the Medical Faculty of the University of Indonesia, addressed "some questions which arise to the Pharmacologist in Indonesia," a lot of which questioned the current relationship among science, the Indonesian state, and society.[119] A chart of Indonesian science showed ominously the "Federal Government" and OSR surrounding all the national scientific institutions.[120] An October 1950 speech by the dean of the Faculty of Technical Sciences in Bandung, which was gloomy about Indonesian science's future in the current political climate, was translated for publication in English, presumably so outsiders could learn of the sorry state of Indonesian science.[121] In 1951 and 1952, as Indonesian scientists came to fully control the OSR, they gave the organization an Indonesian name (Organisasi Penyelidikan Ilmu Pengetahuan Alam), stopped printing criticisms of Indonesian science, and instead began planning for Indonesian science. In 1951 one of the first activities was organizing a small conference in Bogor, as Buitenzorg was now known, in which the future of Indonesian science was discussed.[122] And successive publications of the Organisasi Penyelidikan Ilmu Pengetahuan Alam showed Indonesian science positively again. This led to a reprint in 1952 of Baas Becking's 1949 pamphlet *Buitenzorg Scientific Centre*, of course updated to proudly display national scientific institutions. The institution was now to be known as the Bogor Scientific Centre to coincide with the new Indonesian name of the city.[123] In 1951 the minister of education and culture established a committee charged with investigating the founding of an Indonesian council of sciences, with Sarwono Prawirohardjo, a prominent obstetrician and academic at the University of Indonesia, heading the committee. In the mid-1950s the institution of the former OSR was turned into the Indonesian Council of Sciences, a process examined in the next chapter.

## Conclusion

Van Mook reattached science to the colonial government in the 1940s after science had been cast aside at the beginning of the Depression. His goal had been to centralize scientific decision making near to the governor general's office and thus effect a coordinated scientific administration with power over all science in the colony. Like the earlier and similar attempts at scientific centralization by Treub, there was opposition to these endeavors by both scientists and other colonial officials. And Van Mook's scientific reforms were only partial, even if the institution of the OSR realized some of his goals. Still, after 1945, government science was able to address the usual criticism directed at it—what good is science going to be?—which had been heard as recently as 1940 during the Volksraad debates about the botanical gardens. After 1945 Van Mook positioned science as a way for the state to maintain cultural power and consistency. In federal Indonesia, Dutch scientists were to be empowered as experts with government authority and power. This emphasis on scientific leadership held less credence after the first police action. And Baas Becking, as that new technocrat, was a great disappointment to Van Mook. Nonetheless, the ideal of installing scientists inside the state as technocrats was given institutional life in 1948. With the victory of Indonesian nationalism, the Republic of Indonesia's desire to exert leadership breathed new life into the relationship between science and the state. And it is here that the prewar Indonesian apostles of national enlightenment, and their wish to lead the Indonesian people to progress through science, found an institutional home.

# 7

# Desk Science

In the years after independence, autonomous communities of scientists in Bogor, Bandung, and Jakarta began building new scientific traditions to reflect their Indonesian national identity. First this meant removing the colonial presence from the existing scientific institutions, something the new Indonesian Republic did deliberately and quickly. In 1950 and 1951 the Indonesian Republic removed the Dutch scientific administrators, but that was the extent of government intervention in science at the time. There was a general sense of optimism throughout the Republican government about the possibilities for Indonesian science now that it was no longer shackled by the colonial government and its Dutch administrators, and now had Indonesian scientists in charge of creating the disciplines of Indonesian science. At the botanical gardens in the 1950s, Indonesian biology came into existence under the care of Koesnoto Setyodiwiryo, a prewar graduate in agricultural sciences of the Wageningen University in the Netherlands, who had been one of the few Indonesians to work as a research scientist in Buitenzorg under the Dutch. Koesnoto, as director of the botanical gardens after 1950 (in addition to numerous other teaching and administrative responsibilities), had great freedom to shape the biological discipline as he saw fit. Although initially he imagined this would mean continued collaboration with Dutch scientists, ultimately he decided that professional biology would need to be carried out solely by Indonesian biologists. Unable to find enough Indonesian biologists, he created his own when he founded a biology academy in 1955. From this academy came the first generation of Indonesian biologists, a generation of talented, idealistic, and patriotic scientists who would come to lead Indonesian professional biology after 1966.

The autonomy of Indonesian scientists largely ended in 1959 with Su-karno's authoritarian Guided Democracy. Scientific activities came increasingly under the control of the government. Still, by then the groundwork for Indonesian professional biology in Bogor was in place, at least in terms of personnel. But its larger purpose was still ill-defined. In terms of research, Koesnoto's team tended toward natural history, that is, cataloging and taxonomy, and tried to explain what was Indonesian about Indonesian nature. In this sense the first professional Indonesian biologists were motivated by the Enlightenment ideal of creating useful knowledge, in this case useful to the Indonesian nation. But little thought had been given to working out how professional biology should or would be connected to Indonesian society or culture, and even less was done to integrate professional biology into society. In this way Koesnoto repeated the mistake of his colonial predecessors. Throughout the 1950s, professional biology was an elite occupation, distant from most Indonesians. And after 1959, as the Indonesian government took firmer control over intellectuals, including the biologists in Bogor, the state increasingly gave direction to Indonesian biology. Over the course of the 1960s, biologists and other scientists looked to the state as the principle audience, and patron, of their work. During the 1960s, biologists in Bogor returned to bureaucratic science, where they coupled the idealism of Enlightenment knowledge they had learned a decade earlier with the quick, tangible results expected by government officials. They reorganized themselves around the goals of generating knowledge for the development of Indonesia. They accepted government assignments such as combatting agricultural pests. Yet they still fostered their identity as objective researchers who were part of an international community of biologists. They became what I call desk scientists. From their administrator's desks they juggled a myriad of responsibilities—to the state, the nation, the people of Indonesia, and their local scientific community yet also to international science—and set the agenda for the discipline. These scientific managers' offices became the center, and the norm, of Indonesian science.

Writing recently, the historian Rudolf Mrázek has argued that civil society has been weak because of the sterile office environments that have defined post-independence middle-class society: "The culture of the Indonesian middle class was, and remains, rather than a bourgeois culture an office-like culture. Through colonialism, the Indonesian *Angestellten* [office workers] were robbed of the essential means of production—production of their own material wealth and their own spiritual culture—and, in the post-colonial era, have not gotten them back."[1] In the conclusion to an essay about memories of and nostalgia for the late colonial period, Mrázek is pessimistic about the future, arguing that middle-class Indonesians have been dispossessed, not just by authoritarian governments but by "colonial or global political and economic power structures."[2] I interpret Mrázek's use of *spiritual* as meaning the full engagement of the human intellect. In this chapter I take up Mrázek's suggestion and examine the "spiritual culture" of Indonesian offices, especially during the formative years of the New Order. Science makes a good case study for testing Mrázek's hypothesis: it has had broad middle-class support since independence, was founded in old colonial institutions, and was always linked to the global culture of science and scientists. For many educated Indonesians, the appeal of science was its Enlightenment ideal of revealing the workings of nature in ways that would be useful; ideally, scientific knowledge should transcend any particular political culture. Moreover, scientists have worked largely in offices, many funded by the state. Still, they have also seen themselves as important representatives of civil society, producing a road map to a better future for Indonesians. Viewed from the outside, Indonesian science has been a failure as its research has not been competitive internationally. Indonesian desk scientists, however, have measured their success more by their usefulness to the nation and the creation of a solid intellectual community. This chapter examines how in the last fifty years Indonesian scientists have created a culture of science that balances their Enlightenment ideals with the political logic of authoritarian regimes.

The weakness of civil society during the New Order can be diagnosed as the corollary to an authoritarian regime with roots in the colonial period, as Mrázek and others have done.[3] Historians and other scholars have extensively investigated the reasons for the failure of democratic institutions in the 1950s by debating in particular the motivations and actions of parliament, the military, and President Sukarno. The consensus interpretation now is that Sukarno and the army worked throughout the 1950s to expand their influence and only needed parliament to falter to grab political power.[4]

Fifty years ago Herbert Feith took a different tack in exploring civil society in 1950s Indonesia. He analyzed the country's political history in terms of a political dance between administrators and solidarity makers, between intellectual elites and populists. In tracing the sphere of political activity between 1949 and 1957, he charted the decline of the administrators and the rise of solidarity makers, who could whip up mass political support via campaigning and electioneering.[5] Feith's explanation for the emergence of authoritarianism has not stood the test of scholarship. Nonetheless, I believe he was right about the historical importance of professional administrators. Moreover, their changing political fortunes at the end of the 1950s had an important effect on the functioning of Guided Democracy. After 1957, administrators did not hold legislative or executive power; rather, they were given the task of making the country run. They were subordinated to other state power holders, including army officers, communist leaders, and Sukarno.[6] In justifying the imposition of Guided Democracy, Sukarno was helped by intellectuals and administrators such as Roeslan Abdulgani, Mohammad Yahmin, Djuanda Kartawidjaja, Ki Hadjar Dewantoro (as Soewardi had been known since 1922), and Djokosoetono, then the dean of the Faculty of Law of the University of Indonesia.[7] Djuanda, as an engineering school graduate, a former minister of welfare, and an apolitical administrator, was the prime minister under Sukarno's Guided Democracy from 1959 until his death in 1963. Administrators were willingly and effectively integrated into the authoritarian politics of post-1959 Indonesia.

In the late 1950s, Sukarno and the army pushed an older generation of administrators out of political activity. Administrators as a professional class, however, did not disappear. At the same time Sukarno pursued an alliance with the PKI, whose leaders closely participated in creating and enforcing a revolutionary ideology. Sukarno thus continued consultation with civil society groups that accepted his leadership, including administrators such as Djuanda as well as the communist leadership. Sukarno's Guided Democracy actively sought to enroll civil society leaders in his regime, especially members of compliant political parties, within a variety of new political assemblies and councils. This created a facade of democratic legitimacy but also provided all kinds of opportunities for individuals and institutions not on the wrong side of Sukarno.[8] Leaders of the PKI, the Nahdlatul Ulama party, and the PNI played prominent roles in some of these consultative councils, but on the whole they were comprised of younger elites, including younger professional elites.

Civil society institutions in the era of Sukarno's Guided Democracy have largely been forgotten, and on the whole scholarship is silent about political changes below the highest level of Jakarta politics during the late 1950s and early 1960s. This is a shame. In fact this system for connecting Jakarta politics to local leadership survived Guided Democracy and provided a linkage between the political aspirations of the first generation of Indonesia's elites and the political system of the New Order. The New Order government transformed the assemblies, councils, and other institutions into sterner implements of the state, eventually providing a more efficient means of recruiting administrators and technical experts into the New Order state. But given these institutions' origins outside or on the fringes of the state, its individual members often saw their jobs as transcending the state. The history of science is a particularly poignant case.

A feature of both Guided Democracy and the early New Order was its use of professionals and trained experts, who took up privileged and powerful positions in Sukarno's, and later Suharto's, administrations. In this chapter I will argue that, far from being new, this administrative arrangement had its roots in Van Mook's regime of the 1940s. Technocratic ideals were kept alive in the 1950s, in particular by Feith's administrators. To be sure, within a parliamentary system the administrators were not able to plant their vision within the Indonesian political order. Professionals nurtured their ideas in professional associations and institutions and never found a way to cultivate popular support, a critical hindrance to amassing power in the democratic system of the 1950s. But in the decade after 1957, there was sufficient overlap between the ideals of the scientists and the nationalist ideologies of the authoritarian governments to make them work effectively together.

By the middle of the 1960s, Indonesian professional science was shaped by scholars who began their careers after 1949. Many had trained abroad and returned to work in government institutions. This chapter examines in particular the biologists in Bogor who built Indonesian biology on the ruins of colonial science. They started their scientific careers filled with Enlightenment idealism and enthusiasm in an era when anything still seemed possible for Indonesia. During the 1960s, as scientific institutions moved closer to the state, they were shaped by both the opportunities and the trade-offs of working within the government. Running biology meant balancing these forces, usually within offices tied in to the New Order administration, where scientists began to practice desk science. By following state guidelines, especially those enshrined within five-year plans, such as improving the rice harvest, they became a permanent budget item. And by the early 1970s they had expanded professional biology with the help of the state.

## The First Generation of Indonesian Biologists

In terms of manpower, Indonesian science started nearly from scratch after the revolution. In 1942 there were a mere four hundred Indonesian college graduates, most of them physicians.[9] In the early years of independence, as the Indonesians surveyed what they had inherited from the Dutch, the scientific institutes and universities were seen as valuable assets, especially if they could be staffed with Indonesian scientists. This meant creating the first generation of research scientists; during the Dutch colonial period only five Indonesians had ever worked as scientific researchers in any field.[10] Consequently, after 1950 the hope for science was placed on young Indonesians, such as the new director of the botanical gardens, Koesnoto Setyodiwiryo, who were expected to bridge the distance between Indonesia and the world of science. The groundwork for the discipline of Indonesian science was largely laid in the colonial scientific institutes, although that in itself did not act to constrict new opportunities. The more important legacy of colonialism for Indonesian science, concerning the relationship between the scientists and the state, was an issue that did not confront scientists until the early 1960s. The leaders of the scientific institutions in the 1950s, mostly engineers and doctors educated by the Dutch in the 1920s and 1930s, worked on training the young scientists while also trying to figure out what Indonesian science would be. There was no single controlling body or institution, but on the whole there was broad agreement that an investment in science would benefit all Indonesians while at the same time legitimating Indonesia as a modern civilized nation. In this sense the discipline of Indonesian science was a nationalist endeavor, being both for the Indonesian people and reflecting proudly the Indonesian nation. These ideas were formative for the young generation of scientists trained during the 1950s.

In 1950, Indonesian biology was immediately at the forefront of Indonesian science largely because that is how it had been before independence. Koesnoto became the first head of Indonesian biology when in 1950 he became the director of the botanical gardens, now known as Kebun Raya Indonesia, the Great Indonesian Gardens. Koesnoto was an agricultural engineer with a colonial-era degree from Wageningen University, one of only fourteen Indonesians with a degree from Wageningen in 1945. He had established himself as an expert on drought-resistant plants even prior to the war and under the Japanese occupation had been the chief of the Agricultural Research Bureau. During the revolution he had remained as a senior administrator in Buitenzorg.[11] As the new director of the botanical gardens, he was empowered to oversee the decolonization of Indonesian biology. In 1951 he still saw this in terms of a gentle transition:

> A young nation must maintain respect and standing in the eyes of the world— and Bogor as scientific center is one of the standards used by the international world to judge Indonesia's value. The country should resolutely and without sentiment take precautions to prevent any form and all efforts leading to the decline or dismantling of these institutions. We must squarely face our own deficiencies in the organization and leadership of these universal and scientific-cultural institutions. We will not hesitate to entrust leadership and management of the various institutes to foreign experts who wish to help us.[12]

There was an enthusiasm and excitement about the possibility of participating freely in the world of science. Koesnoto was almost ecstatic about the future of Indonesian science, writing in early 1951 to a Dutch biologist, "It appears that the Indonesian Botanical Gardens shall spread its wings wider than the Botanical Gardens has ever done before."[13] To Koesnoto it felt like the political shackles that had bound science in the colonial period were coming off.

While Indonesian scientists were creating the institutions of an independent Indonesian science, the Dutch staff at the botanical gardens continued working, trying to ignore the end of the Dutch empire. This was possible because Koesnoto was rarely at the gardens, as he had multiple teaching and administrative obligations. In 1955 he held thirty-seven different official jobs.[14] Koesnoto, as he had stated openly since 1951, encouraged the Dutch scientists to stay, as there were few others able to take their place. But this was clearly temporary, as became apparent to Koesnoto in 1955. At a meeting in January, M. A. Donk, the Dutch head of the herbarium, lashed out at the gardens' administration, accusing it of incompetence, which he claimed was dragging down the international reputation of the gardens. Donk then singled out Douglas, the chief in charge of the gardens proper, accusing him of challenging Donk's authority. Donk, as the senior botanist at the gardens and chair of the international nomenclature committee, felt that by all rights the scientific and administrative leadership of the gardens should be his. Koesnoto used this embarrassing incident to clean house, and on January 31, 1955, he fired Donk, effective immediately, and brought in Anwari Dilmy, a young Indonesian botanist from the Forestry Experiment Station in Bogor, as the new herbarium director. The firing of one of the senior Dutch scientists was understandably greeted with concern by other members of Dutch scientific staff, who correctly interpreted this to mean their Indonesian careers were ending.[15]

Under the Dutch, biology had not been seen as an appropriate career for Indonesians. Indonesian applied biologists, educated at the various agricultural colleges, had been trained to serve as liaisons between the colonial government and Indonesian farmers and peasants. A small number of Indonesian agricultural experts, including most prominently Wisaksono Wirjodihardjo and Koesnoto, had risen under Dutch colonialism to become scientific researchers. It was not until the early 1950s, though, that a career in biology was possible. In the 1950s, Indonesian students began studying in the biology departments at Bandung Technical University (Institut Teknologi Bandung, ITB) and the various natural science faculties at the University of Indonesia and Gadjah Mada University. In the optimistic and idealistic few years after Indonesia achieved sovereignty, the idea of becoming a scientist attracted high-achieving high school graduates. Even outside of the fledgling biology departments, there were ad hoc government initiatives for training young scientists. For example, one of the first zoologists of Indonesia, Sampurno Kadarsan, took an apprenticeship at the Department of Fisheries that included formal biological instruction after graduating from high school in 1950. He had gone to a Dutch-language high school on Java and had become interested in evolution after reading about Darwin. He did an internship at the zoological museum in the early 1950s to receive hands-on training.[16] Still, scientific institutions, including the botanical gardens, struggled to find scientific staff throughout the 1950s. The economic and political climate made Dutch scientists less and less tenable. Most newly graduated Indonesian academic biologists were immediately absorbed as teachers by their university biology departments.[17] Koesnoto followed their lead.

Starting in 1955, Koesnoto began to implement his vision of Indonesian participation in the international world of science. For him this meant not jettisoning international contacts but quite the opposite; in a letter to C. G. G. J. van Steenis in the Netherlands, he explained that firing Donk had been necessary in order to maintain institutional strength and continued international cooperation in Bogor. Donk had been undermining this by trying to revive the colonial model of biology in which the herbarium, under his scientific authority, coordinated biology throughout Indonesia.[18] Still, the departure of Donk and other members of the herbarium did create a scientific void, especially as the worsening economic situation meant there was no money to purchase new equipment or even maintain subscriptions to scientific journals. Many of the trained Indonesian support workers were switching to higher-paying jobs in private industry.[19] Few European and American biologists could be lured to

Bogor with an unsure tenure and salary. Koesnoto wanted biologists with loyalty to the Indonesian nation and the botanical gardens, young scientists who would fight to keep the gardens at the center of the biological discipline in Indonesia. When the Dutch no longer filled this role and no support was forthcoming from either the Indonesian government or other Indonesian scientists, he did it himself.

At a general staff meeting on May 28, 1955, Koesnoto announced that the Department of Agriculture, of which the botanical gardens was a part, had decided to establish a Biology Academy attached to the gardens. Recent high school graduates would be immersed in the study of the pure biological sciences. The academy would commence classes in September, and the teachers would be drawn from the staff at the gardens. The rationale behind the academy, according to Koesnoto, was that the quality of academic education was quickly diminishing and the academy could feed students into an international educational exchange program sponsored by the Ford and Rockefeller foundations. English would be an important subject. And anyone receiving an education at this academy would be legally required to work at the gardens. This would solve the pressing problem of manpower. Most of the staff members were taken completely by surprise and worried that precious time was being taken away from scientific research. J. Ruinen, a Dutch citizen and head of the Treub Laboratory, complained that this was a duplication of effort and was being forced on people who had little or no experience teaching. Koesnoto defended the academy, saying that the government had mandated it and he was not forcing the staff to do anything. He desired teachers who would do it for love of the gardens.[20] Ruinen was so disgusted with Koesnoto and the Department of Agriculture for breaking with the principle of free, scientific research that she drafted and sent a letter of resignation the same day.[21] A few days later, suspecting that the academy was nothing more than a vehicle for Koesnoto's ambition, she went to the Department of Agriculture in Jakarta to ask about it. Officials there confirmed that Koesnoto was the driving force behind the academy.[22] But by then the academy would not be stopped.

The academy promised to graduate thirty Indonesian biologists per year and was intended to decolonize Indonesian biology.[23] Koesnoto's two lieutenants, Anwari Dilmy and Soetomo Soerohaldoko, coordinated the work of the planning committee.[24] They sent a delegation to the Ford Foundation to ask for money for a new building. The scientific staff members at the gardens who unequivocally said yes were selected as teachers.[25] Outside examiners were lined up to supervise the final exams of the three-year course.[26] In order to promote national unity, students would be selected from all ethnic groups throughout the archipelago. Places were reserved for women. The government would pay tuition, room, and board.[27] Unlike universities such as ITB, which still used the Dutch system of examinations in which students would schedule classes and exams when they felt ready, the Biology Academy would take

American higher education as its model. All students would follow the same curriculum, and exams for everyone were scheduled at the end of the three-year term. Students were thus required to finish in a fixed amount of time.[28] Koesnoto moved very quickly. In the summer of 1955, announcements in newspapers solicited qualified high school graduates.[29] During the first year students would receive a rigorous introduction to the biological and other sciences, including instruction in botany, taxonomy, cytology, plant geography, zoology, physics, chemistry, and agriculture, as well as English. In the second year students would add marine biology, genetics, ecology, physiology, meteorology, biochemistry, and geology to their schedule.[30] Classes started in early October 1955.

On October 10, 1955, three hundred people attended the opening ceremony at the Cibodas gardens, situated at a temperate 1,200 meters above sea level. On a sunny day, in the gardens where Teysmann had planted the first cinchona tree in 1852, Indonesian biology had its proudest day since independence. Dilmy was chairman of the opening and presided over speeches by Koesnoto as director of the botanical gardens, the minister of agriculture, and Vice President Mohammad Hatta. Numerous government officials, including the prime minister, attended. Comfortably seated under canvas, the crowd listened to Koesnoto's keynote speech about protecting nonrenewable resources. He concluded by predicting that in this nervous, modern age it would be young scientists and their research that would guarantee the continued survival of Indonesian culture. Science, he concluded, would right the wrongs of the recent past.[31] Official photos of the event I saw in Bogor in 2001 show the pomp and circumstance of the opening ceremony. The Dutch biologist M. Jacobs, who did not understand a word of any of the Indonesian speeches, was impressed by the organization and timing. A few weeks later he wrote that as he watched the obviously political spectacle he had a vague sense that the era of European influence in Indonesia was definitively over.[32]

By all accounts the experiences of the first cohort were exceptional. The directors of the school chose thirty students from about nine hundred applicants.[33] In a recent informal poll a number of these students reported that they chose the Biology Academy over a university—where many had also been accepted—because all the students received free room and board at the academy.[34] Most students were from Java, but the Sulawesi, North Sumatra, and Maluku provinces were also represented. The intention had been to establish the school near the botanical gardens, but because time was short and funds nonexistent the students roomed and studied at the Cibodas Botanical Gardens. The teachers would drive up the mountain once a week to lecture. In the visitor's lab, newly rebuilt with funds from the United Nations Education, Scientific, and Cultural Organization (UNESCO), offices were converted into dormitory rooms and the central lab became the lecture room. The five female students slept in the main administrative building a few hundred feet away.[35]

The students worked hard and enthusiastically wrote down every word uttered by their teachers. Anwari Dilmy watched over them like a father, and everyone, faculty, staff, and students, worked toward the success of the school.[36] Probably because of the seclusion of Cibodas, members of the first cohort bonded more than would be usual in a class of thirty students. They did everything together: play, work, and chores.[37] They had little mobility, although on Sundays they could walk the five kilometers to Cipanas to watch a movie. And once a week they would be trucked down (literally!) to Bogor for practicums with proper laboratory equipment.[38] The students' English was marginal, though sometimes better than the lecturers', especially those from Central Europe. Following class with L. M. Olah, a Hungarian cytologist who worked at the Treub Laboratory, students would gather together to decipher the content of the English-language lectures guided only by the illustrations on the chalkboard.[39] In April 1956, after regional rebels from the Indonesian Islamic State movement (Darul Islam) attacked a nearby village, a rumor spread that the rebels were going to "recruit" the female academy students as nurses. The five female students decided they could not stay at Cibodas any longer and sent a spokesperson to inform the school administration. Koesnoto was livid, but he did find temporary housing for the students in the botanical gardens.[40]

The Biology Academy was relocated in late 1956 to a new complex in Ciawi, a suburb of Bogor. At that time the school was renamed the Department of Agriculture Academy. After the move, the Department of Agriculture replaced the broad biological curriculum with one focused on practical agricultural subjects. It quickly became a teacher's college. Almost all of the students after the second cohort became officials in the Department of Agriculture, and many left Ciawi to become teachers in new agricultural high schools that were popping up around the archipelago.[41] In 1968 the school closed after graduating 419 students in ten cohorts.[42]

But when the first cohort of the Biology Academy graduated in 1959, twenty-seven made it to the podium. They were the first large influx of new biologists into Indonesian science, and most of them did not enter the Department of Agriculture hierarchy. Because there would still be no large biology department in the country before the late 1960s (when Bogor Agricultural University opened), this cohort was the last insertion of biological manpower into the botanical gardens for more than a decade.[43] The first batch of Ciawi graduates (Ciawi became the shorthand designation for the school) was the raw talent out of which postcolonial biology would be formed. Throughout the 1960s the botanical gardens and the Indonesian government continued to invest in these students; eighteen of the twenty-seven would receive postgraduate degrees, fourteen of them doctorates.[44] The first cohort of the Ciawi graduates was thrust almost immediately into prominence as scientific leaders. Only a few months after graduation, for example, Didin Sastrapradja was promoted to vice director of the gardens. These young scientists, all in their

early twenties when they started working at the gardens, would be the leaders
of Indonesian biology for the next thirty years. Unlike their Indonesian
superiors, who all had begun their careers in Dutch colonial science, the first
generation of Indonesian scientists had little contact with Dutch scientists;
there is almost no correspondence in the Dutch institutional archives between
them, distinguishing them from Koesnoto, who throughout the 1950s regularly
corresponded with Van Steenis. Most had not attended Dutch-language high
schools, and none went to the Netherlands for graduate training.[45] Consequently, even as they worked inside the old halls of colonial science, they created the discipline of Indonesian biology by filling the colonial institutions in
Bogor with a new vision for Indonesian science.

## Chaos and Opportunity

During the constitutional period of the 1950s, Indonesian elites invented many
of the political and social institutions that were expected to sustain the nation
of Indonesia as an independent entity. Indonesia in the 1950s was built out of
the wreckage of colonial institutions, as well as the organizations and arrangements of the revolution, but still there were few established orthodoxies. Power
was very fluid. Even faced with economic decay, falling cabinets, regional re-
bellions, and a weak state, Indonesian elites vigorously debated their future role
in the country and set in motion the future of Indonesia as a nation. Politicians,
intellectuals, bureaucrats, and army officers gained confidence in their own
national vision.[46] The 1950s were chaotic, pitting different factions of elites
against each other, and sometimes disappointing, particularly due to the failure
to pass a new national constitution.[47] Yet there was opportunity and hope. The
war and revolution of the 1940s had not been the social revolution for which
some had fought. But the vacuum left by the departing Dutch leadership was
an opportunity for a large new set of elites to pursue its own plans and interests.
This was certainly true for the Indonesian scientists. Not only were they young,
ambitious, and experimental, but they also were given every opportunity to
create and lead Indonesian science.

For biology a new generation of Indonesian biologists started setting the
scientific agenda in Bogor. They were idealistic and hopeful and looked for
ways to make Indonesian contributions to the discipline of biology. At the
Ciawi Biology Academy they learned from foreign teachers and visitors about
the latest developments, including the synthesis between evolutionary biology
and natural history research. They were seriously constrained, though. They
largely had to work inside the remains of the colonial institutions with what
equipment and research materials were available. There was little in the way of
government funding. Still, in the herbarium the new director, Anwari Dilmy,
guided and supported by his friend A. J. G. H. Kostermans and the young
Dutch botanist M. Jacobs, brought discipline and purpose back after many

chaotic years.[48] And after 1955 the routine work of dividing, identifying, and storing incoming collections was resumed.[49] Relations with the outside world of science were rekindled. In 1955, Dilmy wrote with confidence to the Dutch botanist Van Steenis as Indonesian scientists took up the same battles of their colonial predecessors, "The battle to obtain money for scientific purposes has not changed since the era of Treub, but for the future, I hold out hope. By also using informal contacts and by making parliament members interested in our institute, we hope that we will not be backed into a corner."[50] Dilmy looked to science outside of Indonesia for opportunities. In 1957 the Ford Foundation sponsored Dilmy on a three-month trip to East Asia, the United States, and Europe, where he laid the groundwork for the exchange of live plants, dried herbarium specimens, scientific personnel, and students. In a report about the trip, Dilmy wrote, "[I]t is now known abroad that science occupies an important place in Indonesia and that the Government puts aside nationalistic sentiment by employing foreign experts."[51] Back home Dilmy's trip promoted the botanical gardens as a credible scientific institution. This credibility was used to secure government money for strengthening the institution; Dilmy thus was able to replace the leaky roof with money from the government, providing shelter for the herbarium collections.[52] By the early 1960s a new research program was put in place when Dilmy and Kostermans began designating which scientists in the herbarium would work on which plant families.[53] By then a first-year Ciawi graduate, Soegeng Reksodihardjo, had been anointed the future leader of the herbarium and was already at Harvard pursuing his doctorate.[54] For the first time ever, Indonesians were controlling the future of the biological discipline.

By the late 1950s, Koesnoto, Dilmy, Soetomo, and a few others had secured institutional survival by training the next generation of biologists and finding work for them at the gardens. What research were they to do? There was as yet no thought of using this science to improve Indonesian society or government pressure to do so. The older generation of scientists continued work begun earlier, which tried to define what was Indonesian about Indonesian nature. For example, Kostermans's and Dilmy's publications at the time were oriented toward the large-scale survey work of the Indonesian flora.[55] This was of less interest to the Ciawi graduates; those at the herbarium received direction from Kostermans, who encouraged taxonomic and ecological research. But most of the others, those stationed at the Treub Laboratory, the zoological museum, and the Ocean Research Center in Jakarta, were left to invent research projects themselves.[56] Some received guidance and advice from one of the few North American scientists who visited Bogor.[57] As a consequence their research interests were closer to the trends in American biology. On the whole they tended to be idealistic about the value of scientific knowledge. They dreamed of practicing science for science's sake and contributing as Indonesians to the Enlightenment dream of using science to reveal the mysteries of nature.[58] To accomplish this they went to Europe and the United States funded by fellow-ships arranged by their government.[59] When Soegeng departed in the summer of 1960, most of his Ciawi cohort saw him off at Kemayoran Airport in Jakarta. One of his friends later wrote that he climbed the steps to board the Pan Am Boeing 707 airliner "à la Robert Mitchum."[60] Great things seemed possible.

## Indonesian Council for Science

While the Ciawi students relished the freedom to pursue their own scientific interests abroad, the leaders of the Indonesian scientific institutions worked to create Indonesian scholarly organizations that would establish what science would mean in independent Indonesia. The Indonesians had inherited the OSR from the Dutch, which in 1951 received an Indonesian name (Organisasi Penyelidikan Ilmu Pengetahuan Alam). This organization kept the building on the Koningsplein Zuid, now renamed Jalan Merdeka Selatan, and in November of 1951 it organized a small workshop in Bogor.[61] After 1950 all scientific and higher-education institutes were to comply with the government's nation-alization regulation, which decreed that institutions must be in compliance with Indonesian national tendencies.[62] In 1951 the minister of education and culture established a committee charged with investigating the founding of an Indonesian council of sciences. Sarwono Prawirohardjo, a prominent obstetrician, headed the committee (Koesnoto was appointed one of nine members in early 1952). By the middle of 1954, the committee had a basic outline of the future council. First, it would be a central organization dedicated to the advancement of Indonesian science, especially research, in the name of national development. The council leadership would coordinate, in theory, all national scientific matters by promoting public understanding of science, providing research grants and fellowships, publishing scientific journals, ensuring cooperation between institutions, coordinating research activities, conducting surveys on the status of Indonesian science, and maintaining international relations. Second, the council would advise the government on problems, projects, and activities related to science. Third, legally the council would be autonomous and would be officially recognized as such through a law signed by the president of Indonesia. Thus the council was to be run by scientists and would not be influenced by either bureaucrats or politicians. Fourth, the OSR would be disbanded (since the early 1950s it had existed in name and building only), but its infrastructure would serve as a building block for the council. By early 1955 the council was effectively in operation, although it would not be until April 1956 that Sukarno signed the Madjelis Ilmu Pengetahuan Indonesia (Indonesian Council for Science, MIPI) into law.[63]

As an institution MIPI embodied the idealism of an autonomous scientific community. And, unlike its colonial predecessors, including the Natuurweten-schappelijke Raad of 1928, which had claimed to aspire to such ideals as well,

MIPI was expected to be more than window dressing for state-run science, as had been the norm under the Dutch. Sarwono and his allies envisioned a future in which Indonesian scientists would generate scientific research about Indonesia but without undue influence from the government. In unveiling MIPI Sarwono criticized the previous, colonial versions of the research council for being "tools of the Government established for the purpose to interfere with the freedom of research for the individual scientist."[64] And, like many Indonesian elites in the 1950s, he and the other leaders of MIPI believed that self-sufficiency, not isolation, would develop Indonesia.[65] Sarwono planned that Indonesian universities would educate most of the country's young scientists to become scientific researchers. Then MIPI would put this trained scientific manpower to work in developing science in Indonesia.[66] It was to play a co-ordinating role in creating this scientific community, for example, by sponsoring a fellowship process that sent the brightest students overseas for advanced training and education.

In January 1957, Indonesian scientists gathered for three days in Bandung to discuss the future of MIPI. The minister of education and culture, the minister of national planning and development, the governor of West Java, and the local UNESCO representative took up the first day with the opening speeches. During the second day the scientists discussed a speech by Sarwono about the future of the council and Koesnoto's comments about the participation of Indonesians in national and international scientific conferences and symposia.[67] On the last day the participants drafted and approved the final findings of the conference. Central to the speeches and discussions was the need to clearly establish how the council would interact with independent scientific institutions. It was felt that MIPI was the only body capable of coordinating scientific endeavors. But the participating scientists did not want the council interfering with the day-to-day operation of scientific institutions. Cooperation between various institutions would be on a voluntary basis. Communication between the Indonesian scientific institutions and the council would take the form of periodic reports. Only in scientific fields hitherto unexplored by Indonesian institutions could MIPI take an active role.[68] Indonesian scientists expressed cautious optimism about the future of the council, which perhaps could have continued in a similarly noninvasive way had the political climate remained unchanged.

As the largest, best-known institution, the botanical gardens played a prominent role in MIPI. The first regular issue of the council's new journal, *Berita M.I.P.I.* (Indonesian Council for Science News), was filled by a review—with two dozen photos—of the gardens.[69] The editor wrote that as the scientific institution that had been engaged in pure and applied research for longer than any other tropical botanical garden it was of great value to the country, as well as to mankind. It would be a model for the future development of science in Indonesia.[70] Funds for biological exploration were made available by MIPI.

Expedition with Russian and French botanists to Sumbawa, 1961 (*left to right*: A. J. G. H. Kostermans, A. Federov, Kuswata Kartiwinata, Wirawan, P. Jauffret, B. Prijanto, D. Soejarto) (Courtesy of the National Herbarium of the Netherlands, Leiden University branch)

In 1961, Kostermans extracted funding for a three-month plant exploration trip to the island of Sumbawa, which included at least one Ciawi graduate (Kuswata Kartiwinata) and was a joint trip with two Russian botanists from Leningrad (A. Federov and N. Kabanov).[71] This and other early initiatives suggested to scientists in Bogor that MIPI promised autonomy and independence from the vagaries of government attention.

## Science under Guided Democracy

In the years following 1959, Sukarno's Guided Democracy—without a parliament and under the powerful executive leadership allowed by the 1945 constitution—pushed science back to its bureaucratic roots. The government expected science to make a direct contribution to the development of the Indonesian nation and participating in international research was no longer sufficient or even appropriate. During the 1950s, when political power was shared among a large number of bureaucratic and political groups, most could agree that an independent and vibrant science would best display the modern achievements of Indonesian scientists.[72] That this was an international endeavor, with foreign contacts and projects, was a point of pride. Although the funding and experience for a grandly conceived Enlightenment study of nature was lacking, in principle most Indonesian elites wanted a thriving culture of scientific research tied into a global network. This changed at the end of the 1950s, when the authoritarian Guided Democracy reined in Indonesian science and its leaders began to think of science as something it should command and direct. The actual work and research of the scientific institutes changed only slowly, but Sukarno's new vision required a close collaboration between science and the state.

Sukarno sought to transform Indonesia into a unified state, by exerting greater executive power. He did not start in a strong position, though, as he had been outside of the day-to-day administration of the country since 1945, and there were few levers of power he could pull directly. Science was a typical case; he had no means of exerting heavy-handed control over science, and even if he had he would not have had an ideological program for scientists to follow. Instead he brought senior scientists closer to the government, and they were expected to create a scientific policy that comported with Sukarno's Guided Democracy. Senior scientists went from directors of institutions to managers of scientific policy. For many scientists this was an excellent opportunity to channel much-needed resources to their programs. Well-established institutions, especially those unconnected to any form of cold war ideology, tended to do best. Some of those institutions, including the botanical gardens, had been designed in the colonial era to thrive under authoritarian conditions and were poised to blossom by the beginning of the New Order.

The ideology of Guided Democracy, referred to by Sukarno as Manipol-USDEK, was revolutionary in nature.[73] Sukarno exhorted Indonesians "to return to the rails of the revolution" by fighting imperialism and restructuring society with the goal of establishing a just and prosperous society.[74] Sukarno wanted to end ten years of argument and stagnation in favor of one all-encompassing ideology. In part this meant stifling dissent. He targeted intellectuals outside of his political control, for example, by shutting down critical newspapers. Iwa Kusumasumantri, a nationalist with associations to the PKI, was minister of higher education and sciences after 1961. He ruthlessly implemented Manipol-USDEK throughout the country's universities, sacking professors and administrators thought to be opposed to Guided Democracy. Students wishing to travel abroad for education could do so only with the minister's personal approval and had to swear allegiance to Manipol-USDEK before leaving.[75] Within the civil service, including the botanical gardens, everyone was threatened with "retooling," the possibility of demotion or transfer for failure to express sufficient revolutionary zeal in their work.[76] This latter action led to a growing sense of distrust among all government employees, who were unsure about what was expected of them. How exactly were they to find the rails of the revolution?

At the end of the 1950s the changing political environment in Indonesia was initially very hard on the botanical gardens' personnel. When in 1958 the Department of Agriculture hijacked the Biology Academy, the leaders of the gardens lost the only consistent means of generating scientific manpower. Moreover, economic decolonization, which had sought to transfer Dutch businesses to Indonesians, reached a crescendo with the seizure of Dutch firms in December of 1957.[77] This not only meant the end of diplomatic ties between Indonesia and the Netherlands, but it also made scientific exchanges with Dutch scientists impossible. The Indonesian government stopped paying the Flora Malesiana bills in 1958.[78] Starting in the late 1950s, Sukarno's international relations were explicitly antagonistic toward the West. He blamed a cabal of global imperialists for continuing to meddle with Indonesia to the detriment of its people. He severed ties with the West and sought alliances with the nonaligned movement. By effectively ending scientific collaboration with the West, which had heretofore made up the bulk of biological contacts, international scientific exchanges became very rare. In Bogor the troubles mounted, and as early as 1959 Dilmy reported to Van Steenis that he and the herbarium staff were feeling "battle weary."[79] Scandal then further disgraced the gardens, when Koesnoto was caught dipping into the official petty cash.[80] At the end of 1959 the Department of Agriculture replaced Koesnoto with Sadikin Soemintawikarta, previously the director of the Biology Academy.[81] The future of Indonesian biology was very uncertain.

Although the initial uncertainty of Guided Democracy destabilized the bureaucracy, by 1962 the restructuring of society had shored up the power of civil servants as opposed to those social groups targeted by Sukarno, the business class in particular. Within the bureaucracy it was the army especially that emerged stronger, but all those who survived "retooling" had greater opportunities than before.[82] Government scientists were among the fortunate ones. Sukarno's targeting of university intellectuals in fact strengthened government and semigovernment organizations such as MIPI. And, although many social scientists, especially economists, were purged from the government ranks, engineers, as well as natural and physical scientists, were on the right side of Sukarno's ideology of self-development. Science and technology were not anathema to Guided Democracy, as Sukarno made clear in a speech to the United Nations in September of 1960:

> Yes, we have learned a lot from Europe and America. We have studied the biographies and lives of the great men from those countries. We have followed their examples; in fact we have tried to exceed them. We speak in their languages and have read their books. We have been inspired by Lincoln and Lenin, by Cromwell and Garibaldi. And indeed there is still much that we must learn from these many, in many fields. But at this time, the fields that we must study are in the area of technology and science, not concepts and movements that are dictated by ideology.[83]

For Sukarno's Indonesia the search for a future free of either American or Soviet dictation was spearheaded by the Dewan Perantjang Nasional (National

Development Council, DEPERNAS), founded in mid-1959. Its most important task was the formulation of the Eight-Year Development Plan, launched in 1961. One of its other responsibilities was securing government science as an institution of Guided Democracy.

Under Guided Democracy the National Development Council transformed MIPI into a bureaucratic institution. As MIPI was well respected within elite political circles, it was chosen to unite Indonesian science.[84] The council, responsible for articulating "Indonesian socialism" under Sukarno's Guided Democracy, began concentrating scientific resources rather than leaving them scattered around the country. In March of 1962 the council transferred command of the botanical gardens (except the library, which to this day remains a part of the Department of Agriculture) to MIPI.[85] After 1962, MIPI was directly responsible to the new Department of National Research. In addition to receiving the botanical gardens and its institutes, now renamed the Lembaga Biologi Nasional (National Biological Institute, LBN), MIPI was charged with setting up six new research institutes dealing with physics, chemistry, geology and mining, social science, metallurgy, and electronics.[86] No longer limited to stimulating scientific development, MIPI started conducting research at its own institutions. After 1962 it became an institution of state science with biology at the botanical gardens as its most important discipline.

In the early 1960s the botanical gardens was rudderless. During the first few years in which it was part of MIPI, the biologists in Bogor really did not know what to do next. Without the powerful Department of Agriculture above them, money was tight and leadership was absent. None of the Ciawi graduates had returned from Europe or the United States yet. Soegeng would not graduate from Harvard until 1964.[87] There was no senior biologist to take over. Instead of appointing an overall director for the new biological institute, Kostermans suggested a council be in charge with a rotating directorship. This plan was adopted, and Soetomo Soerohaldoko, then the head of the Treub Laboratory, became the first director of this five-person council in 1962.[88] Although this power structure was flexible and did not place too much power in the hands of any one person—which had been the problem with Koesnoto— the institute had no direction. Each part of the botanical gardens pursued its own independent course. The overseer of the actual gardens completely neglected regular upkeep, and no one took the trouble to replace the majestic, dying trees in the gardens.[89] Dilmy had lost interest in the herbarium by 1960 after he started a new career as a lecturer in agricultural science at the University of Indonesia. According to Kostermans, Dilmy was actually glad when the Ciawi students left to study abroad because that ended their criticism of his leadership. Dilmy apparently had spent most of his working days gossiping and attending parties.[90] Still, there was some progress. Largely through the intervention of the MIPI head, Sarwono, a new herbarium building was financed in 1962 with government funds, supplemented by a ten-thousand-dollar grant from the U.S. National Science Foundation.[91] But the atmosphere in Bogor was gloomy. In 1963, Kostermans asked Van Steenis (since 1962 the director of the Rijksherbarium) to hold all the Bogor material currently on loan in Leiden as long as possible and to keep the Bogor type-specimens permanently. Kostermans, who recently had become an Indonesian citizen, was so concerned that he wondered if in retrospect "the Dutch were stupid to not take their valuable specimens out" when they had the chance.[92]

Moreover, scientists at the botanical gardens remained uncertain about what was appropriate for Indonesian science under Guided Democracy. How were they to balance international research with development à la Sukarno? This uncertainty led to occasional tension with their superiors in Jakarta, who were more sensitive, at least in deciding what science should not be. In early 1963 the Department of National Research and MIPI startled the biologists in Bogor when they forbade the distribution of the herbarium publication *Reinwardtia*. Sarwono and the minister of research complained that the most recent issue contained unpatriotic content, pointing to the inclusion of Dutch botanists and references to British Malaya, suggesting support for the imperialist Dutch and British. Soetomo, as head of the council running the gardens, was reprimanded at MIPI headquarters and told that future editions of the botanical gardens' journals must be sent to MIPI before publication.[93] In encounters such as this, Indonesian biologists were politically disciplined and learned to create a science that hewed to the nation's needs. The minister of research explained this succinctly in a 1965 speech: "National research has to grow in the fertile ground and environment of Indonesia. If not, it will be an 'intellectual exercise' [English in original] only, completely detached from the Indonesian Revolution. Meaning and direction of national research has to conform to the Indonesian Revolution."[94] As the central government gained confidence in directing science, the scientists' autonomy dissipated. For the scientists to retain a semblance of autonomy after the mid-1960s, and not be overrun by national politics, they had to build a shell protecting them from the government. As became clear to the scientists at the botanical gardens, scientists under Guided Democracy needed to prove that they were sufficiently nationalist. Hence it makes sense that in these years the LBN consciously turned to the task of increasing knowledge of Indonesian flora and fauna by returning to the traditional strength of the botanical gardens: taxonomy and ecology. These fields were Indonesia specific — probably consistent with the aims of the Indonesian revolution — and relatively straightforward in execution considering the collections already housed in Bogor.

In the early 1960s, Sukarno sought to counter the increasing power of the army through his patronage of the PKI. After years of hyperinflation, the social and political structures of the country had become unstable, leading to extreme tension among the political and military elites of Indonesia. This led on September 30, 1965, to a putsch by Lieutenant Colonel Untung in which his group, probably with the cooperation of some members of the Communist Party, murdered six senior army generals. This poorly planned coup was squashed the next day by Suharto, then head of KOSTRAD (Komando Cadangan Strategis Angkatan Darat), the army's strategic reserve, which took control of the army. The army began blaming the communists for the coup, and starting later in October of 1965, killings of suspected communists throughout Indonesia began. Sukarno held on to power but was outmaneuvered by Suharto in March of 1966 when Sukarno gave Suharto the authority to restore order in the country. This was the end of Guided Democracy and the beginning of the New Order.[95]

## From Science Council to Science Institution

Biologists hoped the beginning of the New Order in 1966 would allow them the freedom to expand their institutional authority. Biology, a small professional discipline distant from social and cultural fields, did not suffer from the New Order purge of leftists, and there was the promise of greater funding for science. There were few if any Communist Party members among the Bogor biologists. Among many scientists there was a renewed hope that scientific research would receive greater attention. The editor of *Berita M.I.P.I.* (MIPI News) wrote in 1966 that with the destruction of the Communist Party and its sympathizers academic freedom had returned. No longer did intellectuals need to fear that they would be stigmatized as "textbook thinkers" or that their work would be branded "reactionary." A new "responsible academic freedom" would characterize the future development of Indonesian science.[96] Indeed, the political divisions of the last few years of Sukarno's Guided Democracy had been destabilizing for many scientists. Probably many Indonesian scientists wanted to work for the Indonesian revolution and nation, as Sukarno had asked them to do, but they had done so by increasing their professional credentials and skills, not by political or social agitation. Moreover, the fear of making a mistake and being humiliated by their colleagues meant that most scientists chose to keep a low profile. Scientific organizations such as the botanical gardens and MIPI did likewise. The start of the New Order seemed for many scientists to promise a return to research with less political oversight.

Some of these hopes for expanded opportunities were borne out. But what was less well understood in 1966 was that the New Order government would actually bring science and the state closer together. This process was mirrored in many other intellectual and cultural fields and began almost immediately when in the few years after 1966 the New Order government centralized policy making in the hands of a small number of elites in Jakarta, principally army officers and their chosen technocrats. Scientific managers, that is, those at the top of MIPI and the Department of National Research, became the planners for all scientific research, now in the name of development. This did not mean that they would interfere with research directly, but they did become the principle channel for money. This was institutionalized very early on. In 1967 the Madjelis Permusyawaratan Rakyat Sementara (Provisional Constitutive Assembly, MPRS) replaced both MIPI and the Department of Research with the Lembaga Institut Pengetahuan Indonesia (Indonesian Institute of Sciences, LIPI). Sarwono Prawirohardjo, who transitioned from MIPI to LIPI director, was directly responsible to the president of Indonesia, for "Indonesia needs a potent and competent research organization that can aid—in the field of science and technology—the implementation of planned development for our country. In order to realize this plan, we must firmly coordinate between all research institutions, both government and private."[97] In one of its first regulations, LIPI stipulated that it must approve all research conducted in Indonesia.[98] Institutes formerly controlled by MIPI, such as LBN, continued as before. The goals and tasks of LIPI were similar to those of MIPI, except that LIPI had far greater power "to coordinate, integrate and synchronize all activities in the field of science and technology at the national as well as the regional level."[99] Whereas MIPI had conceived its goal as the formation of an autonomous community of scientists, LIPI saw its role as the scientific manager for the country. Science had become an arm of the state, although the leading Indonesian scientists had not changed.

Led by Otto Soemarwoto after April of 1964, LBN was the most prominent institution directly under LIPI. It was now the government's official institution for leading the biological discipline in Indonesia. Sarwono claimed that biology, while perhaps not as glamorous as some of the other natural sciences, might be the most important science to mankind.[100] Notwithstanding its central position after 1966, the gardens and its staff scientists took some time finding a new direction. In 1965, Soemarwoto wrote that he planned to build Indonesian biology from its existing strengths, in particular the flora and fauna collections in the herbarium and zoology museum.[101] In August of 1967, LBN mounted an exposition meant to show the people of Indonesia the pioneering role biology had played, and would continue to play in the future, in the development of agriculture, forestry, animal husbandry, fishing, and medicine.[102] There were few specifics, although the general outline of Soemarwoto's LBN was an institute where the government made practical the general scientific truths about the natural world. Soemarwoto was never a cipher for New Order technocrats, though. In an English-language article in 1970 he praised the Enlightenment ideal of individual researchers choosing research areas that revealed the fundamental workings of nature.[103] Nonetheless, in official publications, Soemarwoto argued that LBN's research would benefit Indonesia and that the botanical gardens contained an inventory of Indonesia's nature that could be used by government officials, farmers, chemists, industrialists, and even common citizens.[104] At the same time it would serve biologists the world over.[105] Soemarwoto pointed not only to the precedent of Treub's visitors' laboratory but to the more recent collaboration between the marine biologist

Maxwell Doty and LBN (sponsored by grants from the U.S. National Science Foundation, Office of Naval Research, and Atomic Energy Commission), which had resulted in the training of Indonesian students at the University of Hawai'i and had kick-started Indonesian marine exploration.[106]

Ultimately it would not be Soemarwoto who would establish the direction of biology in the New Order. Instead it was the first cohort of Ciawi graduates, many of whom returned from their doctoral work abroad in the late 1960s, that would lead the way. Soemarwoto had great confidence in them and consequently assigned them great administrative responsibility, making them responsible for managing science.[107] Didin Sastrapradja returned from the University of Hawai'i in 1965 and immediately became the assistant director under Otto Soemarwoto. In the same year he was appointed head of the botanical gardens proper. In 1967 he accepted, against Soemarwoto's advice, the position of executive secretary (essentially the vice director) of LIPI.[108] His wife, Setijati Sastrapradja, who was also among the first Ciawi graduates, came back to the Treub Laboratory two years later. She succeeded Soemarwoto as head of LBN in 1973. Soenartono Adisoemarto, after receiving doctoral, master of science, and bachelor of science degrees in Canada, returned to the zoological museum in 1970.[109] He took over Didin's job as assistant director of LBN in 1970 and became the head of the zoological museum in 1977. In 1972, Aprilani Soegiarto (also from the first Ciawi class) became director of the new National Oceanographic Institute, which remained under LIPI but was separated from LBN in 1970.[110] In less than a decade the leadership of virtually all of the biological institutes that formerly made up the Treub biological empire passed into the hands of Ciawi graduates. They would be responsible for implementing professional science.

The difficulty and chaos of finding out how to practice state science was evident at the herbarium. Soegeng since 1965 had held day-to-day responsibility for the herbarium, and, although he was admired for his science, he had failed to bring leadership to the herbarium. In 1968, Soemarwoto, following the advice of Kostermans and Dilmy, promoted Mien Rifai—the only graduate of the third batch of Ciawi students to continue his career in biology—to curator of the herbarium.[111] Rifai would become the head of the herbarium after Dilmy retired in 1970. At that time Dilmy wrote that "the moment of the young, active Indonesians had come."[112] Indeed, the years around 1970 were the watershed for Indonesian biology as a whole as the new generation came into leadership positions. At a 1968 meeting of the herbarium leadership, the vision for the future of the herbarium was laid out. In order to reverse the deterioration of the last ten years, Rifai would first need to take control of the staff and its work. As the only Ciawi graduate trained abroad to return to work in the herbarium, his job was to manage it.[113] The focus was on organization. Materials ordered by Leiden four years earlier had not yet been prepared, and the staff members responsible for entering new collections into tins were six years behind. Discipline would be increased, especially since on any given day a quarter of the staff did not show up for work. Rifai would take special care to protect the 1.6 million sheets of herbarium material for future use by botanists (it was estimated that since the Dutch had left, 5 to 10 percent of the collection had been eaten by insects). With the ten-thousand-dollar grant from the U.S. National Science Foundation used up, the herbarium would need to sharpen its image — especially with publications — if it hoped to receive additional funding from either the Indonesian government or abroad. Research would be scaled back, but regular publications, resulting from increased routine work on the collections, could be guaranteed. A coherent vision of science would turn around the slow liquidation of the herbarium over the past ten years.[114] Scientific research skills were not at a premium, as is evidenced by the process that secured funding to finish the new herbarium building. The building had been under construction since 1962, but with the economy in ruins even Sarwono was unable to secure the money to complete it. When construction stopped in the late 1960s, Kostermans, accompanied by Soegeng and Mien Rifai, had presented himself at the house of one of the top military commanders of the New Order, Major General Hartono Wirjodiprodjo, second deputy (administration) to the commander of the army.[115] Over coffee they pleaded the case of the herbarium. The general was impressed, and after aides checked out the site in Bogor, 30 million rupiah became available.[116] The new building commissioned in 1962 was finished, foreign contacts multiplied (especially with Dutch scientists), and the strong connection to LIPI improved funding. Rifai's success was measured by how well he managed the scientific renaissance of the herbarium.

## Desk Science

Since the 1950s, Indonesian professional scientists have aspired to practice science for science's sake. Solving questions about nature using the best scientific methods has been the guiding ideal. This has meant participating in the transnational scientific community and collaborating with and learning from European, American, Japanese, and Southeast Asian scientists. For example, when Otto Soemarwoto started his biology career in the early 1950s, he — like other young Indonesian scientists — dreamed of becoming a preeminent scientific researcher. While he was at the Treub Laboratory in 1953, he learned about Frits Went's plant hormones research. Like Went, he aspired to examine the underlying principles of growth and to find simple explanations for complex phenomena. Between 1954 and 1959, he did just that at the well-equipped labs at the University of California, Berkeley. When he returned as a professor to Gadjah Mada University, he was heavily burdened with teaching duties, being the only biologist at the university. In any case there were no research facilities at Gadjah Mada, as neither it nor other teaching institutions could afford to spend resources on research facilities.[117] The point of MIPI, to advance independent science, was meant to facilitate cooperative arrangements. But in the 1960s it became clearer and clearer that only the government could realistically fund research. This brought some scientists, including Soemarwoto, out of the universities and into the government and semigovernment institutions linked to MIPI.

Concurrent with the training of Indonesian research scientists, science and the state grew closer. After 1966 the New Order regime kept tight control of the country's critical economic and political functions directly through the military, as well as through the bureaucracy and the New Order parliamentary party Golkar, and indirectly through alliances with businessmen and local political groups.[118] Scientists, too, were brought in as state agents, and by 1970, Indonesian science was a government enterprise. Although the state set the initial parameters, the details of government science were worked out by Indonesian scientists. The state opened an ideological space, proclaiming science to be in the service of the development of Indonesia. Under both Sukarno's and Suharto's regimes, it was expected that Indonesian science would not blindly follow the scientific practice of foreign science but would need to be oriented toward problems specific to Indonesia. Still, balancing Enlightenment ideals with the ideologies of development was not simple. Scientists such as Otto Soemarwoto and his Ciawi graduates spent the 1960s inventing the field of Indonesian biology. Soemarwoto's route linked pure knowledge to national development via the natural collections. Other fields followed different paths, but everywhere science was increasingly bureaucratized.

The result was a practice I call desk science. The generation of scientists that rose to prominence in the 1960s strengthened existing institutions by rationalizing scientific administration. Scientists retooled their disciplines in order to make them appear to be productive tools for Indonesian national development while also trying to preserve the ideals of science. Desk scientists needed to act as conduits between the state and the scientists. They had considerable success. The expansion of desk science after 1970 made possible the growth of Indonesian science, largely because the new policies and goals of science were legible to the Indonesian state and hence funded. Although science was protected, the discipline lost much of its independence.

At LBN this led to a lot of administrative work for the director.[119] Not only did Soemarwoto need to oversee the inexperienced Indonesian biologists but he was also responsible for making their natural knowledge understandable to the government. In Bogor this compromise between government science and scientific research led to the hyperbureaucratization of biology at LBN. The leading biologists spent most or all of their time on administration. Soemarwoto and his lieutenants had so many responsibilities—to bureaucrats, politicians, the universities, the public, and foreign scientists—that they spent their entire working day in offices, busy keeping their various networks intact. This management was done from behind office desks, with scientists linking their diverse networks together with letters, memos, essays, and reports. Science and biology were asked to do so much, for the state, society, and international science also, that the successful administration of scientific knowledge became the real meaning of what was considered good science. Administrators of science earned more, usually far more, than researchers. And as the desk scientists grew in importance, their work became the heart of Indonesian science.

Desk science, compromised as it was, did not appeal to all the young Ciawi biologists. A few years after Soegeng's disappointing homecoming, one of the most promising taxonomists from the first Ciawi class, Engkik Soepadmo, shunned Bogor after receiving his doctorate from Cambridge University in 1966. He did not wish to return to Indonesia, where the pay was poor and the political situation corrupt and chaotic, he wrote. There was no guarantee that he would be employed as a researcher: "There have been many examples before me, when friends of mine did not think more carefully about this matter and they went back straight after their training abroad, thinking that with their M.Sc. or Ph.D. degree they can get a better position at home. Tragically, however, [as] soon [as] they arrived at Bogor they became *'behind the desk botanists.'* One of them even preferred to be appointed as one of the secretaries of Prof. Soemarwoto rather than as a botanist. I do not blame him because as far as the salary is concerned secretary is much better than botanist."[120] Soepadmo would not trade botanical research for desk science. Instead of returning to the Bogor herbarium, he prolonged his stay in Europe. During a visit to Leiden in 1965, Soepadmo had impressed Van Steenis with his work on the oak family, and Van Steenis arranged for a six-month visit in 1967 to work up the family Fagaceae for the Flora Malesiana. The Dutch government extended the grant twice for a total of fifteen months.[121] When Van Steenis formally asked for Soemarwoto's cooperation in arranging this extension, Soepadmo was almost ordered home, as extensions were as a rule denied.[122] Only the quick intervention of Dilmy, Soegeng, and Kostermans prevented the forwarding of this request to LIPI. They did, however, ask Van Steenis to "please impress upon Dr. Soepadmo that this [extension] means at the same time a sacrifice of the staff at the Herbarium Bogoriense, which is overburdened with its task to maintain the collections under the prevailing extremely difficult conditions."[123] Nonetheless, Soepadmo, after his work was done in Leiden, took a three-year contract as a lecturer at the University of Malaya in Kuala Lumpur. When he renewed this contract in 1971 for another three years without consulting anyone in Indonesia, LIPI accused him of breaking the contract he had agreed to when entering the Biology Academy. Didin Sastrapradja, as executive secretary of LIPI, wrote to his old Ciawi classmate Soepadmo promising good pay and a full-time research position if he returned. Soepadmo refused. After Soepadmo ignored this final summons, LIPI instructed the Indonesian embassy in Kuala Lumpur to inform Soepadmo that he was to immediately move to Bogor or lose his Indonesian citizenship. Soepadmo held steadfast in Malaysia. To Sarwono and LIPI, Soepadmo had shirked his duty as an Indonesian scientist. They had his Indonesian citizenship revoked. Eventually Soepadmo became a Malaysian citizen.[124]

One of the most important jobs of the desk scientist was supporting his or her staff on the meager salaries paid to Indonesian civil servants. In the mid-1960s, administrators at LIPI invented a scheme, still widely used, for paying scientists extra if they were attached to a special "project." Kostermans credited the minister of national research, Djoened, with inventing a system for paying the staff twice for the same work: "A project is for example an upcoming expedition (in a given budget); the leader of the project receives 7500 rupiah per month until the project is finished; the leader appoints approximately 10 other people who also receive a monthly stipend. Often the latter do nothing, and are only included on the list to help them."[125] Often these projects were funded by LIPI itself, but ideally they were supported with outside money. In the late 1960s, Soemarwoto conceived of a four- or five-year project about the medicinal values of Indonesian plants. Soegeng, not a real success at the gardens, would be lent to the East-West Center in Hawaii, where he would work on an English translation of K. Heyne's *Nuttige Planten*.[126] By impressing the American biologists, Soegeng could put LBN in a strategic position for the development of Indonesian biology. The Bogor biologists would coordinate the expeditions and fieldwork of interested American scientists. The biologists in Bogor would not only benefit scientifically from the expeditions but would also receive a double salary, half paid by an American grant, while the project was running.[127] These projects did not always produce published results (such as Soegeng's translation work, which fizzled), but they usually strengthened the host institution and supplemented salaries. Educational benefits were often tied to these projects. Long-term projects would often include agreements whereby the American, Japanese, or European field scientist would train young Indonesian biologists at their home university. Dedy Darnaedi, the head of the botanical gardens proper in 2001 and the LIPI sponsor of my research, was educated in such a fashion in Japan during the 1970s.

Soemarwoto was adept at arranging projects for the Bogor scientists, but this nimbleness eventually contributed to his departure from LBN. In 1968, Soemarwoto succeeded in establishing the Southeast Asian Regional Centre for Tropical Biology (BIOTROP), an Indonesian center for tropical biology allied with other such centers in Southeast Asia. Its goal of building "a sound base for the scientific independence of the region and its full participation in the world's scientific community" would be accomplished by training junior scientists and providing opportunities for senior scientists to conduct research. Here was a chance to reinvigorate Indonesian research and perhaps sidestep the usual burdens of government science. The 6.5 million U.S. dollars for 1970–75 came from the government members of the Southeast Asian Ministers of Education Council and the United States government.[128] The primary purpose of BIOTROP was to foster research into biological questions of regional Southeast Asian concern and expand the biology discipline. As such it benefited all Indonesian biologists. But with LBN as host and its location at the botanical gardens, it solidified Bogor's hold on the Indonesian biological discipline. The new BIOTROP building was in the gardens (later it became the administrative headquarters of the gardens), and the lucrative salaries—in dollars—went largely to LBN employees.[129] Furthermore, BIOTROP was separate from LIPI, so Soemarwoto could turn for support, funds, and authority to a scientific network not controlled by Sarwono and LIPI. This benefited the biology discipline in Bogor but also brought Soemarwoto into conflict with Sarwono. Sarwono was jealous of the independent power base held by Soemarwoto, and at the beginning of 1971 this conflict was one of the reasons Soemarwoto resigned as head of both BIOTROP and LBN.[130] Shortly thereafter BIOTROP came under the aegis of the Department of Education and Culture.

During the early 1970s the first cohort of Biology Academy graduates played a leading role in orienting biology as a development science. This was part of a larger shift within LIPI from "Science for Science" to "Science for Development."[131] The goals of LBN were simplified. Its main tasks were now "1) to undertake the exploration and the inventory of the biological resources, 2) to study the living processes of these resources and understand their utilization, and 3) to provide information as required by [*sic*] the scientific communities, the educational sectors, as well as the general public."[132] More than two dozen staff scientists at LBN carried out sustained biological research in plant taxonomy, zoology, microbiology, genetics, ecology, biochemistry, and physiology.[133] But research without focus was no longer enough. In a series of articles in the journal of LIPI, the four leaders of the gardens, Didin Sastrapradja, Setijati Sastrapradja, Soenartono Adisumarto, and Mien Rifai, argued that Indonesian scientific practice had to serve the development of the country. First, this meant convincing Indonesians through educational initiatives that biology, as the science of the environment, had the best chance of improving the welfare of the people.[134] But more important, it meant that scientists needed to use the government's five-year plan as a guide. The New Order government had inaugurated its first five-year plan, Rencana Pembangunan Lima Tahun (Five-Year Development Plan, REPELITA I), on April 1, 1969. According to LBN leaders, one of its two primary goals, to improve local food supplies, should direct the efforts of Indonesian biologists.[135] They berated their colleagues outside LBN for only talking about doing research instead of publishing the results of completed work. No longer was there the excuse of money, as the government was providing extra funds with "projects" sponsored by REPELITA. Universities and scientific institutions could model themselves on LBN. Organized, empirical research would prevent the country from being left farther behind by the West. It was the responsibility of scientists to lay personal interests aside and coordinate their research with the developmental needs of the country. In the 1970s and 1980s, desk scientists in Bogor led the way with financial support from the New Order government. When Setijati Sastrapradja retired as head of LBN in 1986, five times as many scientists were working at the institute than in 1973.[136] During the same period the budget (including income from projects) increased twentyfold from 184 million to 3.6 billion rupiah.[137] In 1980, LIPI built a new home for LBN next to the Agriculture Library, then the highest building in Bogor, five stories tall.[138]

Desk science has become institutionalized across Indonesian institutes of higher education and science. Nowhere is this clearer than in the system for evaluating the productivity of Indonesian scientists known as the cumulative index. The cumulative index started within the universities, managed by the Department of Education, but eventually spread throughout professional science. This system defines scientific achievement and productivity in terms of quantitatively measurable contributions to research, teaching, and service. For example, scientists receive more points for presenting at a conference abroad than at a conference in Jakarta. Advancement and promotion are based on cumulative accomplishments. The index is supposed to be broadly similar among all organizations, but within a university or scientific institute its directors do the regular tallies as part of the yearly performance review, and in this sense the cumulative index is not a true peer review system. It places strong institutional pressures on administrators to base scientists' accomplishments on how well they contribute to the strengthening of their home institution. Service obligations to the institution consume junior staff members' time. Desk science has stifled creativity, as research, even that with dividends for Indonesian development, counts for less and is not as likely to be rewarded with promotion.

## Conclusion

Despite the constraints of desk science, a scientific career has continued to draw Indonesians. It has not been the pay. Salaries have remained low for the research and teaching scientists, although there is still money and power to be gained as a top administrator. Where Indonesian science has broken out of its institutional sterility is in trying to create research relevant to the nation and people of Indonesia. A significant part of the appeal has been the possibility of participating in the creation of useful scientific knowledge about Indonesian nature. This was the course charted by the Ciawi desk scientists who controlled the LBN institutes starting in the early 1970s. In a series of articles in the LIPI trade journal, they sketched the ideal scientific institutions and programs that would allow science to achieve an Indonesian soul.[139] Some of these overlapped clearly with state goals—for example, scientists' projected role in the green revolution—while others were about explaining how scientific research could help Indonesia. There is no evidence that these projects were prescribed by the state. Scientific administrators made certain that the day-to-day workings of science adhered to New Order ideology. In this way they retained control.

Indonesian science has grown. Since 1970, LIPI has expanded to run a dozen of its own scientific institutes and has even spawned a rival, the Pusat

Penelitian, Ilmu Pengetahuan, dan Teknologi (Center for Science, Technology, and Research, PUSPITEK), founded in the late 1970s by B. J. Habibie. Moreover, scientists have moved into teaching positions in the expanding number of universities and colleges. There is much greater public awareness that Indonesia is a scientific nation. Increasing numbers of Indonesians are susceptible to the claims of science through vectors such as public school instruction. Scientific knowledge of the nation seems to be a valuable and patriotic goal to more than just professional scientists. Celia Lowe, in her ethnography of Indonesian environmentalism in the Togean islands, argues the Indonesian nation "has successfully cleared a space for Indonesian scientific expertise."[140] As this chapter has shown, because of the spread of Indonesian nationalism, science encompasses more than the professional scientists' institutions. But even with the spread of science as a credible feature of nationalism, the elite nature of professional science has remained intact.

The new professional scientific institutions created after 1970 by the state, even ones with thoroughly Indonesian national goals, have not been shaped with civic leadership in mind. These research sites are distant from the public eye. As a result scientists have not forged their own leadership of civil society, and professional science has remained an elite affair. This can be illustrated by tracking the relationship between LBN and the botanical gardens proper. The Kebun Raya of Bogor, that is, the gardens itself, remains well known to Indonesians. It is probably the best-known natural monument in the country and is visited by hundreds of Indonesian tourists and schoolchildren every day. But its connection to biology and science is tenuous. The trend in the twentieth century, starting with Treub's Department of Agriculture, was to decouple the activities of the biologists from those of the people maintaining the gardens, and this trend became more pronounced after 1950. With the growth of LBN and Indonesian professional biology, the gardens and its grounds became less and less a part of Indonesian biology. In 1980, with construction of the new LBN building, professional biology moved across the street from the gardens. And in subsequent years, professional science has moved farther away from engaging with the Indonesian visitors to the gardens. This has especially been the case since the headquarters of LIPI biology (now Pusat Penelitian Biologi) moved to the Cibinong Science Center. This state-of-the-art zoology research facility, completed in 2000, replaced the old zoology museum dating from Treub's reign. But unlike the old zoology museum, it is not open to casual visitors. It is located about ten miles north of Bogor in a lightly populated area between the old and new Jakarta-Bogor highways. With its improved research facility, which includes, for example, DNA sequencing equipment, it is more capable of engaging in world-class research. But, like most other Indonesian scientific institutions, it is located far from public view, and few Indonesians are even aware of its existence.

# Conclusion

Indonesian professional science, with its long-standing ties to the state, has hardly been unique in colonial and postcolonial Asia and Africa. Scholars of colonial and postcolonial politics have shown convincingly that the state sought out any discipline that could provide theories or practices that simplified the complex world they aimed to control. The resulting scholarship has deepened our understanding of the role of science in aiding those in power. For example, Charles Keyes, in his 2002 presidential address to the Association for Asian Studies, argued that many Asian governments have drawn on the anthropological theories of ethnologists to gain knowledge and control of their populations.[1] James C. Scott's *Seeing Like a State* is probably the definitive examination of how centralizing regimes the world over saw science's pretensions to universality and what he calls high-modernist ideology as a way to mold people and land to their wills. He demonstrates not only the repeated failure of this high-modernist ideology to make sense of the complexity of the natural and social worlds but also its destructive results in the hands of modern authoritarian regimes. Scott offers relentless and thorough testimony of how the state has used science to bad ends. In the process he constructs a careful indictment of science under authoritarian regimes.[2] But Scott never asks why science would be such an easy tool for the state to use. Where did these scientific theories come from, who created them, and why did the government draw on science for its modernist ideology?

In another camp, historians of science have been more attuned to examining the constraints and opportunities for scientists, particularly in colonial contexts. They have been singularly adept at showing how science developed in scientific enclaves, often by maintaining complex relationships with scientific communities in Europe. Much has been gained by understanding the varied practices of science, which led to traditions and disciplines unlike those in Europe. This scholarship has demonstrated that the narrative of a simple transfer of science from Europe to the colonies is incomplete at best, and often patently false. At the peripheries of empire, scientists, mostly of European origin but often with the help of local inhabitants, created independent traditions with their own institutions, their own theories, and their own bases of support.[3] Thus no world history of science is complete without an understanding of these traditions, which are not just derivative of European science as previously thought.[4] Still, in celebrating the diversity of scientific traditions, historians of science have not often assessed the legacy of colonial science on postcolonial societies. Moreover, that literature has not adequately addressed the question of colonial, and also postcolonial science's closeness to the state. Historians of colonial science have been hesitant to examine how their protagonists, and their scientific work, were readily available for the colonial and postcolonial state's use.[5] In short they have not judged the work of science done under authoritarian conditions.

This book explains how it is that science has developed into a tool of the Indonesian state. I have sought out moments when ambitious naturalists attempted to bring something meaningful into their community by fostering the common use of scientific knowledge. At their most successful these naturalists created scientific disciplines and institutions where they could work on their ideas, research, and knowledge about the natural world around them. These institutions, however, invariably fell under the sponsorship and control of the state, and naturalists became floracrats, state experts of nature.

It is possible to imagine a different outcome if the naturalists had been able to build permanent relationships with civil society and maintain some independence for their institutions. In the early 1850s, Piet Bleeker and his colleagues attempted to spur economic and cultural innovation through the 1853 Batavian Exhibition. But this was a onetime event, and its champions ended up working for the colonial government, which did not expand popular Enlightenment. In the first decade of the twentieth century, Melchior Treub believed he could improve native agriculture with scientific research. But his Department of Agriculture was designed to ignore the agricultural practices of Javanese peasants and in the end was a mechanism for the state taking control of economic development. And in the 1950s, young Indonesian biologists tried to create a biological discipline that served the new Indonesian nation by immersing themselves in Western biology yet staying clear of the debates and conflicts about what Indonesian democracy would look like. Thus, even while keeping themselves detached from the political process, they built up a national science to bolster their nation and consequently whoever ruled it. In these endeavors, instead of spreading Enlightenment skills and knowledge or otherwise fomenting a popular Enlightenment, generations of scientists have created government-funded institutions of professional science and thus placed the onus of enlightening people on the state. State use of scientific knowledge results from enrolling willing scientists to help advance their political or administrative agendas. To be sure, the Indonesian biological discipline was not created by government officials but by scientists working under the rules, constraints, and opportunities they found in Leviathan's garden. And while this dynamic began under Dutch colonial rule, its legacy would be felt by independent Indonesian scientists as well.

What would be necessary for Indonesian scientists to establish scientific institutions outside of state control? A useful comparison can be made to science in the neighboring countries of India and Australia. Historians of science have analyzed the varied ways Indian and Australian scientists rose to local as well as international prominence, even under colonialism, in contrast to what happened in Indonesia. As late as the 1840s the practice of Western science was similar in colonial Indonesia, India, and Australia. Natural history, particularly botany, loomed larger than other sciences in that it was a popular European pursuit, was promoted by European scientific enterprises interested in learning about distant floras and faunas, and was practically suited to European exploration and settlement. Across these colonies European naturalists consciously tried to create local scientific organizations and institutions based on European models, including in particular botanical gardens and academic societies. And, finally, by 1850 the colonial administrations had begun to employ, if erratically, naturalists, geologists, surveyors, and foresters in mapping and administering the colonial territories. In short, in colonial Indonesia, India, and Australia the professionalization of natural history had begun, however haphazardly.[6] It is after 1850 that the scientific cultures of the three colonies diverge.

In the Australian colonies the gold rush and increasing economic opportunities fueled European immigration in the 1850s, and by the end of that decade Australia was a true settler colony with over one million inhabitants. This boom increased support of educational and academic institutions serving local communities, and a number of these institutions, including Sydney University and the University of Melbourne after 1852 and 1855, provided a permanent presence for professional scientists. At the same time, the colonial governments increased their investment in natural history museums, geological surveys, and botanical gardens. Perhaps most important, though, popular and sizable interest supported scientific education and advancement. By the end of the nineteenth century, professional science was anchored in the various universities and other professional scientific societies, and this came to include disciplines outside of natural history, such as physics.[7] Like in the Netherlands East Indies, government science was organized to aid agriculture, but because there were three Australian governments (Victoria, New South Wales, and South Australia) investing in scientific institutes there was much greater diversity and variety of scientific organizations even among those backed by the government.[8] In this way the Australian apostles of enlightenment were far more accomplished than their Indonesian counterparts at establishing connections to local society and in creating research and education programs that drew broad support.

The scientific networks of Australian scientists were much more tightly integrated into metropolitan science than was the case in the Dutch empire. There was a long history of subordinating Australian science to the needs of the British Empire, and into the twentieth century, Australian scientists were still trained, appointed, and directed from England. Yet after 1900 these close connections would serve Australian scientists at home in that they allowed them to participate in imperial, and hence international, science. And under certain circumstances a shared sense of creating science for the empire meant that Australian scientists were seen as the equals of British scientists, at least in the eyes of local Australian governments and populations.[9] Moreover, the extension of self-government in some of the Australian colonies in the mid- to late nineteenth century meant that by the time Australia united federally in 1901 a diverse collection of professional science institutes was well established. Thus, when in 1926 the federal Council for Scientific and Industrial Research (reorganized in 1949 as the Commonwealth Scientific and Industrial Research Organization [CSIRO]) united state science, scientists therein retained considerable autonomy in organizing their research agendas despite increasing bureaucratic and government control.[10] Among its other achievements, Australia in the postwar years played a leading role in pioneering new astronomical research, much of it done under the aegis of CSIRO.[11]

The similarities between colonial India and Indonesia might suggest that science would have had a similar trajectory in the two countries. And in fact there are many points of overlap. After India became a Crown possession in 1858, it was the colonial administration that facilitated the professionalization of science, beginning with botany at the Calcutta Botanic Garden. Subsequent institutions, including educational institutes, were created not just from above but with colonial interests paramount.[12] In British India, as in the Netherlands East Indies, the colonial regimes used science to systematically control colonial nature and society.[13] What, then, explains the difference between Indian and Indonesian science? Serious comparative scholarship of the two empires has been rare, but studies of the two colonial administrations suggest that, although each colonial regime faced similar challenges, they adopted different administrative strategies. The colonial official and theorist J. S. Furnivall made the famous distinction between the "magistrate" of the British Raj and the "policeman" of the Dutch colonial administration.[14] A more recent examination of the two systems by the historians C. Fasseur and D. H. A. Kolff argues that the Dutch were fixated on regulating and policing colonial society while the British were content as long as the law was upheld.[15] This difference also begins to explain their divergent histories of science. In India science did not operate exclusively as an extension of the colonial state. In the nineteenth century, Indian elites took to science as a means of creating a modernity separate from British colonialism, drawing on precolonial scientific and cultural traditions. British colonialism largely ignored these movements.[16] Starting at the end of the nineteenth century, Indians built an infrastructure of professional science free of state supervision out of research and teaching institutes dedicated to astronomy, physics, botany, geology, and medicine, all intended to serve an Indian nation and people.[17] In this way science was always more than an instrument of the state. And as a result Indian national science developed with a rich and varied history and competing visions of science as modernity.[18] Not so in Indonesia, where there were no universities or independent institutions of science. It is not that British or Indian civilization was superior or more dynamic. As I argue in this book, since the 1840s, different sets of elites in Indonesia have repeatedly and variously attempted to tap into the ideals of science as a means of creating a vibrant and modern colony and nation. The colonial and postcolonial state, however, carefully policed these movements; it co-opted those it believed were useful and pliable and pushed others aside. Unlike in Australia and India, no alternative set of institutions of science took shape. In Indonesia, science and scientists have found it very hard to escape from Leviathan's garden.

Is there a way now for Indonesian science to break away from this dynamic of state-sponsored knowledge? Scott addresses this issue more generally in his conclusion to *Seeing Like a State*. He suggests that intellectuals can stay clear of becoming entangled with large-scale state planning by working at nonstate, decentralized institutions of practical and local knowledge.[19] He presents this more as a moral counterpoint to his indictment of state science than as a proven argument. In light of my study of Indonesian state science, I think an important question would be what can scientists accomplish working at nongovernment institutions in countries with a history of authoritarian regimes? The experience of one prominent Indonesian biologist, Otto Soemarwoto, suggests that the answer for Indonesia is that there remain significant barriers to advancing science outside of the state.

The late Otto Soemarwoto was an exceptional Indonesian scientist, not least in that in 1971 he moved from the heart of Leviathan's garden, where he then held the directorship of the LBN, and joined Padjadjaran University in Bandung, one of Indonesia's first private universities (founded in 1957). There he became a leading scientific voice outside of the government, trying to establish a new role for higher education as a domain for independent research and teaching.[20] Among other accomplishments, he wrote the standard Indonesian textbook on human ecology.[21] In an interview with me in 2001, Soemarwoto expressed concern over Indonesia's contemporary culture of science, in which established scientists were only interested in advancing their careers and pay, in particular by seeking government work. He suggested that greater scientific autonomy for science, including higher salaries for scientists, would be necessary before Indonesian science could blossom.[22] But without an alternative system of scientific institutions, paid for with private funds, scientists were

unlikely to be able to command higher salaries. And there is considerable entrenched opposition from government science to the creation of an alternative,
as Soemarwoto himself experienced. He told me that a major reason he left the
LBN leadership post in 1971 was a conflict with the LIPI leadership over his
control of BIOTROP, an organization that since 1968 had fostered research
about the Southeast Asian flora and fauna largely funded by the U.S. government. Notably, the BIOTROP employees, including those in Indonesia, were
paid in U.S. currency.[23]

At Padjadjaran University, Soemarwoto became a leading scientific voice
of Indonesian environmentalism. He continued a century-long tradition of
naturalists and biologists as environmental activists that stretches back to S. H.
Koorders and K. W. Dammerman, who were active in nature conservation
programs in the 1910s and 1920s.[24] In his 2001 book *Atur-Diri-Sendiri: Paradigma Baru Pengeloloaan Lingkungan Hidup* (Self-Regulation: A new paradigm
for managing the living environment), Soemarwoto made the argument that
Indonesian civil society must solve Indonesia's environmental problems. He
began by laying out the environmental crises enveloping Indonesians, ranging
from the epidemic of forest fires during the El Niño years of 1991, 1994, and
1997 to the heaping pollution of Indonesia's cities. He linked these problems to
the top-down environmental policies developed under Suharto's New Order:
"During the New Order when the development of the economy was strongly
pushed, natural resources such as oil, mined ores, and forests became the spine
of the development economy. Unfortunately, this meant the natural riches
needed for improving people's prosperity, as instructed in the 1945 constitution,
were ignored by the government."[25] In particular he pointed to the five-year
plans of the New Order, which brought Indonesia's nature under a command-
and-control environmental regulatory system. As the state and its agents came
to control Indonesia's environment, they did not place the citizens' needs and
interests first.

The demise of the New Order was an opportunity, Soemarwoto wrote, for
civil society to manage environmental policy. Through a careful study of the
international environmental policy literature, he argued that better environmental outcomes occur when businesses self-regulate their environmental im-
pacts through voluntary environmental practice codes. And businesses would
adopt these codes, he suggested, if the Indonesian people were to exert pressure
on them to adopt environmentally friendly policies. Weakening government
regulation and an increased dialogue between informed citizens and business
would lead to productive and successful environmental policies for both business and those impacted by its operations.[26] The critical change, though, would
be minimizing the government's power by debureaucratizing environmental
control while erecting a bottom-up social process that would democratize environmental management.[27] Soemarwoto singled out universities, businesses,
and nongovernmental organizations (NGOs) as the leading institutions of civil

society. They would coordinate discussions about environmental codes, transmit knowledge to the people, and propose national policies and laws. The Indonesian government, when appropriate, would be subjected to the will of civil society.

In his book, Soemarwoto identified a new role to be played by intellectuals, and professional scientists such as himself, in the Indonesian democracy after the resignation of Suharto in 1998. Under Suharto's regime little doubt existed that the regime was in firm control of the political process, including setting environmental policy, despite the continued existence of intellectuals and others advocating political reform.[28] Soemarwoto was optimistic that the political context had changed and intellectuals would play a greater role after Suharto's fall. Nonetheless, the ability of intellectuals to guide the political process has been less than he hoped. Their role in ending the New Order was limited, and civil society leadership in the era of after Suharto's demise has been building from what now looks like rudimentary beginnings.[29] And Soemarwoto himself did not directly explain how science and scientists might take control of the environmental debate. He had many suggestions for what environmental policies should be pursued — mostly addressed to other Indonesian elites and intellectuals — but he only very cursorily addressed how in a democracy scientists would persuade the Indonesian population to trust science's authority.[30]

The question, it seems to me, is whether the Indonesian people would have enough confidence in science and scientists to force the Indonesian government to adopt environmentally self-sustaining practices. Soemarwoto couched the book's analysis and proposals as scientifically appropriate, citing scientific studies and evidence from the West (he pointed to Rachel Carson's 1962 *Silent Spring* as an inspiration). He suggested that future Indonesian governments should simply step back from authoritatively directing critical decisions in areas such as environmental policy and leave it to civil society institutions. But can his words carry more weight than the government's long history of environmental regulation? To do so, I think, would require a fundamental realignment of civil society leadership, which would be harder to effect than Soemarwoto implied. Three things strike me as necessary: first, environmental policy makers at universities, businesses, and NGOs would need to decouple from the state; second, they would need to establish their own leadership of civil society by at the very least increasing the popular authority of professional science; and, third, an alternative source of funding would need to support these Indonesian scientists. These processes may be beginning, but, for scientists at least, who according to Soemarwoto would be critical players in crafting new environmental policy for Indonesia, this would necessitate serious changes to their work practices. Constitutional changes to Indonesia's political system since 2000 have led to greater checks and balances inside the state, including ceding power to representative assemblies. While this may have improved the authority of the Indonesian government and expanded the notion of the rule of law, the relationship between the state and civil society has changed less. And these constitutional changes have done little to change the culture of science. There are more private universities and NGOs than there were fifteen years ago, and they employ an increasing number of Indonesian scientists. But these scientific communities, many of which maintain ties to government science, have not yet spawned a true alternative to desk science.

Like so many professional scientists before him, Soemarwoto underestimated the strength of the Indonesian state. His suggestion that the state should step aside and allow civil society, guided by scientists, to set environmental policy was politically naive. The Indonesian state remains committed to regulating its nation's environment, as environmental policies are seen as a means of controlling resources. While the environment may constitutionally belong to the nation's inhabitants, in practice it belongs to the state and its officials. That is not to say that environmental activists have been mute or powerless. At times they have prodded the state to decide environmental protection was necessary and appropriate. Starting a century ago, Dutch environmentalists promoted conservation by pressuring the state to establish territories where wildlife would be protected, which ultimately led to a handful of nature preserves and parks.[31] Nonetheless, the Dutch environmentalists, a number of whom were professional scientists based in Buitenzorg, only proposed, and it was the Dutch colonial state that implemented. After independence the Dutch environmental policies were maintained by the Republic of Indonesia. A new set of environmental protection policies after 1975 were instigated by the Indonesian government in an effort to control, and even slow down, the logging of the tropical rain forests. The first environmental NGO, Yayasan Indonesia Hijau (Green Indonesia Foundation), founded in Bogor in 1978, was closely allied with the government.[32] Since the 1990s a separate Ministry of the Environment has managed conservation for the state. The politics of the environment tended to be marked by intragovernmental contests, although environmental NGOs were important in raising concerns and acting as a conduit between Indonesians and the government.[33] Recent research about the politics of natural resources after the fall of Suharto suggests that while some of the actors have changed—in particular local governments have much more power to decide environmental policies now—Indonesian government officials are still the principal agents of conservation and natural resource management.[34]

Soemarwoto's book came out of a long tradition of scholarship by Indies and Indonesian apostles of enlightenment. He proposed that Indonesian society adopt enlightened and scientific processes to solve local problems. Notwithstanding my sympathy for Soemarwoto's arguments, I hold little hope this can be implemented. The institutions in a position to back Soemarwoto's decentralized environmentalism are in fact associated with the Indonesian government. Environmental policy continues to be driven by the Indonesian government, aided by both international agencies such as the World Bank and scientific institutes in Indonesia such as the Center for International Forestry Research in Bogor.[35] And while Otto Soemarwoto was not personally co-opted by the Indonesian government, other Indonesian environmentalists have been. As generation after generation of apostles of enlightenment have found, the easiest way to implement some form of their ideas is as government floracrats.

# Notes

## Introduction

1. For a map of Linneaus's global reach, see Lisbet Koerner, *Linnaeus: Nature and Nation* (Cambridge, MA: Harvard University Press, 1999), 116-17.

2. This led Darwin to publish his famous theory, by then already long in the works. For a fine introduction to Wallace and evolution, see Peter Raby, *Alfred Russel Wallace: A Life* (Princeton, NJ: Princeton University Press, 2001). For an overview of the role of Wallace in the history of Darwinism, see Peter J. Bowler, *Evolution: The History of an Idea*, rev. ed. (Berkeley: University of California Press, 1989), 184-86.

3. Alfred Wallace, *The Malay Archipelago* (New York: Harper, 1869).

4. A famous instance was the Siboga expedition, a yearlong maritime exploration of the archipelago by Dutch scientists. Weber-van Bosse's travelogue has recently been republished in the Netherlands as A. Weber-van Bosse, *Een jaar aan boord H. M. Siboga* (Amsterdam: Atlas, [1904] 2000).

5. For the original article describing the skeleton, see M. J. Morwood et al., "Archaeology and Age of a New Hominin from Flores in Eastern Indonesia," *Nature* 431 (2004): 1087-91. Other articles in the October 28, 2004, issue of *Nature* explore how this skeleton contradicts the consensus theory that hominids had principally evolved in Africa, and then dispersed across Eurasia. This new species suggests that following dispersal from Africa hominids continued to evolve under conditions of isolation. The first big paleontology discovery in Indonesia was of *Pithecanthropus erectus*, so-called Java man, in the early 1890s by Eugène Dubois. Dubois had come to the Dutch colony for the express purpose of searching for the origins of man. Bert Theunissen, *Eugène Dubois en de aapmens van Java: Een bijdrage tot de geschiedenis van de paleoantropologie* (Amsterdam: Rodopi, 1985).

6. G. E. Rumphius, *The Ambonese Curiosity Cabinet*, trans. E. M. Beekman (New Haven, CT: Yale University Press, 1999).

7. M. J. Sirks, *Indisch natuuronderzoek* (Amsterdam: Ellerman/Harms, 1915); L. M. R. Rutten, ed., *Science in the Netherlands East Indies* (Amsterdam: Koninklijke Akademie van Wetenschappen, 1929); Pieter Honig and Frans Verdoorn, eds., *Science and Scientists in the Netherlands Indies* (New York: Board for the Netherlands Indies, Surinam and Curaçao, 1945); B. J. O. Schrieke, ed., *Report of the Scientific Work Done in the Netherlands on Behalf of the Dutch Overseas Territories during the Period between Approximately 1918 and 1943* (Amsterdam: North-Holland, 1948); H. H. Zeijlstra, *Melchior Treub: Pioneer of a New Era in the History of the Malay Archipelago* (Amsterdam: KIT, 1959); Lewis Pyenson, *Empire of Reason: Exact Sciences in Indonesia, 1840-1940* (Leiden:

Brill, 1989); Sampurno Kadarsan et al., *Satu Abad Museum Zoologi Bogor, 1894–1994* (Bogor: Pusat Penelitian dan Pengembangan Biologi, LIPI, 1994); *Sejarah Penelitian Pertanian di Indonesia* (Jakarta: Badan Penelitian dan Pengembangan Pertanian, 1995).

8. Lewis Pyenson, "Assimilation and Innovation in Indonesian Science," in "Beyond Joseph Needham: Science, Technology, and Medicine in East and Southeast Asia," ed. Morris F. Low, *Osiris* 13 (1998): 46.

9. The biologist Didin Sastrapradja told me in August 2001 that collaborations have markedly decreased since the 1980s.

10. C. G. G. J. van Steenis, "Treub, Melchior," in *Dictionary of Scientific Biography*, ed. Charles Gillispie (New York: Scribner's, 1976), 13:458–60; J. A. Prins, "Clay, Jacob," in Gillispie, *Dictionary of Scientific Biography*, 3:312–13. A number of other scientists who spent only portions of their careers in Indonesia, including C. Eijkman, Alfred Wallace, and Stamford Raffles, do receive entries. The twentieth-century-centric supplement 2, compiled during the 1980s, did not add any new colonial or postcolonial Indonesian scientists.

11. John Pemberton, *On the Subject of "Java"* (Ithaca, NY: Cornell University Press, 1994); Laurie Sears, *Shadows of Empire: Colonial Discourse and Javanese Tales* (Durham, NC: Duke University Press, 1996); Ann Laura Stoler, *Carnal Knowledge and Imperial Power: Race and the Intimate in Colonial Rule* (Berkeley: University of California Press, 2002); Eric Tagliacozzo, *Secret Trades, Porous Borders: Smuggling and States along a Southeast Asian Frontier, 1865–1915* (New Haven, CT: Yale University Press, 2005).

12. Nancy Lee Peluso, *Rich Forests, Poor People: Resource Control and Resistance in Java* (Berkeley: University of California Press, 1992), 44–78.

13. Wim van der Schoor, "Biologie and Landbouw: F. A. F. C. Went en de Indische Proefstations," *Gewina* 17 (1994): 145–61.

14. Suzanne Moon, *Technology and Ethical Idealism: A History of Development in the Netherlands East Indies* (Leiden: CNWS, 2007); Robert Cribb, "Development Policy in the Early 20th Century," in *Development and Social Welfare: Indonesia's Experiences under the New Order*, ed. Frans Hüsken, Mario Rutten, and Jan-Paul Dirkse (Leiden: KITLV Press, 1993), 225–45.

15. Bowler, *Evolution*. See also Robert E. Kohler, *Landscapes and Labscapes: Exploring the Lab-Field Border in Biology* (Chicago: University of Chicago Press, 2002).

16. David Philip Miller and Peter Hanns Reill, eds., *Visions of Empire: Voyages, Botany, and Representations of Nature* (Cambridge: Cambridge University Press, 1996); B. C. Sliggers and M. H. Besselink, eds., *Het verdwenen museum: Natuurhistorische verzamelingen, 1750–1850* (Blaricum: V+K Publishing, 2000).

17. E.C. Spary, *Utopia's Garden: French Natural History from Old Regime to Revolution* (Chicago: University of Chicago Press, 2000); Michael A. Osborne, *Nature, the Exotic, and the Science of French Colonialism* (Bloomington: Indiana University Press, 1994).

18. Richard Drayton, *Nature's Government: Science, Imperial Britain, and the "Improvement" of the World* (New Haven, CT: Yale University Press, 2000); Lucille Brockway, *Science and Colonial Expansion: The Role of the British Royal Botanic Gardens* (New York: Academic Press, 1979).

19. Christophe Bonneuil, "The Manufacture of Species: Kew Gardens, the Empire and the Standardisation of Taxonomic Practices in Late Nineteenth-Century Botany," in *Instruments, Travel and Science: Itineraries of Precision from the Seventeenth to*

*the Twentieth Century*, ed. Marie-Noëlle Bourguet, Christian Liccope, and H. Otto Sibum (London: Routledge, 2002), 189–215.

20. Richard H. Grove, *Green Imperialism: Colonial Expansion, Tropical Island Edens, and the Origins of Environmentalism, 1600–1860* (Cambridge: Cambridge University Press, 1995); Eugene Cittadino, *Nature as the Laboratory: Darwinian Plant Ecology in the German Empire, 1880–1900* (Cambridge: Cambridge University Press, 1990).

21. Even Kew Gardens did not have full control over British botanical collecting, as British naturalists in China at the end of the nineteenth century were under local British institutional authority. See Fa-ti Fan, *British Naturalists in Qing China: Science, Empire, and Cultural Encounter* (Cambridge, MA: Harvard University Press, 2004).

22. Zoology and botany were taught as part of the curriculum at the medical schools in Batavia and Surabaya, as well as at the various agricultural colleges. After Indonesian independence in 1949, biological departments were created at the University of Indonesia in Jakarta and at Gadjah Mada University in Yogyakarta.

23. Peter Boomgaard, "Oriental Nature, Its Friends and Its Enemies: Conservation of Nature in Late-Colonial Indonesia, 1889–1949," *Environment and History* 5 (1999): 257–92; Hans Pols, "European Physicians and Botanists, Indigenous Herbal Medicine in the Dutch East Indies, and Colonial Networks of Mediation," *East Asia Science, Technology, and Society: An International Journal* 3 (2009): 173–208.

24. Lynn K. Nyhart, *Modern Nature: The Rise of the Biological Perspective in Germany* (Chicago: University of Chicago Press, 2009).

25. Pyenson, *Empire of Reason*, 56–77; Bambang Hidayat, "Under a Tropical Sky: A History of Astronomy in Indonesia," *Journal of Astronomical History and Heritage* 3, no. 1 (2000): 45–58.

26. Anne Secord, "Science in the Pub: Artisan Botanists in Early Nineteenth-century Lancashire," *History of Science* 32 (1994): 269–315.

27. Ann B. Shteir, *Cultivating Women, Cultivating Science: Flora's Daughters and Botany in England, 1760 to 1860* (Baltimore: Johns Hopkins University Press, 1996). See also David Allen, *The Naturalist in Britain: A Social History* (London: A. Lane, 1976).

28. For a sampling, see Gerard Termorshuizen, ed., *In de binnenlanden van Java: Vier negentiende-eeuwse reisverhalen* (Leiden: KITLV Press, 1993).

29. D. F. van Slooten, "De Nederlandsch Indische Natuur-Historische Vereeniging en *De Tropische Natuur*," in "Jubileum Uitgave," special issue, *De Tropische Natuur*, 1936: 3–8.

30. William Clark, Jan Golinski, and Simon Schaffer, eds., *The Sciences in Enlightened Europe* (Chicago: University of Chicago Press, 1999).

31. Bert Theunissen, *"Nut en nog eens nut": Wetenschapsbeelden van Nederlandse natuuronderzoekers, 1800–1900* (Hilversum: Verloren, 2000).

32. Huib J. Zuidervaart and Rob H. Van Gent, "'A Bare Outpost of Learned European Culture on the Edge of the Jungles of Java': Johan Maurits Mohr (1716–1775) and the Emergence of Instrumental and Institutional Science in Dutch Colonial Indonesia," *Isis* 95, no. 1 (2004): 1–33.

33. J. M. Mohr to Thomas Hope, November 2, 1768, quoted in Zuidervaart and Van Gent, "'A Bare Outpost,'" 13–14.

34. R. E. Elson, *The Idea of Indonesia: A History* (Cambridge: Cambridge University Press, 2008).

## Chapter 1. Apostles of Enlightenment

1.   Ann Laura Stoler, *Along the Archival Grain: Epistemic Anxieties and Colonial Common Sense* (Princeton, NJ: Princeton University Press, 2008), 73–102; C. Fasseur, *De Indologen: Ambtenaaren voor de Oost, 1825–1950* (Amsterdam: Bert Bakker, 1994), 120–24. See also Paul van 't Veer, "Een Revolutiejaar, Indische Stijl: Wolter Robert baron van Hoëvell, 1812–1879," in *Geen blad voor de mond: Vijf radicalen uit de negentiende eeuw* (Amsterdam: Arbeiderspers, 1963), 101–44; Herman Stapelkamp, "De rol van Van Hoëvell in de Bataviase Mei-Beweging van 1848," *Jambatan* 4, no. 3 (1986): 11–20.

2.   W. R. van Hoëvell, "Geschiedkunding Overzigt van de beofening van Kunsten en Wetenschappen in Neêrland's Indië," *Tijdschrift voor Neêrlands Indië* 2, no. 1 (1839): 112–14.

3.   E. E. van Delden, *Klein repertorium: Index op tijdschriftartikelen met betrekking tot voormalig Nederlands-Indië*, vol. 5, *Tijdschrift voor Nederlandsch Indië, 1838–1866* (Amsterdam: KIT, 1990).

4.   C. F. Winter, "Romo: Een Javaansch gedicht, naar de bewerking van Joso Dhipoero," *Verhandelingen van het Bataviaasch Genootschap* 21, no. 2 (1846–47): 1–596.

5.   For the history of the early Batavian Society, see Hans Groot, *Van Batavia naar Weltevreden: Het Bataviaasch Genootschap van Kunsten en Wetenschappen, 1778–1867* (Leiden: KITLV Press, 2009). See also Jean Gelman Taylor, *The Social World of Batavia: European and Eurasian in Dutch Asia* (Madison: University of Wisconsin Press, 1983), 78–96.

6.   For the central role of the Batavian Society in printing these journals of the 1840s, see P. Bleeker, "Overzigt der geschiedenis van het Bataviaasch Genootschap van Kunsten en Wetenschappen van 1778–1853," *Verhandelingen van het Bataviaasch Genootschap* 25 (1853): 9–11; and Groot, *Van Batavia naar Weltevreden*, 289–347. For an index and short introduction to some of these journals, see E. E. van Delden, *Klein Repertorium: Index op tijdschriftartikelen met betrekking tot voormalig Nederlands-Indië*, vol. 7, *Acht algemene tijdschriften, 1834–1864* (Amsterdam: KIT, 1993).

7.   C. Fasseur, *The Politics of Colonial Exploitation: Java, the Dutch, and the Cultivation System* (Ithaca, NY: Cornell University Southeast Asia Program, 1992), 17.

8.   Max Havelaar, and presumably the novel's author, E. Douwes Dekker, thought *De Kopiïst* sufficiently important to merit an anecdote and a poem about renewing his subscription. Multatuli, *Max Havelaar* (London: Penguin, [1860] 1987), 177–78.

9.   F. Junghuhn, "Oproeping en beleefd verzoek aan Nederlansch-Indië's ingezetenen," *De Kopiïst* 2 (1843): 358–62.

10.   Ian F. McNeely, *The Emancipation of Writing: German Civil Society in the Making, 1790s–1820s* (Berkeley: University of California Press, 2003).

11.   H. W. van den Doel, *De stille macht: Het Europese binnenlands bestuur op Java en Madoera, 1808–1942* (Amsterdam: Bert Bakker, 1994), 66.

12.   Robert van Niel, "Government Policy and the Civil Administration in Java during the Early Years of the Cultivation System," in *Java under the Cultivation System: Collected Writings* (Leiden: KITLV, 1992), 87–106.

13.   Rudolf Mrázek, *Engineers of Happy Land: Technology and Nationalism in a Colony* (Princeton, NJ: Princeton University Press, 2002), 5.

14.   Fasseur, *The Politics of Colonial Exploitation*.

15.   Paul van 't Veer, *Het leven van Multatuli* (Amsterdam: Arbeiderspers, 1979), 197–98.

16. C. S. W. Graaf van Hogendorp was acting governor general until early 1841. P. Merkus succeeded him as acting governor general until early 1843, when he was formally appointed by The Hague. After Merkus died in August of 1844, J. C. Reynst became acting governor general until he was replaced by J. J. Rochussen in September of 1845. Graaf van Hogendorp, Merkus, and Reynst all came from local families that had been powerful since the VOC era.

17. Paul van 't Veer, "Een Revolutiejaar, Indische Stijl," 126. See also C. Fasseur, "Indische persperikelen, 1847–1860," *Bijdragen en Mededelingen betreffende de Geschiedenis der Nederlanden* 91, no. 1 (1976): 56–75.

18. Rochussen to Baud, February 28, 1846/44, file 4565, Archief Ministerie van Koloniën, 1819–1850, Nationaal Archief, The Hague, quoted in Fasseur, *The Politics of Colonial Exploitation*, 77.

19. C. Fasseur, "Een koloniale paradox: De Nederlandse expansie in de Indische archipel (1830–1870)," in *De weg naar het paradijs en andere Indische geschiedenissen* (Amsterdam: Bert Bakker, 1995), 62.

20. Malcolm Nicolson, "Historical Introduction," in Alexander von Humboldt, *Personal Narrative of a Journey to the Equinoctial Regions of the New Continent* (London: Penguin, 1995), ix–xxxiv; Susan Faye Cannon, "Humboldtian Science," in *Science in Culture: The Early Victorian Period* (New York: Dawson, 1978), 73–110.

21. Andreas Daum, "Science, Politics, and Religion: Humboldtian Thinking and the Transformation of Civil Society in Germany, 1830–1870," in "Science and Civil Society," ed. Lynn K. Nyhart and Thomas H. Broman, *Osiris* 17 (2002): 107–40.

22. Max C. P. Schmidt, *Franz Junghuhn: Biographische Beiträge zur hundersten Wiederkehr Seines Geburtstages* (Leipzig: Verlag der Duerr'schen Buchhandlung, 1909). For further biographical material about Junghuhn, see Rob Nieuwenhuys and Frits Jaquet, *Java's Onuitputtelijke Natuur: Reisverhalen, tekeningen en fotografieen van Franz Wilhelm Junghuhn* (Alphen aan de Rijn: Sijthoff, 1980); Paul van 't Veer, "In de Schaduw van de Kinaboom: Franz Wilhelm Junghuhn, 1809–1864," in *Geen blad voor de mond: Vijf radicalen uit de negentiende eeuw* (Amsterdam: Arbeiderspers, 1958), 54–100; *Gedenkboek Franz Junghuhn, 1809–1909* (The Hague: Nijhoff, 1910); and F. Guenst, "Levensschets van dr. Franz Wilhelm Junghuhn," in F. W. Junghuhn, *Licht- en schaduwbeelden uit de binnenlanden van Java*, 6th ed. (Amsterdam: Guenst, 1867).

23. Franz Junghuhn, *Java, deszelfs gedaante, bekleeding, en inwendige structuur*, 3 vols. (Amsterdam: Van Kampen, 1850–53). *Java* was well received in Europe and immediately saw a second, improved edition, as well as a German translation. Junghuhn to Pahud, January 10, 1852, Exh. January 12, 1852/24, file 149, Archief Ministerie van Koloniën, 1850–1900, Nationaal Archief, The Hague.

24. Junghuhn, *Java*, 84–87.

25. On Humboldt and the importance of measurement, see Michael Dettelbach, "Humboldtian Science," in *Cultures of Natural History*, ed. N. Jardine, J. A. Secord, and E. C. Spary (Cambridge: Cambridge University Press, 1996), 287–304; and Michael Dettelbach, "Global Physics and Aesthetic Empire: Humboldt's Physical Portrait of the Tropics," in *Visions of Empire: Voyages, Botany, and Representations of Nature*, ed. David Philip Miller and Peter Hanns Reill (Cambridge: Cambridge University Press, 1996), 258–92.

26. Junghuhn, *Licht- en schaduwbeelden uit de binnenlanden van Java*, 6th ed. (Amsterdam: Guenst, 1867), 75. The book was quickly embraced for its antireligious qualities.

See Peter Sep, "De receptie van *Licht- en schaduwbeelden uit de binnenlanden van Java* van F. W. Junghuhn," *Indische Letteren* 2 (1987): 53–64.

27. Junghuhn, *Licht- en schaduwbeelden*, 77–78.

28. Ibid., 192.

29. Ibid., 97.

30. Ibid., 260–61.

31. Sep, "De receptie."

32. Mary Louise Pratt, *Imperial Eyes: Travel Writing and Transculturation* (London: Routledge, 1992).

33. When Junghuhn returned to the Netherlands in 1849, Blume, in his capacity as chief botanist of the country and its colonies, made a bid for all of Junghuhn's botanical specimens, arguing that they were the state's property. Junghuhn refused, setting off a public quarrel in which Blume tried to paint Junghuhn as an amateur in need of close supervision and Junghuhn responded with a nasty smear campaign. (Junghuhn alluded to the "stiff, dumb, pedantic professor" Blume as being one of those "sanctimonious hypocrites, scientific liars and sophists, vain Xanthippes, stiff aristocrats, money-grubbers, whose pockets weigh more than their heads.") Blume lost not only the argument with Junghuhn but all his colonial prerogatives when the new political regime of liberals led by Thorbecke dismantled the Natural History Commission in 1850 (by royal decree) and, thinking in particular of Junghuhn, passed the right to appoint naturalists on to Parliament. For the end of the Natural History Commission, see Nieuwenhuys and Jaquet, *De onuitputtelijke Natuur*, 127–33; and H. J. Veth, *Overzicht van hetgeen, in het bijzonder door Nederlands, gedaan is voor de kennis der fauna van Nederlandsch-Indië* (Leiden: Doesburgh, 1879), 120–23. For the mutual accusations of Blume and Junghuhn, see C. L. Blume, *Rumphia sive commentationes botanicae de plantis Indiae orientalis* 3 (1849): 219–20; F. Junghuhn, "Een woord over den boom *Sambinoer* op Sumatra, betrekkelijk deszelfs botanische bepalingen volgens C. L. Blume," *Nederlandsch Kruidkundig Archief* 2 (1850): 261–75; C. L. Blume, *Antwoord op den Heer W. H. de Vriese* (Leiden: Couvée, 1850); F. Junghuhn, "Inlichtingen, aangeboden aan het publiek over zeker geschrift van den Heer C. L. Blume en antwoord aan dien Heer," *Algemeene Konst- en Letterbode*, October 11, 1850, 232–40; and C. L. Blume, *Opheldering van de inlichtingen van den Heer Fr. Junghuhn* (Leiden: Couvée, 1850). The insulting description of Blume is in F. Junghuhn, *Terugreis van Java naar Europa* (Zalt-Bommel: Joh Noman en Zoon, 1851), 127.

34. Gerard Termorshuizen, *Journalisten en heethoofden: Een geschiedenis van de Indisch-Nederlandse dagbladpers, 1744–1905* (Amsterdam: Nijgh and Van Ditmar, 2001), 68–73, 461–64.

35. S. D. Schiff, "Cirkulaire van de kommissie to het beheer der Tentoonstelling, te houden te Batavia in de maand September van het jaar 1853," *Natuurkundig Tijdschrift voor Nederlandsch Indië* 3 (1852): 128–31.

36. "Programma voor de tentoonstelling," *Natuurkundig Tijdschrift voor Nederlandsch Indië* 3 (1852): 648.

37. "Tentoonstelling," *Natuurkundig Tijdschrift voor Nederlandsch Indië* 4 (1853): 432.

38. "Bataviasche Tentoonstelling," *Java-Bode*, October 19, 1853.

39. Nieuwenhuys and Jaquet, *Java's Onuitputtelijke Natuur*, 130.

40. "Prospectus van den natuurkundig tijdschrift voor Nederlandsch Indië," *Natuurkundig Tijdschrift voor Nederlandsch Indië* 1 (1850): 1–3. Piet Bleeker became the first

president of the organization. He had been a military doctor in Batavia since 1842. When he arrived in Batavia in March of 1842, he lived for a few months in a house with his childhood friend E. Douwes Dekker. He quickly became part of the community of naturalists who were members of the Batavian Society. The best source for Bleeker's life is his autobiography, P. Bleeker, "Levensbericht van P. Bleeker," in *Jaarboek van de Koninklijke Akademie van Wetenschappen* (1877): 11–158. On his friendship with Douwes Dekker, see Veer, *Het Leven van Multatuli*, 94.

41. P. Bleeker, "Algemeen Verslag der Werkzaamheden van de Natuurkundige Vereeniging in Nederlandsch Indië," *Natuurkundig Tijdschrift voor Nederlandsch Indië* 2 (1851): 15.

42. Ibid., 13.

43. Ibid., 23–34.

44. P. Bleeker, "Algemeen Verslag der Werkzaamheden van de Natuurkundige Vereeniging in Nederlandsch Indië over het jaar 1851," *Natuurkundig Tijdschrift voor Nederlandsch Indië* 3 (1852): 3.

45. Regeeringsbesluit, February 25, 1852, no. 3, quoted in *Natuurkundig Tijdschrift voor Nederlandsch Indië* 3 (1852): 126–28.

46. "Tentoonstelling te Batavia, te houden in 1853," *Natuurkundig Tijdschrift voor Nederlandsch Indië* 3 (1852): 647.

47. Ibid., 649–52.

48. P., "Residentie Batavia," *Java-Bode*, October 29, 1853.

49. "Vervolg der beschouwingen over de Tentoonstelling, Residentie Cheribon," *Java-Bode*, November 2, 1853.

50. "Vervolg der beschouwingen over de Tentoonstelling, Residentie Soerakarta," *Java-Bode*, November 9, 1853.

51. *Katalogus der tentoonstelling van produkten der natuur en der industrie van den Indischen archipel, te houden te Batavia in Oktober en November 1853* (Batavia: Lange, 1853).

52. *Natuurkundig Tijdschrift voor Nederlandsch Indië* 5 (1853): 250–64.

53. Between October 29, 1853, and December 3, 1853, the author P. wrote a series of reviews in the *Java-Bode* summarizing each residency's submissions.

54. Duymaer van Twist to Pahud, June 14, 1852/118, file 6529, Archief Ministerie van Koloniën, 1851–1899, Semi-Officieele Missiven, Nationaal Archief, The Hague.

55. Duymaer van Twist to Pahud, November 7, 1853/135, file 6529, Archief Ministerie van Koloniën, 1851–1899, Semi-Officieele Missiven, Nationaal Archief, The Hague.

56. H. C. D. de Wit, "De K. N. V. en de botanie in Indië," in *Een eeuw natuurwetenschap in Indonesië, 1850–1950: Gedenkboek Koninklijke Natuurkundige Vereeniging*, ed. P. J. Willikes MacDonald (Bandung: KNV, 1950), 167.

57. Taylor, *The Social World of Batavia*, 78–134.

58. A. W. P. Weitzel, *Batavia in 1858: Schetsen en beelden uit de hoofdstad van Nederlandsch Oost Indië* (Gorinchem: Noorduijn, 1860), 196–97.

59. Ibid., 146.

60. Ibid., 204–5. Weitzel greatly admired Bleeker and his associates for the organization of the 1853 exhibition. But he was critical of the exhibition's successor, the Maatschappij tot Bevordering der Nijverheid in Indië (Society for the Advancement of Industry in the Indies) for failing to accomplish any of its goals. He argued that industry, especially advanced industry, could only flourish in a settler colony where the inhabitants

had a communal stake in their future. In his view Bleeker and his colleagues had no popular appeal because there was no educated population to whom they could speak.

61. Tony Bennett, *The Birth of the Museum: History, Theory, Politics* (London: Routledge, 1995).

62. Paul Greenhalgh, *Ephemeral Vistas: The* Expositions Universelles, *Great Exhibitions, and World's Fairs, 1851–1939* (Manchester: Manchester University Press, 1988).

63. The 1853 exhibition has not attracted the interest of either Indonesian historians or historians of expositions. A recent comprehensive survey of world's fairs and expositions makes no mention of it. John E. Findling and Kimberly Pelle, eds., *Historical Dictionary of World's Fairs and Expositions, 1851–1988* (New York: Greenwood Press, 1990).

64. I. J. Brugmans, *Geschiedenis van het onderwijs in Nederlandsch-Indië* (Groningen: Wolters, 1938), 285.

65. Duymaer van Twist to Pahud, September 19, 1853/134, file 6529, Archief Ministerie van Koloniën, 1851–1899, Semi-Officieele Missiven, Nationaal Archief, The Hague.

66. *Natuurkundig Tijdschrift voor Nederlandsch-Indië* 5 (1853): 269.

67. "Koninklijk Besluit," January 15, 1815, quoted in W. H. de Vriese, ed., *Reis naar het Oostelijk Gedeelte van den Indischen Archipel in het jaar 1821, door C. G. C. Reinwardt* (Amsterdam: Frederick Muller, 1858), 33–48.

68. M. J. Sirks, *Indisch natuuronderzoek* (Amsterdam: Ellermans/Harms, 1915), 86–140; Veth, *Fauna van Nederlandsch-Indië.*

69. Directeur van het Kabinet des Konings to Baud, October 27, 1844/9, V. November 2, 1844/1, file 1599, Archief Ministerie van Koloniën, 1815–1849, Nationaal Archief, The Hague.

70. C. Fasseur and D. H. A. Kolff, "Some Remarks on the Development of Colonial Bureaucracies in India and Indonesia," *Itinerario* 10, no. 1 (1986): 31–55. See also Tony Day, *Fluid Iron: State Formation in Southeast Asia* (Honolulu: University of Hawai'i Press, 2002), 207–23.

71. Niel, "Government Policy and the Civil Administration," 95; Fasseur, *The Politics of Colonial Exploitation,* 57, 71.

72. Von Siebold and Blume to Baud, November 16, 1840, Exh. November 25, 1840/3, file 1338, Archief Ministerie van Koloniën, 1815–1849, Nationaal Archief, The Hague.

73. Von Siebold and Blume to Baud, February 9, 1843, V. February 16, 1843/1, file 1490, Archief Ministerie van Koloniën, 1815–1849, Nationaal Archief, The Hague.

74. Teysmann to Fisscher, June 28, 1843, V. November 15, 1843/2, no. 1538, Archief Ministerie van Koloniën, 1815–1849, Nationaal Archief, The Hague.

75. Teysmann was instructed, per earlier agreement, that he should mail requested plants only to Blume "except with special permission from the Governor General." Bt. 15 August., 1843/16, quoted in Melchior Treub, *Geschiedenis van 's Lands Plantentuin: Eerste gedeelte* (Batavia: Landsdrukkerij, 1889), 67.

76. J. K. Hasskarl, *Tweede Catalogus der in 's Lands Plantentuin te Buitenzorg gekweekte gewassen* (Batavia: Landsdrukkerij, 1844).

77. In 1845 the library held only thirty-nine natural history books, one-third of which belonged to the Natural History Commission. Intendant van 's Gouvernement Hotels to Rochussen, September 8, 1845/171, Exh. September 28, 1846/37, file 1736, Archief Ministerie van Koloniën, 1815–1849, Nationaal Archief, The Hague.

78. Ibid.

79. W. H. de Vriese, "Vorstel tot bevordering van den bloei van den Akademie tuin der Leydensche Hoogeschool," January 19, 1847, V. March 9, 1847/6, file 1768, Archief Ministerie van Koloniën, 1815-1849, Nationaal Archief, The Hague.

80. De Vriese's protégé S. Binnendijk became Teysmann's assistant in 1850. Sirks, *Indisch natuuronderzoek*, 133-34.

81. Teysmann to Intendant van 's Gouvernements Hotels, January 31, 1849, Exh. August 1, 1849/23, no. 1951, Archief Ministerie van Koloniën, 1815-1849, Nationaal Archief, The Hague.

82. J. E. Teysmann, "'s Lands Plantentuin te Buitenzorg in 1850," in *Verslag van het beheer en den staat der Nederlandsche Bezittingen en Kolonien, 1850* (Utrecht: Kemink, 1858), 93-97.

83. J. E. Teysmann, "Botanische reis van den heer Teijsmann," *Natuurkundig Tijdschrift voor Nederlandsch-Indië* 4 (1853): 206.

84. Treub, *Geschiedenis van 's Lands Plantentuin*, 70-72. See also D. F. van Slooten, "Biologisch en landbouwkundig werk in Nederlandsch Indië: Het Herbarium en Museum voor Systematische Botanie van 's Lands Plantentuin," *Vakblad voor Biologen* 14, no. 9 (1933): 162-63.

85. Teysmann and Junghuhn had sparred bitterly and publicly as early as 1843.

86. Teysmann to Merkus, August 28, 1842, quoted in Treub, *Geschiedenis van 's Lands Plantentuin*, 53-54.

87. The point of the catalog was to organize the twenty-two hundred plants according to Endlicher's naturally ordered system. Treub, *Geschiedenis van 's Lands Plantentuin*, 77. Hasskarl later wrote that his most important contribution to the botanical gardens was instilling European taxonomic classification in Buitenzorg. Hasskarl to Van Gorkom, quoted in K. W. van Gorkom, "Levensbericht van Rudolph Herman Christiaan Carel Scheffer," *Jaarboek van de Koninklijke Akademie van Wetenschappen* (1880): 1-21.

88. Baud to Minister van Binnenlandse Zaken, V. March 9, 1847/6, file 1768, Archief Ministerie van Koloniën, 1815-1849, Nationaal Archief, The Hague.

89. Until 1868 Teysmann's immediate supervisor was the military intendant of the palace grounds.

90. For reprints of some of Teysmann's official letters, see Treub, *Geschiedenis van 's Lands Plantentuin*.

91. Wallace faulted the botanical gardens for its artificial organization. He was more interested in wild nature and had no use for gardens organized family by family. Alfred Wallace, *The Malay Archipelago* (New York: Harper, 1869), 121.

92. Multatuli, *Max Havelaar*, 124.

93. Stoler, *Along the Archival Grain*, 94.

## Chapter 2. Quinine Science

1. Siep Stuurman, *Wacht op onze daden: Het liberalisme en de vernieuwing van de Nederlandse staat* (Amsterdam: Bert Bakker, 1992); E. H. Kossman, *De Lage Landen, 1780-1940: Anderhalve eeuw Nederland en België* (Amsterdam: Elsevier, 1976).

2. Kossmann, *De Lage Landen*, 192-96.

3. C. Fasseur, *The Politics of Colonial Exploitation: Java, the Dutch, and the Cultivation System* (Ithaca, NY: Cornell University Southeast Asia Program, 1992).

4. Richard Drayton, *Nature's Government: Science, Imperial Britain, and the "Improvement" of the World* (New Haven, CT: Yale University Press, 2000).

5. Mulder to Minister van Koloniën, undated, file 30, De Vriese Papers, Nationaal Archief, The Hague. Mulder was instrumental in reorienting Dutch academic science toward practical, useful knowledge. Bert Theunissen, *"Nut en nog eens nut": Wetenschapsbeelden van Nederlandse natuuronderzoekers, 1800–1900* (Hilversum: Verloren, 2000), 80–97.

6. For a short introduction to this trip, see A. M. Tempelaars, *Inventaris van het archief van Prof. Dr. W. H. de Vriese (1806–1862) betreffende zijn onderzoek naar de kultures in Nederlands-Indië, 1857–1862, met retroacta vanaf 1817* (Schaarsbergen: Algemeen Rijksarchief, 1977).

7. De Vriese died before completing, or really even starting, this report, but he did leave behind the archive of his research. The social historian Peter Boomgaard has utilized De Vriese's archives in constructing a demographic history of mid-nineteenth-century Java. Peter Boomgaard, *Children of the Colonial State: Population Growth and Economic Development in Java, 1795–1880* (Amsterdam: Free University Press, 1989).

8. De Vriese to Pahud, January 30, 1859/258, file 5, De Vriese Papers, Nationaal Archief, The Hague.

9. Pahud to De Vriese, November 12, 1857/1, file 1, De Vriese Papers, Nationaal Archief, The Hague. De Vriese did dispense occasional advice such as when he advised that Teysmann look into growing cotton on Java. De Vriese to Teysmann, March 14, 1858/14, file 23, De Vriese Papers, Nationaal Archief, The Hague.

10. Fasseur, *The Politics of Colonial Exploitation.*

11. P. D. Curtin, "'The White Man's Grave': Image and Reality, 1780–1850," *Journal of British Studies* 1 (1961): 107. For a narrative discussion of quinine as a tool of British expansion in Africa, see Daniel Headrick, *The Tools of Empire: Technology and European Imperialism in the Nineteenth Century* (New York: Oxford University Press, 1981), 58–79.

12. For the history of acclimatization, see Michael A. Osborne, "Acclimatizing the World: A History of the Paradigmatic Colonial Science," in "Nature and Empire: Science and the Colonial Enterprise," ed. Roy MacLeod, *Osiris* 15 (2001): 135–51; and Daniel Headrick, *The Tentacles of Progress: Technology Transfer in the Age of Imperialism, 1850–1940* (New York: Oxford University Press, 1988), 209–58.

13. Kavita Philip, *Civilizing Natures: Race, Resources, and Modernity in Colonial South India* (New Brunswick, NJ: Rutgers University Press, 2004), 171–95.

14. One of the cinchona trees from this French attempt had been traded with W. H. de Vriese for some rare tropical plants. De Vriese dispatched this tree to Java in 1851. Although the tree did not survive the trip, Teysmann took a cutting off the dying tree and planted it above the Puncak Pass. Neither this tree nor its descendents produced quinine-rich bark, but it was the first attempt at cinchona acclimatized on Java. See P. van Leersum, "Kina," in *Dr. K. W. van Gorkom's Oost-Indische Cultures,* ed. H. C. Prinsen Geerligs (Amsterdam: De Bussy, 1919), 3:175.

15. For descriptions of this trip, see W. H. de Vriese, *De Kina-Boom uit Zuid-Amerika overgebragt naar Java onder de Regering van Koning Willem III* (The Hague: Mieling, 1855); Carl Müller, "Die Verpflauzung des Chinabaumes und seine Cultur," *Unsere Zeit: Deutsche Revue der Gegenwart* 9, no. 2 (1873): 62–74, 186–215, 258–73; and K. W. van Gorkom, *A Handbook of Cinchona Culture* (Amsterdam: J. H. Bussy, 1883), 38–62.

16. M. Gr., "Schets van Hasskarl's leven en werken: 1811–1894," *Indische Gids* 16, no. 1 (1894): 294.

17. De Vrij to Junghuhn, October 9, 1857, file 33, Junghuhn Papers, KITLV archive, Leiden.

18. Pahud had a particular interest in science. Before leaving for his post in the Netherlands East Indies, he visited Alexander von Humboldt in Berlin, where they talked of plans to tie Java into a worldwide network of surveying stations. Lewis Pyenson, *Empire of Reason: Exact Sciences in Indonesia, 1840–1940* (Leiden: Brill, 1989), 83.

19. Fasseur, *The Politics of Colonial Exploitation*, 145.

20. Quoted in H. W. van den Doel, *De stille macht: Het Europese binnenlands bestuur op Java en Madoera, 1808–1942* (Amsterdam: Bert Bakker, 1994), 61.

21. F. Junghuhn, "Catalogues en Inventaris der boeken, instrumenten and andere voorwerpen bij het Natuurkundig Onderzoek in Neerlandsch Indie," 1859, file 35, Junghuhn Papers, KITLV archive, Leiden.

22. F. Junghuhn, "Verantwoording van f.4200 's jaars, bij Art. 2 van het Gouv Besluit 28 Feb. 1856, no. 18 behouden nadere verantwoording en overeenkomstig GS besluit van den 4 Maart 1858, w.g. Art. 2 ter beschikking gesteld van den Inspecteur voor het Natuurkundig Onderzoek in Neerlandsch Indië," 1860, file 36, Junghuhn Papers, KITLV archive, Leiden.

23. De Vriese to Pahud, May 21, 1858/56, file 23, De Vriese Papers, Nationaal Archief, The Hague.

24. H. A. M. Snelders, "Gerrit Jan Mulders Bemoeienissen met het Natuurwetenschappelijk Onderzoek in Nederlands Indië," *Tijdschrift voor de Geschiedenis der Geneeskunde, Natuurwetenschappen, Wiskunde en Techniek* 13, no. 4 (1990): 260.

25. F. Junghuhn, "Toestand der aangekweekte kinaboomen op het eiland Java tijdens het bezoek van zijne Excellentie den Gouverneur Generaal Chs. F. Pahud in het laatst der maand Julij 1857," *Natuurkundig Tijdschrift voor Nederlandsch-Indië* 15 (1858): 23–24.

26. J. E. Teysmann, "Bijdrage tot de Geschiedenis der Kina-Kultuur op Java," *Natuurkundig Tijdschrift voor Nederlandsch Indië* 25 (1863): 65.

27. Junghuhn, "Toestand der aangekweekte Kinaboomen," 86.

28. F. Junghuhn and J. E. de Vrij, "De Kinakultuur op Java op het einde van het jaar 1859," *Natuurkundig Tijdschrift voor Nederlandsch-Indië* 21 (1860): 179–275.

29. He wrote full-length quarterly reports augmented by month-to-month updates. For a sample of a report in De Vriese's collection, see file 138, De Vriese Papers, Nationaal Archief, the Netherlands.

30. Junghuhn and De Vrij, "De Kinakultuur op Java op het einde van het jaar 1859," 180.

31. Teysmann to Miquel, April 15, 1861, Miquel Papers, Handschrift (hereafter HS) 1873, Utrecht University Library.

32. Teysmann, "Bijdrage tot de Geschiedenis der Kina-Kultuur op Java."

33. J. E. Teysmann, "Kritische Opmerkingen op de Bijdragen van Doctor Junghuhn," *Tijdschrift voor Neerlandsch-Indië* 5, no. 1 (1843): 486–508.

34. Teysmann, "Bijdrage tot de Geschiedenis der Kina-Kultuur op Java."

35. Markham had a parallel career to Junghuhn's as an apostle of enlightenment in the British Empire. He began his colonial career as a junior clerk in the India Office in London and traveled to Peru on the cinchona foray in the mid-1850s. He was careful to

select a wide variety of different cinchona species by sending collectors to different regions of the Andes. It was his plants and seeds that launched the British cinchona initiatives. Clements Markham, *Travels in Peru and India While Superintending the Collections of Chinchona Plants and Seeds in South America and Their Introduction into India* (London: John Murray, 1862). For a discussion of the British cinchona projects, see Philip, *Civilizing Natures*; and Lucille Brockway, *Science and Colonial Expansion: The Role of the British Royal Botanic Gardens* (New York: Academic Press, 1979), 112–17.

36. In the decades following his death, though, planters found that successful cultivation of cinchona required some shade. Furthermore, the bark of *C. pahudiana* was later found to contain more quinine alkaloids than many other species. P. van Leersum, "Junghuhn and Cinchona Cultivation," in *Science and Scientists in the Netherlands East Indies*, ed. Pieter Honig and Frans Verdoorn (New York: Board for the Netherlands Indies, Surinam and Curaçao, 1945), 195–96.

37. C. Fasseur, *De Indologen: Ambtenaren voor het Oost, 1825–1950* (Amsterdam: Bert Bakker, 1993), 189–92.

38. P. Bleeker, "Levensbericht van P. Bleeker," *Jaarboek van de Koninklijke Akademie van wetenschappen* (1877): 43.

39. Pahud to Junghuhn, August 14, 1861, file 38, Junghuhn Papers, KITLV archive, Leiden.

40. Mulder and Miquel to Uhlenbeck, April 17, 1862, V. April 23, 1862/20, file 1173, Archief Ministerie van Koloniën, 1850-1900, Nationaal Archief, The Hague.

41. Fasseur, *The Politics of Colonial Exploitation*, 238.

42. Uhlenbeck to Sloet van de Beele, V. April 23, 1862/20, file 1173, Archief Ministerie van Koloniën, 1850-1900, Nationaal Archief, The Hague.

43. Pahud to Junghuhn, April 8, 1863, file 38, Junghuhn Papers, KITLV archive, Leiden.

44. Sloet van de Beele to Uhlenbeck, August 13, 1862/232/K7, V. November 24, 1862/18, file 1267, Archief Ministerie van Koloniën, 1850-1900, Nationaal Archief, The Hague.

45. Junghuhn to Uhlenbeck, August 8, 1862, V. November 24, 1862/18, file 1267, Archief Ministerie van Koloniën, 1850-1900, Nationaal Archief, The Hague. For Uhlenbeck's decision to back down on privatization, refer to Uhlenbeck to Sloet van de Beele, V. November 24, 1862/18.

46. F. Junghuhn and J. E. de Vrij, "Kritiek," October 21, 1862, V. February 14, 1863/33, file 1299, Archief Ministerie van Koloniën, 1850-1900, Nationaal Archief, The Hague.

47. K. W. van Gorkom, "Cinchona Cultivation after Junghuhn's Death," in *Science and Scientists in the Netherlands East Indies*, ed. Pieter Honig and Frans Verdoorn (New York: Board for the Netherlands Indies, Surinam and Curaçao, 1945), 198.

48. Sloet van de Beele to Fransen van de Putte, March 10, 1864/222/20, V. June 23, 1864/7, file 1485, Archief Ministerie van Koloniën, 1850-1900, Nationaal Archief, The Hague.

49. *Handelingen van de Beide Kamers der Staten-Generaal: Zitting, 1861-1862*, May 27, 1862, 811–12.

50. Uhlenbeck to Sloet van de Beele, V. November 24, 1862/18.

51. *Handelingen van de Beide Kamers der Staten-Generaal: Zitting, 1862-1863*, Bijlagen, 1213–18.

52. Ibid., 1317-25.

53. W. R. van Hoëvell, "De kina-kultuur op Java beoordeeld in den vreemde," *Tijdschrift voor Nederlandsch-Indië* 24, no. 2 (1862): 360-74. Junghuhn's reply to Van Hoëvell in the *Java-Bode* is translated in Van Leersum, "Junghuhn and Cinchona Cultivation," 190-96.

54. *Handelingen van de Beide Kamers der Staten-Generaal: Zitting, 1862-1863, Bijlagen,* 1317-25.

55. C. Fassuer, "Purse or Principle: Dutch Colonial Policy in the 1860s and the Decline of the Cultivation System," *Modern Asian Studies* 25, no. 1 (1991): 33-52.

56. Doel, *De stille macht,* 88.

57. "Besluit, December 15, 1866, no. 149," in *Staatsblad van Nederlandsch-Indië* (Batavia: Landsdrukkerij, 1866).

58. Fasseur, "Purse or Principle."

59. J. A. A. van Doorn, *De laatste eeuw van Indië: Ontwikkeling en ondergang van een koloniaal project* (Amsterdam: Ooievaar, 1996).

60. Fasseur, "Purse or Principle," 41-42.

61. Van Gorkom, since being snubbed by Junghuhn, had worked at an agricultural chemical lab in Buitenzorg. Snelders, "Gerrit Jan Mulders Bemoeienissen," 260-61.

62. Quoted in Norman Taylor, "Chapters in the History of Cinchona: Modern Developments," in *Science and Scientists in the Netherlands East Indies,* ed. Pieter Honig and Frans Verdoorn (New York: Board for the Netherlands Indies, Surinam and Curaçao, 1945), 203.

63. See, for example, Van Gorkom's gloomy review of 1868 published in a colonial natural history journal. K. W. van Gorkom, "Verslag nopens de Kina-Kultuur op Java over het jaar 1868," *Natuurkundig Tijdschrift voor Nederlandsch-Indië* 31 (1870): 147-62.

64. Van Gorkom, "Cinchona Cultivation after Junghuhn's Death," 200.

65. Van Gorkom to Miquel, June 29, 1866, Miquel Papers, HS 1873, Utrecht University Library.

66. Ibid.

67. Van Gorkom to Miquel, June 24, 1865, Miquel Papers, HS 1873, Utrecht University Library. In 1869 Miquel published his monograph on cinchona species, and his taxonomy was immediately adopted by Van Gorkom, Teysmann, and Scheffer. Miquel confirmed that *C. calisaya* was the best species and *C. pahudiana* was in fact *C. carabayensis,* a previously known species. F. A. W. Miquel, *De Cinchonae speciebus quibusdam, adiectis iis quae in Java coluntur* (Amsterdam: Van der Post, 1869).

68. Van Gorkom, " Verslag nopens de Kina-Kultuur 1868," 150.

69. Van Gorkom, "Cinchona Cultivation after Junghuhn's Death," 202.

70. J. C. Bernelot Moens, "Onderzoek van eenige Kina-basten van Java," *Natuurkundig Tijdschrift voor Nederlandsch Indië* 31 (1870): 176.

71. Teysmann to Miquel, May 13, 1865, Miquel Papers, HS 1873, Utrecht University Library.

72. Van Gorkom to Miquel, undated, Miquel Papers, HS 1873, Utrecht University Library.

73. For Van Gorkom's political obstacles in the 1860s, see Van Gorkom, *Handbook,* 69-85.

74. K. W. van Gorkom, "Scheikundige Bijdragen tot de Kennis der Java-kina, 1872-1907" (1908), quoted in M. Kerbosch, "Cinchona Culture in Java: Its History and

Development," in *Proceedings of the Celebration of the Three Hundredth Anniversary of the First Recognized Use of Cinchona* (Saint Louis: Missouri Botanical Gardens, 1931), 190.

75. Van Gorkom, *Handbook*, 83–86.

76. A. Groothoff, *De Kinacultuur*, 2nd ed. (Haarlem: Tjeenk Willink, 1915), 100.

77. Gabriele Gramiccia, *The Life of Charles Ledger (1819–1905): Alpacas and Quinine* (London: Macmillan, 1988), 123–28.

78. Miquel had been involved with the Dutch cinchona initiatives since the early 1850s. See F. A. Stafleu, *F. A. W. Miquel, Netherlands Botanist* (Utrecht: Botanisch Museum en Herbarium, 1966), 38–39.

79. Groothoff, *De Kinacultuur*, 15–16. When a year later the seeds had germinated into twenty thousand plants, the government paid Ledger an additional f.500.

80. For a narrative of the Dutch exploitation of Ledger's seeds with a different emphasis, see Gramiccia, *The Life of Charles Ledger*, 153–65.

81. K. W. van Gorkom, "Verslag nopens de Kina-Kultuur op Java over het jaar 1869," *Natuurkundig Tijdschrift voor Nederlandsch Indië* 32 (1873): 2.

82. Arnold Groothoff, *Rationeele exploitatie van Kina-Plantsoenen* (Haarlem: Tjeenk Willink, 1919), 7–8.

83. Kerbosch, "Cinchona Culture in Java," 189.

84. Groothoff, *De Kinacultuur*, 18.

85. Van Gorkom, *Handbook*, 87.

86. C. E. Ridsdale, "Hasskarl's Cinchona Barks, Historical Review," *Reinwardtia* 10, no. 2 (1985): 255.

87. Kerbosch, "Cinchona Culture in Java," 191–92; Norman Taylor, *Cinchona in Java: The Story of Quinine* (New York: Greenberg, 1945), 58–59.

88. For more about trees no. 23 and no. 38 and their descendents, see Van Gorkom, *Handbook*, 137–38; Kerbosch, "Cinchona Culture in Java," 194–95; Taylor, *Cinchona in Java*, 63; and Van Leersum, "Kina," 241–42.

89. K. W. van Gorkom, "Cinchona in Java from 1872 to 1907," *Agricultural Ledger* 17, no. 4 (1912): 48.

90. Kerbosch, "Cinchona Cultivation in Java," 188–89.

91. On Van Loon, see Taylor, *Cinchona in Java*, 64.

92. Van Gorkom, *Handbook*, 144–45.

93. J. C. Bernelot Moens, *De Kinacultuur in Azië, 1854 t/m 1882* (Batavia: Ernst, 1882).

94. Groothoff, *De Kinacultuur*, 100.

95. Anton Hogstad, "The Three Hundredth Anniversary of the First Recognized Use of Cinchona," in *Proceedings of the Celebration of the Three Hundredth Anniversary of the First Recognized Use of Cinchona* (Saint Louis: Missouri Botanical Gardens, 1931), 2.

96. Kerbosch, "Cinchona Culture in Java," 198.

97. W. H. de Vriese, *Wetenschap en beschaving, de grondslagen van welvaart der landen en volken van den Indischen archipel* (Leiden: Hazenberg, 1861), 10–11.

98. R. H. C. C. Scheffer, "Sur quelques Palmiers du groupe des Arécinées," *Natuurkundig Tijdschrift van Nederlandsch-Indië* 32 (1873): 149–93; R. H. C. C. Scheffer, "Sur deux espèces du genre *Genocaryum* Miq.," *Annales du Jardin Botanique de Buitenzorg* 1 (1876): 96–102.

99. Scheffer to Miquel, January 12, 1868, and Scheffer to Miquel, August 6, 1869, Miquel Papers, HS 1873, Utrecht University Library.

100. *Verslag omtrent den staat van 's lands plantentuin te Buitenzorg, 1878* (Batavia: Landsdrukkerij, 1879); K. W. van Gorkom, "Levensbericht van Rudolph Herman Christiaan Carel Scheffer," *Jaarboek van de Koninklijke Akademie van Wetenschappen* (1880): 14–19.

101. Groothoff, *Rationeele exploitatie.*

102. Wim van der Schoor, "Biologie en landbouw: F. A. F. C. Went en de Indische Proefstations," *Gewina* 17 (1994): 227–34.

## Chapter 3. Treub's Beautiful Science

1. P. Bleeker, February 14, 1860, quoted in H. C. de Wit, "De K. N. V. en de botanie in Indië," in *Een eeuw natuurwetenschap in Indonesië, 1850–1950: Gedenkboek Koninklijke Natuurkundige Vereeniging*, ed. P. J. Willikes MacDonald (Bandung: KNV, 1950), 167.

2. Eugene Cittadino, *Nature as the Laboratory: Darwinian Plant Ecology in the German Empire, 1880–1900* (Cambridge: Cambridge University Press, 1990), 110.

3. Gottlieb Haberlandt, *Eine Botanische Tropenreise: Indo-Malayische Vegetations-bilder und Reiseskizzen*, 2nd ed. (Leipzig: Verlag von Wilhelm Engelmann, 1910), 73.

4. For recent scholarship about Melchior Treub and the Department of Agriculture, see Suzanne Moon, *Technology and Ethical Idealism: A History of Development in the Netherlands East Indies* (Leiden: CNWS, 2007); Harro Maat, *Science Cultivating Practice: A History of Agricultural Science in the Netherlands and Its Colonies, 1863–1986* (Dordrecht: Kluwer, 2001), 53–67; H. W. van den Doel, "Practical Agricultural Education in the Netherlands East Indies: The Transfer of Agricultural Knowledge to the Indigenous Population of Java, 1875–1920," *Journal of the Japan-Netherlands Institute* 6 (1996): 78–94; and H. W. van den Doel, *De stille macht: Het Europese binnenlands bestuur op Java en Madoera, 1808–1942* (Amsterdam: Bert Bakker, 1994), 227–34.

5. H. H. Zeijlstra, *Melchior Treub: Pioneer of a New Era in the History of the Malay Archipelago* (Amsterdam: KIT, 1959).

6. A. W. P. Weitzel, *Batavia in 1858: Schetsen en beelden uit de hoofdstad van Nederlandsch Oost Indië* (Gorinchem: Noorduijn, 1860), 39.

7. Intendant der Gouvernements-Hotels to Teysmann, July 15, 1865/Secret, Miquel Papers, HS 1873, Utrecht University Library.

8. Kenji Tsuchiya, "Kartini's Image of Java's Landscape," *East Asian Cultural Studies* 25, nos. 1–4 (1986): 59–86.

9. Doel, *De stille macht.*

10. M. Treub, "Korte Geschiedenis van 's Lands Plantentuin," in *'s Lands Plantentuin te Buitenzorg, 18 Mei 1817–18 Mei 1892* (Batavia: Landsdrukkerij, 1892), 38.

11. Teysmann to Miquel, May 13, 1865, Miquel Papers, HS 1873, Utrecht University Library.

12. Binnendijk to H. Witte, July 25, 1864, in "Correspondentie hoofdzakelijk van Teysmann en Binnendrijk voorts Wigman, Lovink, Treub, enz. met de Hortulanus H. Witte te Leiden, 1847–1885," ed. H. C. D. de Wit and C. G. G. J. van Steenis, December 1948, Nationaal Herbarium Nederland (hereafter NHN), Leiden branch library.

13. Scheffer's successor found both in chaos. *Verslag omtrent den staat van 's Lands Plantentuin te Buitenzorg, 1881* (Batavia: Landsdrukkerij, 1882), 5–7.

14. The minister of colonies closed the agricultural school in mid-1884 (no new students had been taken on since 1882). *Verslag omtrent den staat van 's lands plantentuin te Buitenzorg, 1884* (Batavia: Landsdrukkerij, 1885), 5.

15. Treub, unlike Junghuhn, Van Gorkom, or Scheffer, had already established a name for himself in the Netherlands; in 1879, at the age of twenty-seven, he was elected to the Royal Academy of Sciences in Amsterdam on the strength of his research on the development of the embryo in orchids. Zeijlstra, *Melchior Treub*, 14–15.

16. The appointment to Java, still early in his career, may have been temporary, and in the first few years he would have taken a professorship in Europe if one had been offered. J. P. Lotsy, "Levensbericht van Melchior Treub," *Levensberichten der afgestorven medeleden van de Maatschappij der Nederlandsche Letterkunde te Leiden* (1911–12): 16.

17. Treub to Went, December 22, 1888, Treub correspondence archive, Boerhaave Museum, Leiden.

18. Bert Theunissen, *"Nut en nog eens nut": Wetenschapsbeelden van Nederlandse natuuronderzoekers, 1800–1900* (Hilversum: Verloren, 2000), 168.

19. Cittadino, *Nature as the Laboratory*, 76–81.

20. W. Burck, "Het Herbarium en Museum van 's Lands Plantentuin," in *'s Lands Plantentuin te Buitenzorg, 18 Mei 1817–18 Mei 1892*, 218–33.

21. *Verslag 's Lands Plantentuin, 1881*, 9.

22. *Verslag 's Lands Plantentuin, 1882* (Batavia: Landsdrukkerij, 1883), 6.

23. *Verslag Omtrent den Staat van 's Lands Plantentuin te Buitenzorg, 1884*, 9.

24. *Verslag 's Lands Plantentuin 1882*, 12. For more about Treub's upgrades to the gardens proper, see Theo F. Rijnberg, *'s Lands Plantentuin, Buitenzorg, 1817–1992* (Enschede: Johanna Oskamp, 1992).

25. M. Treub, "Études sur les Lycopodiacées," *Annales du Jardin Botanique de Buitenzorg* 4 (1884): 107–37; M. Treub, "Études sur les Lycopodiacées," *Annales du Jardin Botanique de Buitenzorg* 5 (1886): 87–139.

26. M. Treub, "Some Words on the Life-History of Lycopods," *Annals of Botany* 1, no. 2 (1887): 119.

27. M. Treub, "Études sur les Lycopodiacées," *Annales du Jardin Botanique de Buitenzorg* 7 (1888): 141–50.

28. W. Burck, "Sur l'organisation florale chez quelques Rubiacées," *Annales du Jardin Botanique de Buitenzorg* 3 (1883): 105–19.

29. W. Burck, "Sur les diptérocarpées des Indes Néerlandaises," *Annales du Jardin Botanique de Buitenzorg* 6 (1887): 145–48; W. Burck, "Sur les Sapotacées des Indes Néerlandaises et les origins botaniques de la gutta-percha," *Annales du Jardin Botanique de Buitenzorg* 5 (1886): 1–86.

30. M. J. Sirks, *Indisch natuuronderzoek* (Amsterdam: Ellerman/Harms, 1915), 272–73.

31. *Verslag Omtrent den Staat van 's Lands Plantentuin te Buitenzorg, 1886* (Batavia: Landsdrukkerij, 1887), 8.

32. M. Treub, "Un jardin botanique tropical," *Revue des Deux Mondes* 97 (1890): 162–83, translated as M. Treub, "A Tropical Botanic Garden," in *Annual Report of the Board of Regents of the Smithsonian Institution to July, 1890* (Washington, DC: Government Printing Office, 1891), 389–406.

33. Treub, "A Tropical Botanic Garden," 402–3.

34. M. Treub, *Geschiedenis van 's Lands Plantentuin: Eerste gedeelte* (Batavia: Lands-drukkerij, 1889).

35. *'s Lands Plantentuin te Buitenzorg, 18 Mei 1817-18 Mei 1892* (Batavia: Lands-drukkerij, 1892).

36. A German edition printed a year later in Leipzig likely reached a wider European audience. M. Treub, *Der botanische Garten: "'s Lands Plantentuin" zu Buitenzorg auf Java—Festschrift zur Feier seines 75, jährigen bestehens (1817-1892)* (Leipzig: Engelmann, 1893).

37. Treub, September 22, 1909, Treub correspondence archive, Boerhaave Museum, Leiden. This letter was probably written to F. A. F. C. Went.

38. Doel, *De stille macht*, 227.

39. K. W. Dammerman, "History of the Visitors' Laboratory ('Treub Laborato-rium') of the Botanic Gardens, Buitenzorg, 1884-1934," in *Science and Scientists in the Netherlands Indies*, ed. Pieter Honig and Frans Verdoorn (New York: Board for the Netherlands Indies, Surinam and Curaçao, 1945), 59.

40. *Verslag Omtrent den Staat van 's Lands Plantentuin te Buitenzorg, 1883* (Batavia: Landsdrukkerij, 1885), 11-12.

41. *Verslag 's Lands Plantentuin 1884*, 11.

42. Treub, "A Tropical Botanic Garden," 398.

43. This figure does not include Dutch scientists from either the Netherlands or the Netherlands East Indies. Dammerman, "History of the Visitors' Laboratory."

44. When in that year the mountain bungalow of the governor general at Cipanas was torn down to make way for a new building, Treub secured permission to use the old building's material, and a small research lab with a few guest rooms was built at the Cibodas gardens. Zeijlstra, *Melchior Treub*, 54.

45. *Verslag Omtrent den Staat van 's Lands Plantentuin te Buitenzorg, 1885* (Batavia: Landsdrukkerij, 1886), 8.

46. *Verslag Omtrent den Staat van 's Lands Plantentuin te Buitenzorg, 1888* (Batavia: Landsdrukkerij, 1889), 58.

47. Treub to Engler, 20 July, 1886/28 October, 1889/July 2, 1888/August 24, 1891/December 26, 1895/September 29, 1896/November 3, 1897/September 27, 1906, Lb 1875, Handschriftsabteilung, Staatsbibliothek, Berlin.

48. Dirk R. Walters and David J. Keil, *Vascular Plant Taxonomy*, 4th ed. (Dubuque, IA: Kendall/Hunt, 1996), 469-70.

49. Cittadino, *Nature as the Laboratory*, 73-74.

50. Zeijlstra, *Melchior Treub*, 67-68; Sirks, *Indische Natuuronderzoek*, 160-61.

51. Bert Theunissen, *Eugène Dubois en de aapmens van Java: Een bijdrage tot de ge-schiedenis van de paleoantropologie* (Amsterdam: Rodopi, 1985), 55-56, 73.

52. Zeijlstra, *Melchior Treub*, 68-69.

53. Eliza Scidmore, *Java: The Garden of the East* (Singapore: Oxford University Press, [1897] 1984), 306-7.

54. Tsuchiya, "Kartini's Image of Java's Landscape," 62.

55. Haberlandt, *Eine Botanische Tropenreise*, 46.

56. Goebel was in Buitenzorg during the winter of 1885-86. Cittadino, *Nature as the Laboratory*, 1.

57. Haberlandt, *Eine Botanische Tropenreise*, 17, 59.

58. Ibid., 62.

59. Cittadino, *Nature as the Laboratory*, 79–81.

60. See, for example, Haberlandt, *Eine Botanische Tropenreise*, 73.

61. David Fairchild, "An American Plant Hunter in the Netherlands Indies: Buitenzorg and Doctor Treub," in *Science and Scientists in the Netherlands Indies*, ed. Pieter Honig and Frans Verdoorn (New York: Board for the Netherlands Indies, Surinam and Curaçao, 1945), 80.

62. Ibid., 81–82.

63. Ibid., 85.

64. For the idea of modular colonialism, see Ann Laura Stoler and Frederick Cooper, "Between Metropole and Colony: Rethinking a Research Agenda," in *Tensions of Empire: Colonial Cultures in a Bourgeois World*, ed. Frederick Cooper and Ann Laura Stoler (Berkeley: University of California Press, 1997), 13.

65. Elmer D. Merrill, *Report on Investigations Made in Java in the Year 1902* (Manila: Bureau of Public Printing, 1903), 7–9. For more on the Java Forest Flora, see chapter 4.

66. Ibid., 70.

67. Ibid., 72–73.

68. Ibid., 56–69.

69. William J. Robbins, "Elmer Drew Merrill (1876–1956)," *Biographical Memoirs* 32 (1958): 273–333.

70. For Haeckel's work at Jena, see Lynn K. Nyhart, *Biology Takes Form: Animal Morphology and the German Universities, 1800–1900* (Chicago: University of Chicago Press, 1995), 143–67.

71. Ernst Haeckel, *A Visit to Ceylon* (New York: Peter Eckler, n.d.). The quotation is from Haeckel to Franziska von Altenhausen, January 24, 1901, in *The Love Letters of Ernst Haeckel*, ed. Johannes Werner (New York: Harper, 1930), 219–20.

72. Haeckel to Franziska von Altenhausen, November 21, 1900, in Werner, *The Love Letters of Ernst Haeckel*, 212–16.

73. J. C. Koningsberger, "Herinneringen aan Melchior Treub," 1945, KIT library, Amsterdam.

74. Docters van Leeuwen to Went, August 10, 1925, Docters van Leeuwen correspondence archive, Boerhaave Museum, Leiden.

75. Quoted in Zeijlstra, *Melchior Treub*, 112.

76. Sartono Kartodirdjo, *The Peasant Revolt of Banten in 1888* (The Hague: Nijhoff, 1966).

77. Martin Bossenbroek, "Joannes Benedictus van Heutsz en de leer van het functionele geweld," in *In de vaart der volken: Nederlanders rond 1900*, ed. Herman Beliën, Martin Bossenbroek, and Gert Jan van Setten (Amsterdam: Bert Bakker, 1998), 87–96.

78. See Louis Couperus, *The Hidden Force* (Amherst: University of Massachusetts Press, [1900] 1985).

79. Doel, *De stille macht*.

80. Claire Holt, *Art in Indonesia: Continuities and Change* (Ithaca, NY: Cornell University Press, 1967), 194.

81. Rudolf Mrázek, *Engineers of Happy Land: Technology and Nationalism in a Colony* (Princeton, NJ: Princeton University Press, 2002), 113.

## Chapter 4. Ethical Professionals

1. Robert Cribb, "Development Policy in the Early 20th Century," in *Development and Social Welfare: Indonesia's Experiences under the New Order*, ed. Frans Hüsken, Mario Rutten, and Jan-Paul Dirkse (Leiden: KITLV Press, 1993), 225–45. See also Suzanne Moon, *Technology and Ethical Idealism: A History of Development in the Netherlands East Indies* (Leiden: CNWS, 2007), 17–23; and Elsbeth Locher-Scholten, *Ethiek in fragmenten: Vijf studies over koloniaal denken en doen van Nederlanders in de Indonesische Archipel, 1877–1942* (Utrecht: Hes Publishers, 1981).

2. Ann Laura Stoler, *Along the Archival Grain: Epistemic Anxieties and Colonial Common Sense* (Princeton, NJ: Princeton University Press, 2008), 155–72.

3. Cribb, "Development Policy," 225.

4. For more about the colonial bureaucracy after 1866, see H. W. van den Doel, *De stille macht: Het Europese binnenlands bestuur op Java en Madoera, 1808–1942* (Amsterdam: Bert Bakker, 1994), 80–104.

5. J. Thomas Lindblad, "The Late Colonial State and Economic Expansion, 1900–1930s," in *The Emergence of a National Economy: An Economic History of Indonesia, 1800–2000*, ed. Howard Dick (Honolulu: University of Hawai'i Press, 2002), 123.

6. Kees van Dijk, *The Netherlands Indies and the Great War, 1914–1918* (Leiden: KITLV Press, 2007), 1–18.

7. Anne Booth, *The Indonesian Economy in the Nineteenth and Twentieth Centuries: A History of Missed Opportunities* (London: Macmillan, 1998), 149.

8. M. Treub, "A Tropical Botanic Garden," in *Annual Report of the Board of Regents of the Smithsonian Institution to July, 1890* (Washington, DC: Government Printing Office, 1891), 400.

9. This short pamphlet was printed in-house with a limited circulation. Historians of Treub and the botanical gardens (except Zeijlstra, who quotes only from the first page) have ignored this important document. M. Treub, *Over de taak en den werkkring van 's Lands Plantentuin te Buitenzorg* (Buitenzorg: 's Lands Plantentuin, 1899).

10. Ibid., 10.

11. Ibid., 21.

12. Ibid., 24 (emphasis in original).

13. Ibid., 27.

14. Melchior Treub, *Geschiedenis van 's Lands Plantentuin: Eerste gedeelte* (Batavia: Landsdrukkerij, 1889).

15. Gouvernement Besluit (hereafter Bt.) July 30, 1868, no. 38, in *Staatsblad van Nederlandsch-Indië 1868* (Batavia: Landsdrukkerij, 1869).

16. Bt. May 22, 1890, no. 6, Archief Algemeene Secretarie, 1817–1890, Arsip Nasional Republik Indonesia (hereafter ANRI), Jakarta. This is the first instance in which a matter concerning Treub's botanical gardens reached the governor general's office.

17. Van der Kemp, April 17, 1890/3471, Bt. May 22, 1890, no. 6, Archief Algemeene Secretarie, 1817–1890, ANRI, Jakarta.

18. Memo, May 14, 1890, and "Advies van den raad van Nederlandsch-Indië 2 Mei 1890," in Bt. May 22, 1890, no. 6, Archief Algemeene Secretarie, 1817–1890, ANRI, Jakarta.

19. *Verslag Omtrent den Staat van's Lands Plantentuin te Buitenzorg, 1888* (Batavia: Landsdrukkerij, 1889), 10–11.

20. *Verslag Omtrent den Staat van's Lands Plantentuin te Buitenzorg, 1889* (Batavia: Landsdrukkerij, 1890), 7–8.

21. During a sabbatical in the Netherlands in 1887, Treub collected f.16,000 from private hands to serve as the capital to fund a Dutch botanist's trip to Buitenzorg every other year. The minister of interior added a f.700 yearly subsidy. An additional f.2,400 was raised so that Boerlage could immediately accompany Treub back to Buitenzorg in early 1888. *Verslag's Lands Plantentuin 1888*, 1–20.

22. K. W. Dammerman, "History of the Visitors' Laboratory ('Treub Laboratorium') of the Botanic Gardens, Buitenzorg, 1884-1934," in *Science and Scientists in the Netherlands Indies*, ed. Pieter Honig and Frans Verdoorn (New York: Board for the Netherlands Indies, Surinam and Curaçao, 1945), 63.

23. W. Burck, "Het Herbarium en Museum van's Lands Plantentuin," in *'s Lands Plantentuin te Buitenzorg, 18 Mei 1817–18 Mei 1892* (Batavia: Landsdrukkerij, 1892), 224.

24. Treub to Van der Kemp, June 15, 1893, no. 620/0, file 8, Van Steenis Papers, Nationaal Archief, The Hague.

25. M. Treub, "Voorwoord," in J. G. Boerlage, *Handleiding tot de kennis der Flora van Nederlandsch Indië* (Leiden: Brill, 1890-1900), 1:i–ii.

26. Nancy Lee Peluso, *Rich Forests, Poor People: Resource Control and Resistance in Java* (Berkeley: University of California Press, 1992), 44–78; Peter Boomgaard, "Colonial Forest Policy in Java in Transition, 1865-1916," in *The Late Colonial State in Indonesia: Political and Economic Foundations of the Netherlands Indies, 1880-1942*, ed. Robert Cribb (Leiden: KITLV Press, 1994), 117–37.

27. S. H. Koorders and Th. Valeton, *Bijdragen tot de kennis der boomsoorten van Java*, 13 vols. (Buitenzorg: 's Lands Plantentuin/Departement van Landbouw, 1894-1910).

28. Kroesen to Rooseboom, December 30, 1902/1043/1, and Treub to Rooseboom, July 25, 1904, file 15, Indische Comité voor Wetenschappelijke Onderzoekingen (hereafter ICWO) Collection, Nationaal Archief, The Hague.

29. Sampurno Kadarsan et al., *Satu Abad Museum Zoologi Bogor, 1894-1994* (Bogor: Pusat Penelitian dan Pengembangan Biologi, LIPI, 1994), 4–8. See also M. A. Lieftinck and A. C. V. van Bemmel, "The Development of the Zoological Museum at Buitenzorg," in *Science and Scientists in the Netherlands Indies*, ed. Pieter Honig and Frans Verdoorn (New York: Board for the Netherlands Indies, Surinam and Curaçao, 1945), 226–31.

30. H. H. Zeijlstra, *Melchior Treub: Pioneer of a New Era in the History of the Malay Archipelago* (Amsterdam: KIT, 1959), 74–79.

31. M. J. Sirks, *Indisch natuuronderzoek* (Amsterdam: Ellerman/Harms, 1915), 267–71.

32. See Wim van der Schoor, "Pure Science and Colonial Agriculture: The Case of the Private Java Sugar Experiment Stations (1885-1940)," *Journal of the Japan-Netherlands Institute* 6 (1996): 68–77.

33. For a sympathetic sketch of Van Kol, see Paul van't Veer, "Het Einde van een Eeuw: Ir. Henri Hubertus van Kol, 1852-1925," in *Geen Blad voor de Mond: Vijf Radicalen uit de Negentiende Eeuw* (Amsterdam: Arbeiderspers, 1958), 183–217.

34. H. van Kol, *Uit onze Koloniën* (Leiden: Sijthoff, 1903), 574.

35. Malangsche Vereeniging van Koffie Planters to Van der Wijck, September 19, 1899, Bt. July 31, 1903/48, Archief Algemeene Secretarie, 1891-1942, ANRI, Jakarta.

36. Treub to Rooseboom, October 3, 1899/2429/0, Bt. July 31, 1903/48, Archief Algemeene Secretarie, 1891-1942, ANRI, Jakarta.

37. Departement van Onderwijs, Eeredienst, en Nijverheid, February 12, 1900/ 2544, Bt. July 31, 1903/48, Archief Algemeene Secretarie, 1891-1942, ANRI, Jakarta.

38. In his 1901 proposal to locate the school at his institution, Treub noted that in 1891 and again in 1896 he had advocated reopening the school (which had been moth-balled in 1884) but had been stonewalled by his superiors. Melchior Treub, "Nota over de oprichting eener Landbouwschool in Nederlandsch Indië," July 13, 1901, Bt. July 31, 1903/48, Archief Algemeene Secretarie, 1891-1942, ANRI, Jakarta.

39. Abendanon to Rooseboom, August 3, 1901/9641, and Abendanon to Roose-boom, June 30, 1902/12486, Bt. July 31, 1903/48, Archief Algemeene Secretarie, 1891-1942, ANRI, Jakarta. For more on Abendanon as the director of education, religion, and industry, see Hans van Miert, *Bevlogenheid en onvermogen: Mr. J. H. Abendanon en de ethische richting in het Nederlandse kolonialisme* (Leiden: KITLV Press, 1991).

40. "Reglement voor den Landbouwschool bij 's Lands Plantentuin te Buiten-zorg," Bt. July 31, 1903/48, Archief Algemeene Secretarie, 1891-1942, ANRI, Jakarta.

41. Mgs. to Treub, July 15, 1903, and Bt. July 4, 1903/1, in Bt. July 31, 1903/48, Archief Algemeene Secretarie, 1891-1942, ANRI, Jakarta.

42. H. W. van den Doel, "Practical Agricultural Education in the Netherlands East Indies: The Transfer of Agricultural Knowledge to the Indigenous Population of Java, 1875-1920," *Journal of the Japan-Netherlands Institute* 6 (1996): 82-83.

43. Suzanne Moon, "Constructing 'Native Development': Technological Change and the Politics of Colonization in the Netherlands East Indies, 1905-1930" (PhD diss., Cornell University, 2000), 50.

44. Rooseboom to Van Asch van Wijck, April 14, 1902/3/C/Secret, in *Het ekono-misch beleid in Nederlandsch-Indië: Capita selecta—een bronnenpublicatie*, ed. P. Creutz-berg (Groningen: Wolters-Noordhoff, 1972), 1:279. In this letter to the minister of col-onies, Governor General Rooseboom took credit for coming up with the idea of an agricultural department. Zeijlstra, in his biography, claims that the idea came from Treub, though without citation (Zeijlstra, *Melchior Treub*, 87–88). Zeijlstra worked as Treub's assistant in 1908 and 1909 and may have been drawing on personal recollections of stories he heard at that time about the founding of the Department of Agriculture. A letter written by Zeijlstra in 1959 suggests that he indeed used personal memories to draw together some of the material in the biography. Zeijlstra to Rooseboom, Novem-ber 10, 1959, Treub correspondence archive, Boerhaave Museum, Leiden. Other circumstantial evidence lends credence to Zeijlstra's argument. Governor General Rooseboom's initial request from 1901 or 1902, and Treub's answer, left Treub's superior, Director of Education, Religion, and Industry Abendanon, out of the loop, and this im-plies that the matter had been discussed beforehand and they had intentionally circum-vented Abendanon.

45. Melchior Treub, "Schematisch nota over de oprichting een agricultuur departe-ment in Nederlandsch Indië," January 30, 1902, V. August 7, 1902/74, file 136, Archief Ministerie van Koloniën, 1901-1953, Nationaal Archief, The Hague. The basic descrip-tion of the creation of the Department of Agriculture is in W. M. F. Mansvelt, "De Om-zetting van 's Lands Plantentuin tot Departement van Landbouw," *Koloniale Studiën* 21, no. 1 (1937): 115-33. More recently historians have analyzed the creation of the Depart-ment of Agriculture in the broader context of the ethical policy. See Moon, *Technology*

*and Ethical Idealism,* 25–43; Doel, *De stille macht,* 227–34; Doel, "Practical Agricultural Education"; and Harro Maat, *Science Cultivating Practice: A History of Agricultural Science in the Netherlands and Its Colonies, 1863–1986* (Dordrecht: Kluwer, 2001), 55–60.

46. Treub, "Schematische nota." A later budget addendum suggested that this would be a cheap reorganization. Melchior Treub, "Gegevens ter verkrijging van een algemeen denkbeeld der finantieele consequenties eener oprichting van een agricultuur departement, als in de 'Schematsich nota' bedoeld," March 6, 1902, V. August 7, 1902/74, file 136, Archief Ministerie van Koloniën, 1901–1953, Nationaal Archief, The Hague.

47. "Advies van de raad van Nederlansch-Indië," December 19, 1902/XVIII, in Creutzberg, *Het ekonomisch beleid in Nederlandsch-Indië,* 1:286–88.

48. Vehement opposition to Treub and the department came from the few colonial officials who were practically oriented agriculturalists, such as foresters, who objected to a laboratory scientist being installed as agricultural chief. J. S. van Braam, *Een landbouwdepartement in Indië* (Batavia: Kolff, 1903).

49. Neither Director of Education, Religion, and Industry Abendanon nor Director of Civil Administration P. C. Arends accepted Treub's notion of a technical department without the usual bureaucratic rules. Abendanon to Rooseboom, July 13, 1903/58/Secret, Bt. September 23, 1904/20, Algemeene Secreatrië, 1891–1942, ANRI; Arends to Rooseboom, September 27, 1902/276/A, in Creutzberg, *Het ekonomisch beleid in Nederlandsch-Indië,* 1:283–86.

50. Treub to Idenburg, March 16, 1903, Bt. September 23, 1904/20, Archief Algemeene Secretarie, 1891–1942, ANRI, Jakarta (emphasis in original).

51. This point is forcefully made in Idenburg to Queen Wilhelmina, January 28, 1904/50, Bt. September 23, 1904/20, Archief Algemeene Secretarie, 1891–1942, ANRI, Jakarta.

52. Idenburg, "Memorie van voorlichting: Voorgeslagen Department van Landbouw in Nederlandsch-Indië," in *Ontwerp-Begrooting voor het dienstjaar 1904,* Bt. September 23, 1904/20, Archief Algemeene Secretarie, 1891–1942, ANRI, Jakarta, reprinted in A. W. F. Idenburg, "Oprichting van een Landbouw departement in Ned.-Indië," *Tijdschrift voor het Binnenlandsch Bestuur* 26 (1904): 394–417.

53. "92ste Vergadering van Donderdag 2 Juni 1904," *Handelingen der Staten-Generaal 1903–1904 II,* Bt. September 23, 1904/20, Archief Algemeene Secretarie, 1891–1942, ANRI, Jakarta. The debates in the Dutch Parliament about the department are reprinted in "Kamerdebatten over de oprichting van een Departement van Landbouw in Nederlandsch-Indië," *Tijdschrift voor het Binnenlandsch Bestuur* 27 (1904): 151–73, 259–84, 367–431.

54. C. Pijnacker Hordijk (a past governor general), D. Fock (a future minister of colonies), and J. Th. Cremer (a past minister of colonies) spoke in favor of the proposal. "93ste Vergadering van Vrijdag 3 Juni 1904," *Handelingen der Staten-Generaal 1903–1904 II,* Bt. September 23, 1904/20, Archief Algemeene Secretarie, 1891–1942, ANRI, Jakarta.

55. "96ste Vergadering van Donderdag 9 Juni 1904," *Handelingen der Staten-Generaal 1903–1904 II,* Bt. September 23, 1904/20, Archief Algemeene Secretarie, 1891–1942, ANRI, Jakarta.

56. "31ste Vergardering van Vrijdag 15 Juli 1904," *Handelingen der Staten-Generaal 1903–1904 I,* Bt. September 23, 1904/20, Archief Algemeene Secretarie, 1891–1942, ANRI, Jakarta.

57. See the Department of Agriculture's first yearly report in *Jaarboek van het Departement van Landbouw in Nederlandsch-Indië 1906* (Batavia: Kolff, 1907).

58. M. Treub, "Inleiding," in *Jaarboek van het Departement van Landbouw in Nederlandsch-Indië 1907* (Batavia: Landsdrukkerij, 1908), viii.

59. It went poorly with the European component of the school, however, and no European students started the course in 1906.

60. J. C. Koningsberger, "Landbouwschool en Cultuurtuin," in *Jaarboek van het Departement van Landbouw in Nederlandsch-Indië 1907* (Batavia: Landsdrukkerij, 1908), 62–69.

61. Moon, *Technology and Ethical Idealism*, 35–38; Doel, "Practical Agricultural Education."

62. M. Treub, *Landbouw, Januari 1905–October 1909* (Amsterdam: Scheltema and Holkema, 1910), 29.

63. Ibid., 76.

64. Treub to Went, December 11, 1908, Treub correspondence archive, Boerhaave Museum, Leiden.

65. Doel, "Practical Agricultural Education," 87.

66. Moon, *Technology and Ethical Idealism*, 40.

67. "Treub and his staff should have demonstrated, not just for a few cases but it needed to radiate from everything they did, that they also had, in addition to a thorough knowledge of agricultural technology, clear insight into the economics of agriculture." H. H. Rijkens, "Nog eens: Over het slibbezwaar van eenige rivieren in het Serajoedal en daarmede in verband staande onderzoekingen door Dr. E. C. Jul. Mohr, 1908, Batavia, G. Kolff and Co., Mededeelingen, uitgaande van het Dep. v.d. Landbouw, no. 5, 1908," *Landbouwkundig Tijdschrift* 22 (1910): 271.

68. Quoted in Maat, *Science Cultivating Practice*, 59.

69. J. C. Koningsberger, "Herinneringen aan 1907–1923," November 1929, KIT library, Amsterdam.

70. Treub to Went, July 2, 1909, Treub correspondence archive, Boerhaave Museum, Leiden.

71. Moon, *Technology and Ethical Idealism*, 41.

72. *Jaarboek van het Departement van Landbouw, Nijverheid en Handel 1911* (Batavia: Landsdrukkerij, 1912).

73. Maat, *Science Cultivating Practice*, 59

74. Moon, *Technology and Ethical Idealism*, 44–69.

75. Immediately after Treub heard the news about Lovink, he reportedly tele-grammed Idenburg, who was still in the Netherlands at that time, writing, "Choice worst reaction and grave misjudgment [of] meaning natural research for agronomy." Treub, September 22, 1909, Treub correspondence archive, Boerhaave Museum, Leiden.

76. Treub, *Landbouw*. His brothers Hector and M. W. F. Treub published this book posthumously.

77. For recent scholarship on the two ends of these poles, see Wim Ravensteijn and Jan Kop, eds., *For Profit and Prosperity: The Contributions Made by Dutch Engineers to Public Works in Indonesia, 1800–2000* (Leiden: KITLV Press, 2008); and Adrian Vickers, *A History of Modern Indonesia* (Cambridge: Cambridge University Press, 2005).

78. Harry Benda, "The Pattern of Reforms in the Closing Years of Dutch Rule in Indonesia," *Journal of Asian Studies* 25, no. 4 (1966): 589–605.

79. Cribb, "Development Policy in the Early 20th Century"; Peter Boomgaard, "The Welfare Services in Indonesia, 1900–1942," *Itinerario* 10, no. 1 (1986): 57–81.

## Chapter 5. The Nationalists' Enlightenment

1. For excellent studies about radical nationalism and Indonesian social revolutionaries, see Takashi Shiraishi, *An Age in Motion: Popular Radicalism in Java, 1912–1926* (Ithaca, NY: Cornell University Press, 1990); John Ingleson, *Road to Exile: The Indonesian Nationalist Movement, 1927–1934* (Singapore: Heinemann, 1979); and Ruth McVey, *The Rise of Indonesian Communism* (Ithaca, NY: Cornell University Press, 1965).

2. The definitive treatment of Boedi Oetomo is Akira Nagazumi, *The Dawn of Indonesian Nationalism: The Early Years of Budi Utomo, 1908–1918* (Tokyo: Institute of Developing Economies, 1972). See also Robert van Niel, *The Emergence of the Modern Indonesian Elite* (The Hague: Van Hoeve, 1960).

3. "The beteekenis van de woorden Boedi Oetama," *Indische Gids* 32, no. 1 (1910): 514–15.

4. During World War I Boedi Oetomo did become a political party, and a few of its members joined the elected parliament, the Volksraad, in 1918. It cast itself as a moderate voice, trying to position itself between the colonial government and the social revolutionaries. Hans van Miert, "Benepen stemmen in het zwembassin: Boedi Oetomo in de eerste Volksraad, 1918–1921," *Jambatan* 11, nos. 1–2 (1993): 3–35.

5. Ruth McVey, "Taman Siswa and the Indonesian National Awakening," *Indonesia* 4 (1967): 128–49.

6. Tjipto Mangoenkoesoemo, "De vrees voor Demos," *Het Tijdschrift* 5 (1911): 154–58.

7. Djaja Pranata, "Orgaan Doentoekkan Kemadjoean Hindia dan Anak Boemi," *Djawa-Moeda* 1, no. 1 (1916): 1.

8. Previously these schools had added Dutch as a subject starting in the third grade. Dwidjo Sewojo, "Boedi Oetomo: Rede gehouden voor de Defltsche Studentenvereeniging 'Onze Koloniën' op 2 Mei 1917," *Nederlandsch-Indië Oud en Nieuw* 2 (1917–18): 67–72.

9. S. L. van der Wal, *De opkomst van de nationalistische beweging in Nederlands-Indië* (Groningen: Wolters, 1967), 60–62.

10. Nagazumi, *Dawn of Indonesian Nationalism*, 58.

11. For Tjipto's quote, see "Boedi Oetomo en het volks onderwijs," *Indische Gids* 32, no. 2 (1910): 1390–91, reprinted in Pitot Soeharto and A. Zainoel Ihsan, eds., *Cahaya di Kegelapan* (Jakarta: Jayasakti, 1981), 83–88. The Dutch editorial writer mocked what he saw as Tjipto's conceit of native scientific leadership. See also "Het tweede congres van Boedi Oetomo," *Indische Gids* 32, no. 1 (1910): 94–96.

12. E. F. E. Douwes Dekker, "Onderwijs," *Het Tijdschrift* 1, no. 9 (1912): 312–16.

13. S. S. J. Ratoe Langie, "Waarom geen strijd?" *Hindia Poetra* 1, no. 4 (1916): 89–91.

14. Soewardi was the force behind *Hindia Poetra*, and he wrote a short note at the bottom of Ratoe Langie's article, cautioning Indies students in the Netherlands to do more for their development than simply acquire a technical education. Soewardi made his nationalist debut in 1913 with an incisive parody of Dutch colonial ideology, "If I Were a Dutchman," which is analyzed inter alia in Benedict Anderson, *Imagined Communities: Reflections on the Origin and Spread of Nationalism*, rev. ed. (London: Verso, 1991), 116–18.

15. F. X. Prawirotaroemo, in "Soembangsih: Gedenkboek Boedi-Oetomo, 1908–20 Mei 1918," ed. Noto Soeroto, special issue, *Nederlandsch Indië Oud en Nieuw* (1918): 134.

16. See W. Drees and N. Ant. Jansen, *Eerste koloniaal onderwijscongres: Stenografisch verslag* (The Hague: N.p., 1916), 62–67, for Soewardi's comments at the 1916 conference, which included the claim that for the Indies, Dutch was the "key to the moral culture [*geestelijke beschaving*] of the West."

17. McVey, "Taman Siswa," 129–30; Shiraishi, *An Age in Motion*, 41–69.

18. Shiraishi, *An Age in Motion*, 94–96.

19. Takashi Shiraishi, "The Disputes between Tjipto Mangoenkoesoemo and Soetatmo Soeriokoesoemo: Satria vs. Pandita," *Indonesia* 32 (1981): 93–108. Also see Kenji Tsuchiya, *Democracy and Leadership: The Rise of the Taman Siswa Movement in Indonesia* (Honolulu: University of Hawai'i Press, 1987), 35–37.

20. Th. Thomas, "Een hoogeschool voor Nederlandsch-Indië," pts. 1 and 2, *Jong Indië* 1, no. 49 (1909): 537–38, and 1, no. 50 (1909): 549–51; J. E. Stokvis, "Geen universitair, wel hooger vakonderwijs," *Het Tijdschrift* 1, no. 9 (1912): 276–78.

21. Noto Kworo, "Oude en Nieuwe Methoden in de Verloskundige Hulp op Java," in "Soembangsih: Gedenkboek Boedi-Oetomo 1908–20 Mei 1918," edited by Noto Soeroto, special issue, *Nederlandsch Indië Oud en Nieuw* (1918): 66.

22. This advice went against the by then well-established European medical community's claim that only male doctors with university training could use forceps. Londa Schiebinger, *The Mind Has No Sex? Women in the Origins of Modern Science* (Cambridge, MA: Harvard University Press, 1989), 102–12.

23. Comments by Soewardi Soerjaningrat, quoted in Drees and Ant. Jansen, *Eerste koloniaal onderwijscongres*, 119.

24. Radjiman, June 19, 1918, in *Handelingen van de Volksraad (eerste gewone zitting in het jaar 1918)*, 153–58.

25. R. Sastrawidjono, June 20, 1918, in *Handelingen van de Volksraad (eerste gewone zitting in het jaar 1918)*, 169–71.

26. M. Aboekasan Atmodirono, June 20, 1918, in *Handelingen van de Volksraad (eerste gewone zitting in het jaar 1918)*, 193–95.

27. These institutions included, for example, many of the schools run by Muhammadiyah, an Islamic organization that promoted modernist Islamic thought. Deliar Noer, *The Modernist Muslim Movement in Indonesia, 1900–1942* (Singapore: Oxford University Press, 1973).

28. Tsuchiya, *Democracy and Leadership*, 126–41.

29. Quoted in Tsuchiya, *Democracy and Leadership*, 127.

30. R. Sastrawidjono, November 18, 1918, in *Handelingen van de Volksraad (eerste gewone zitting in het jaar 1918)*, 236–37.

31. Noer, *The Modernist Muslim Movement in Indonesia*, 202.

32. Niel, *Emergence*, 224–26; Benedict Anderson, "A Time of Darkness and a Time of Light: Transpositions in Early Indonesian Nationalist Thought," in *Language and Power: Exploring Political Cultures in Indonesia* (Ithaca, NY: Cornell University Press, 1990), 241–70.

33. An organization of teachers in north Sulawesi, in an effort to secure political control of a local school board, published a weekly newspaper, *Soerat Chabar Boelanan oentoek Kemadjoean dan Keamanan Kaoem dan Negeri*, May 31, 1926.

34. B. Simoen D., "Kewadjiban Nederland dan Jang Haroes Diterima Hindia," *Kemadjoean-Hindia*, April 18, 1923.

35. Harry Benda, "Christiaan Snouck Hurgronje and the Foundations of Dutch Islamic Policy in Indonesia," *Journal of Modern History* 30, no. 4 (1958): 338–47. The Dutch strategy of using native elites to indirectly rule colonial society had been practiced since the early nineteenth century. After 1830 the colonial state began actively cementing alliances with native elites, who would serve as links between the state and civil society. See Heather Sutherland, *The Making of a Bureaucratic Elite: The Colonial Transformation of the Javanese Priyayi* (Singapore: Heinemann, 1979); and Taufik Abdullah, "The Beginning of the Padri Movement," in *Papers of the Dutch-Indonesian Historical Conference*, ed. A. B. Lapian and Anthony Day (Leiden: Bureau of Indonesian Studies, 1978), 143–53.

36. Hans van Miert, *Bevlogenheid en onvermogen: Mr. J. H. Abendanon en de ethische richting in het Nederlandse kolonialisme* (Leiden: KITLV Press, 1991). Since the 1850s, Dutch officials had been using official schools to educate and train native elites with the ultimate goal of recruiting them into the colonial service. With the official ethical policy, however, the colonial state expanded schooling opportunities with an emphasis on preparing colonial inhabitants, European, Chinese, and natives, for colonial careers in trade, agriculture, and industry. I. J. Brugmans, *Geschiedenis van het onderwijs in Nederlandsch-Indië* (Groningen: Wolters, 1938), 340–42.

37. Shiraishi, *An Age in Motion*, 117–248; McVey, *The Rise of Indonesian Communism*.

38. P. J. Drooglever, *De Vaderlandse Club: Totoks en de Indische politiek* (Franeker: T. Wever, 1980).

39. Sutherland, *The Making of a Bureaucratic Elite*, 113–29.

40. Niel, *Emergence*, 244–51.

41. Ibid., 241–42.

42. Anderson, *Imagined Communities*, 83–111.

43. R. E. Elson, *The Idea of Indonesia: A History* (Cambridge: Cambridge University Press, 2008), 92–94.

44. H. Colijn, *Koloniale vraagstukken van heden en morgen* (Amsterdam: De Standaard, 1928).

45. Paul W. van der Veur, "The Social and Geographical Origins of Dutch-Educated Indonesians," in *Education and Social Change in Indonesia*, ed. Paul W. van der Veur (Athens: Ohio University Center for International Studies, 1969), 18–49.

46. Paul W. van der Veur, "Progress and Procrastination in Education Prior to World War II," in *Educational and Social Change in Indonesia*, ed. Paul W. van der Veur (Athens: Ohio University Center for International Studies, 1969), 1–17.

47. Lewis Pyenson, *Empire of Reason: Exact Sciences in Indonesia, 1840–1940* (Leiden: Brill, 1989), 133–35.

48. Bernard Dahm, *Sukarno and the Struggle for Indonesian Independence* (Ithaca, NY: Cornell University Press, 1969), 43–45.

49. H. J. Lam, "Conspectus of Institutions of Pure and Applied Science in or Concerning the Netherlands East Indies," in *Science in the Netherlands East Indies*, ed. L. M. R. Rutten (Amsterdam: De Bussy, 1929), 383–432.

50. Brugmans, *Geschiedenis van het onderwijs*, 342–43.

51. Suzanne Moon, *Technology and Ethical Idealism: A History of Development in the Netherlands East Indies* (Leiden: CNWS, 2007), 57–58.

52. *Middelbare Landbouwschool te Buitenzorg, programma voor het leerjaar, 1915–1916* (Buitenzorg: Departement van Landbouw, Nijverheid, en Handel [hereafter LNH], 1915), 2.

53.  W. G. Boorsma, "Toespraak, door den Directeur der school gehouden bij de opening van het nieuwe gebouw op 3 Juni 1916," in *Middelbare Landbouwschool te Buitenzorg, programma voor het leerjaar, 1916–1917* (Buitenzorg: LNH, 1916), 36–42.

54.  *Middelbare Landbouwschool, 1915–1916*, 17–22.

55.  Boorsma, "Toespraak," 39. Boorsma further noted that this was the case despite the school's efforts to recruit Europeans. In 1920 three students with European names appear on the roster of students. *Middelbare Landbouwschool te Buitenzorg, programma voor het leerjaar, 1920–1921* (Buitenzorg: LNH, 1920), 32. The names of the students enrolled in 1915–16 suggest that nearly all were from Java. *Middelbare Landbouwschool, 1916–1917*, 34–35.

56.  "Pendidikan Adjunct-Landbouw Consulent," *Pandji Poestaka* 5, no. 38 (1927): 611.

57.  Moon, *Technology and Ethical Idealism*, 44–69.

58.  In the 1930s the leader of the education informational service itemized all the possibilities of agricultural and medical education in the colony in D. J. W. J. Kluiver, *Het Landbouw en Medisch Onderwijs in Nederlandsch-Indië* (Batavia: Volkslectuur, 1935).

59.  Fourteen native graduates of Wageningen University were still in Indonesia in 1945. See Zainal Mahmud et al., eds., *Sejarah Penelitian Pertanian Indonesia* (Jakarta: Badan Penelitian dan Pengembangan Pertanian, 1996), 33–34.

60.  Dahm, *Sukarno*, 59–126.

61.  J. D. Legge, *Sukarno: A Political Biography* (Sydney: Allen and Unwin, 1972), 89–96.

62.  Participants pledged to forge national unity on the basis of "one land, one people, and one language." Ibid., 97–98.

63.  W. F. Wertheim, *Indonesian Society in Transition: A Study of Social Change*, 2nd ed. (The Hague: Van Hoeve, 1964), 321–25.

64.  Susan Abeyasekere, *One Hand Clapping: Indonesian Nationalists and the Dutch, 1939–1942* (Clayton, Vic.: Monash University Centre of Southeast Asian Studies, 1976).

65.  Education remained the one area where the functional elite felt able to challenge the Dutch colonial government. In the 1930s, so-called wild schools (schools without an approved state curriculum) multiplied, including especially Taman Siswa schools. Reacting to this development, the colonial government passed an ordinance in 1932 forcing all schools without government funding to obtain a state permit to operate. Colonywide protests, led by westernized educators who had been associated with Boedi Oetomo (such as Soewardi, by then known as Ki Hadjar Dewantara, and Mohammad Sjafei in Sumatra), led to the ordinance being rescinded.

66.  Elson, *The Idea of Indonesia*, 60–61.

67.  The Soetardjo petition called on the colonial government to convene a conference that would create a framework for extending dominion status to the colony in ten years. Susan Abeyasekere, "The Soetardjo Petition," *Indonesia* 15 (1973): 81–107.

68.  Arimbo, "Volkshoofden en z.g. intellectueele leiders," *Soeloeh Indonesia* 2, no. 8 (August 1927): 10.

69.  Shiraishi, *An Age in Motion*, 249–98; Tsuchiya, *Democracy and Leadership*, 140–48; Paul W. van der Veur, "Introduction," in Soetomo, *Toward a Glorious Indonesia: Reminiscences and Observations of Dr. Soetomo*, ed. Paul W. van der Veur (Athens: Ohio University Center for International Studies, 1987), xv–lxii.

70.  John Ingleson, "Sutomo, the Indonesian Study Club, and Organised Labour in Late Colonial Surabaya," *Journal of Southeast Asian Studies* 39, no. 1 (2008): 56.

71. The apostles of enlightenment shifted from Dutch to Malay- and Indonesian-language publications during the 1920s.

72. For Ratoe Langie's career as a Minahasan nationalist and politician, see David Henley, *Nationalism and Regionalism in a Colonial Context: Minahasa in the Dutch East Indies* (Leiden: KITLV Press, 1996).

73. "Madame Curie," *Penindjauan* 1, no. 32 (1934): 755.

74. "Ilmoe Djiwa," *Penindjauan* 1, no. 4 (1934): 83.

75. Pak Bilalang, "Sampai Kemana?" *Penindjauan* 1, no. 22 (1934): 521. The author, writing under a pseudonym, was probably Ratoe Langie.

76. Pak Bilalang, "Sampai Kemana?" *Penindjuaan* 1, no. 23 (1934): 545.

77. Another cultural journal from this period, the 1937 *Bangoen* (Awakening), was novel in addressing itself to an international audience in English. The English articles petered out after the first few issues, and it quickly became a more conventional journal (in Dutch) reporting on the achievements of the functional elite. See "Introduction: Indonesia, to Our Readers Abroad," *Bangoen* 1, no. 1 (1937): 4.

78. "No. 3000," *Penindjauan* 1, no. 32 (1934): 750.

79. "Perhimpoenan Akademisi Indonesia," *Penindjauan* 1, no. 3 (1934): 68–69.

80. R. M. Hardjokoesoemo, "Perbaikilah Keada'an, Kedoedoekan dan Penghidoepan Toean!" *Madjallah Boelanan Oentoek Kemadjoean Ra'jat* 1, no. 1 (1937): 3.

81. R. M. Hardjokoesoemo, "Madjallah untuk Kemadjoean Ra'jat," *Madjallah Boelanan Oentoek Kemadjoean Ra'jat* 1, no. 12 (1937): 178.

82. "Perkawinan dari Prinses Juliana dan Prins Bernhard," *Madjallah Boelanan Oentoek Kemadjoean Ra'jat* 1, no. 1 (1938): 4.

83. "Pemandangan Doenia," *Madjallah Boelanan Oentoek Kemadjoean Ra'jat* 1, no. 3 (1937): 36.

84. "Pengetahoean tentang Mendidik Anak-Anak," *Madjallah Boelanan Oentoek Kemadjoean Ra'jat* 1, no. 1 (1937): 8.

85. "Ilmoe Chewan," *Madjallah Boelanan Oentoek Kemadjoean Ra'jat* 1, no. 1 (1937): 10.

86. "Ilmoe Chewan," *Madjallah Boelanan Oentoek Kemadjoean Ra'jat* 1, no. 11 (1937): 172.

87. "Madjallah oentoek Kemadjoean Ra'jat," *Madjallah Boelanan Oentoek Kemadjoean Ra'jat* 1, no. 12 (1937): 178.

88. "Pengetahoean tentang Hoekoem Negeri," *Madjallah Boelanan Oentoek Kemadjoean Ra'jat* 2, no. 2 (1938): 31.

89. American audiences came to know Kartini through an abridged English language version of *From Darkness to Light*. Raden Adjeng Kartini, *Letters of a Javanese Princess* (New York: Norton, 1964).

90. "Orang Prempoean dan Kemadjoean Djaman," *Madjallah Boelanan Oentoek Kemadjoean Ra'jat* 1, no. 1 (1937): 14.

91. "Faedahnja Pendidikan dan Pengetahoean Jang Loeas," *Madjallah Boelanan Oentoek Kemadjoean Ra'jat* 2, no. 9 (1938): 163.

92. See, for example, "Riwajat Orang Pandai, Tersohor: Benjamin Franklin (7)," *Madjallah Boelanan Oentoek Kemadjoean Ra'jat* 2, no. 1 (1938): 8.

93. James Siegel, *Fetish, Recognition, Revolution* (Princeton, NJ: Princeton University Press, 1997), 93.

94. Rudolf Mrázek, *Engineers of Happy Land: Technology and Nationalism in a Colony* (Princeton, NJ: Princeton University Press, 2002).

95. "Bagaimana Kita Dapat Memateri Sendiri?" *Madjallah Boelanan Oentoek Kemadjoean Ra'jat* 2, no. 9 (1938): 164–66.

96. "Madjallah oentoek Kemadjoean Ra'jat," 178.

97. "Pembahagian dari isinja," *Madjallah Boelanan Oentoek Kemadjoean Ra'jat* 2, no. 24 (1938): 474–76.

98. Dr. L. P., "Een Britisch Indische Natuurkundige benoemd tot Nederlandsche Hoogleraar," *Nationale Commentaren* 1, no. 13 (1938): 212. Shortly after this article appeared, the official budget for 1939 included funds for expanding higher education by adding a Faculty of Eastern Literature to the Law School in Batavia. This was the beginning of an expansion of university education, which in 1940 included a proposal for an agricultural faculty.

99. G. S. S. J. Ratoe Langie, "De Indonesische Universiteit," pt. 1, *Nationale Commentaren* 1, no. 19 (1938): 323–25.

100. G. S. S. J. Ratoe Langie, "De Indonesische Universiteit," pt. 2, *Nationale Commentaren* 1, no. 20 (1938): 340–42.

## Chatper 6. Technocratic Dreams

1. J. A. A. van Doorn, *De laatste eeuw van Indië: Ontwikkeling en ondergang van een koloniaal project* (Amsterdam: Ooievaar, 1996), 105–65.

2. Rudolf Mrázek, *Engineers of Happy Land: Technology and Nationalism in a Colony* (Princeton, NJ: Princeton University Press, 2002), xvii, 59.

3. Bt., December 22, 1933/509, *Staatsblad van Nederlands-Indië, 1933.*

4. E. P. Wellenstein had previously been a senior official in the Department of Finance and since 1926 had been chair of the Native Welfare Commission (Commissie tot het nagaan en verzamelen van gegevens betrekking hebbende op de welvaart der Inheemsche Bevolking). In 1931 he published a book, *Nederlandsch-Indië in de wereldcrisis* (The Hague: Martinus Nijhoff, 1931).

5. For a list of agricultural and general crisis measures taken in 1933 and 1934 by the Department of Economic Affairs, see "Departement van Economische Zaken," in *Regeerings Almanak voor Nederlandsch Indië 1935* (Batavia: Landsdrukkerij, 1935), 1:243–46.

6. For a discussion of the Depression era economy in Indonesia, see J. S. Furnivall, *Netherlands India: A Study of Plural Economy* (Cambridge: Cambridge University Press, 1944), 428–45; Anne Booth, "The Evolution of Fiscal Policy and the Role of Government in the Colonial Economy," in *Indonesian Economic History in the Dutch Colonial Era*, ed. Anne Booth, W. J. O'Malley, and Anna Weidemann (New Haven, CT: Yale University Program in Southeast Asian Studies, 1990), 210–43.

7. Suzanne Moon, *Technology and Ethical Idealism: A History of Development in the Netherlands East Indies* (Leiden: CNWS, 2007), 92–108. For more about the dual economy thesis, see *Indonesian Economics: The Concept of Dualism in Theory and Policy* (The Hague: Van Hoeve, 1961).

8. E. P. Wellenstein, "Remarks on 'Dualistic Economics,'" in *Indonesian Economics: The Concept of Dualism in Theory and Policy* (The Hague: Van Hoeve, 1961), 213.

9. J. van Gelderen, *The Recent Development of Economic Foreign Policy in the Netherlands East Indies* (London: Longmans, 1939), 10–11.

10. H. W. van den Doel, *De stille macht: Het Europese binnenlands bestuur op Java en Madoera, 1808–1942* (Amsterdam: Bert Bakker, 1994), 412–14.

11. Moon, *Technology and Ethical Idealism*, 126–27.

12. Harry Benda, "The Pattern of Administrative Reforms in the Closing Years of Dutch Rule in Indonesia," *Journal of Asian Studies* 25, no. 4 (1966): 591.

13. Doel, *De stille macht*, 417.

14. *Statistisch Jaaroverzicht van Nederlandsch-Indië 1929* (Buitenzorg: Statistisch Kantoor, 1930).

15. Andrew Goss, "Decent Colonialism? Pure Science and Colonial Ideology in the Netherlands East Indies, 1910–1929," *Journal of Southeast Asian Studies* 40, no. 1 (2009): 187–214.

16. Docters van Leeuwen to Went, December 19, 1931, Docters van Leeuwen correspondence archive, Boerhaave Museum, Leiden.

17. K. W. Dammerman, "Algemeene Beschouwingen," 1932, file 188, ICWO Collection, Nationaal Archief, The Hague.

18. Ibid.; K. W. Dammerman, "Natuurwetenschappelijke Instituten," 1932, file 190, ICWO Collection, Nationaal Archief, The Hague. For similar arguments, see R. Wind, "Landbouwkundige Instituten," 1932, file 191, ICWO Collection, Nationaal Archief, The Hague.

19. Docters van Leeuwen to Went, May 4, 1932, Docters van Leeuwen correspondence archive, Boerhaave Museum, Leiden.

20. R. C., "Bezuiniging op 's Lands Plantentuin: Een ander oordeel," *Bataviaasch Nieuwsblad*, January 22, 1932, file 250, Van Steenis Papers, Nationaal Archief, The Hague. Based on information Kees van Steenis received from the editor of the *Bataviaasch Nieuwsblad*, Van Steenis speculated that the author R. C. was in fact the zoologist H. J. V. Sody, then a teacher at the Secondary Agricultural School in Buitenzorg, who was still bitter about being rejected for a job at the gardens a few years earlier. Van Steenis to Polak, February 3, 1932, file 250, Van Steenis Papers, Nationaal Archief, The Hague. For the conflict between Sody and Dammerman after 1930, see J. H. Becking, *Henri Jacob Victor Sody (1892–1959), His Life and Work: A Biographical and Bibliographical Study* (Leiden: Brill, 1989), 15–17.

21. Docters van Leeuwen to F. A. F. C. Went, January 25, 1932, Docters van Leeuwen correspondence archive, Boerhaave Museum, Leiden.

22. David Fairchild and Thomas Barbour, "The Crisis at Buitenzorg," *Science* 80 (1934): 33–34.

23. V. J. Koningsberger, "Het werk van Melchior Treub na 25 jaren," *Koloniale Studiën* 18, no. 1 (1934): 253.

24. Ibid., 257.

25. K. W. Dammerman, "'Het werk van Melchior Treub na 25 jaren,' een antwoord aan Dr. V. J. Koningsberger," *Koloniale Studiën* 18, no. 2 (1934): 402–12; K. W. Dammerman, "'s Lands Plantentuin te Buitenzorg," *Koloniale Studiën* 22, no. 1 (1938): 33–57.

26. W. M. F. Mansvelt, "De omzetting van 's Lands Plantentuin tot Department van Landbouw," *Koloniale Studiën* 21, no. 2 (1937): 115–33; D. Tollenaar, "Ontwikkeling en toekomst van het natuurwetenschappelijk onderzoek voor Nederlandsch-Indië," *Koloniale Studiën* 21, no. 3 (1937): 237–53.

27. In the 1935 issue of the official almanac of the colonial government, the section about the Department of Economic Affairs starts with an extensive description of all the activities and policies related to the economic crisis. All the policies listed concerned

the regulation of export crops and consumer goods. "Departement van Economische Zaken," 1:243–46.

28. Dakkus to Lam, July 23, 1934, Buitenzorg correspondence archive, 1933–1945, NHN, Leiden.

29. Ibid.

30. Dammerman's restaurant was derided by colleagues. Backer to Van Steenis, Louwmaand 2, 1937, file 4, Van Steenis Papers, Nationaal Archief, The Hague. The restaurant stands to this day and is one of the nicest spots in which to eat lunch in greater Jakarta.

31. Van Steenis to Danser, December 21, 1938, file 3d, Van Steenis Papers, Nationaal Archief, The Hague.

32. Van Steenis to Pulle, June 20, 1936, file 4, Van Steenis Papers, Nationaal Archief, The Hague.

33. This progressive lobby argued in its writings that the continued implementation of Western civilization in the colony should be coupled with greater autonomy for the Indies, expanded development opportunities, and increased political leadership by Indonesians. *De Stuw* was short-lived and was shut down in 1933 by Governor General De Jonge. Elsbeth Locher-Scholten, *Ethiek in fragmenten: Vijf studies over koloniaal denken en doen van Nederlanders in de Indonesische Archipel, 1877–1942* (Utrecht: Hes Publishers, 1981), 118–49.

34. H. J. van Mook, *De organisatie van de Indische Regeering* (Batavia: Albrecht, 1932), 34–35.

35. Ibid., 41.

36. Doel, *De stille macht*, 43–44.

37. H. J. van Mook, "Vragenlijst," Buitenzorg correspondence archive, 1933–1945, NHN, Leiden.

38. L. G. M. Baas Becking, "De beteekenis van 's Lands Plantentuin . . . ," 1936, V. J. Koningsberger to Van Mook, December 4, 1936, Lam to Van Mook, December 9, 1936/1179, Buitenzorg correspondence archive, 1933–1945, NHN, Leiden.

39. Onghokham, *Runtuhnya Hindia Belanda* (Jakarta: Gramedia, 1987), 74.

40. Baas Becking started his career in the 1920s as a professor of biology at Stanford University. In 1930 he was appointed a professor at Leiden University and thoroughly modernized the biology discipline in his first few years there. For biographical details, see J. Ruinen, "L. G. M. Baas Becking in Memoriam," *Vakblad voor biologen* 43, no. 3 (1963): 41–47.

41. On the retirement, see Bt. January 26, 1939/24, Archief Algemeene Secretarie, 1891–1942, ANRI, Jakarta.

42. Bt. February 16, 1939/21, Archief Algemeene Secretarie, 1891–1942, ANRI, Jakarta.

43. Van Steenis to Verdoorn, September 20, 1939, Van Steenis file, Institute for History and Foundations of Science, Utrecht University.

44. Van Steenis to Verdoorn, June 12, 1939, Van Steenis file, Institute for History and Foundations of Science, Utrecht University.

45. Baas Becking to Lam, September 27, 1939, Buitenzorg correspondence archive, 1933–1945, NHN, Leiden.

46. Van Mook to Tjarda van Starkenborgh Stachouwer, December 28, 1939/4614 Geheim, Geheim Verbaal, March 30, 1940/Y19, file 580, Ministerie van Koloniën Geheim Archief, 1901–1940, Nationaal Archief, The Hague.

47. Ibid.

48. Lieftinck to Boschma, April 20, 1939, Lieftinck to Boschma, March 3, 1940, Lieftinck correspondence archive, Naturalis Museum, Leiden; Van Steenis to Verdoorn, May 22, 1939, Van Steenis to Verdoorn, June 12, 1939, Van Steenis to Verdoorn, September 20, 1939, Van Steenis file, Institute for History and Foundations of Science, Utrecht University.

49. Van Mook, "Memorie van Inlichting," Volksraad, zittingsjaar, 1939–1940, "Zevende aanvullende begrooting betreffende afdeeling VI der begrooting van Nederlandsch-Indie voor 1940," Onderwerp 126, February 1940, Geheim Verbaal, March 30, 1940, no. Y19, file 580, Ministerie van Koloniën Geheim Archief, 1901–1940, Nationaal Archief, The Hague. This was a reworking of the memo to the governor general cited above in note 46, with many phrases verbatim but also with some points, including the one cited here, expanded. There was much less about the economic benefits.

50. J. A. van Helsdingen et al., "Afdeelingsverslag," February 15, 1940, Volksraad, zittingsjaar, 1939–1940, "Zevende aanvullende begrooting betreffende afdeeling VI der begrooting van Nederlandsch-Indie voor 1940," Onderwerp 126, February 1940, Geheim Verbaal, March 30, 1940/Y19, file 580, Ministerie van Koloniën Geheim Archief, 1901–1940, Nationaal Archief, The Hague.

51. For speeches made by Volksraad members, mostly against Van Mook's plan, see *Handelingen van de Volksraad 1939–40*, 2:1873–1884, 1953–1959.

52. Ibid., 1963. See also Lieftinck to Boschma, March 3, 1940, Lieftinck correspondence archive, Naturalis Museum, Leiden.

53. Shigeru Sato, *War, Nationalism, and Peasants: Java under the Japanese Occupation, 1942–1945* (Armonk, NY: M. E. Sharpe, 1994), 22–25.

54. Prasenjit Duara, *Sovereignty and Authenticity: Manchukuo and the East Asian Modern* (Lanham, MD: Rowman and Littlefield, 2003), 48.

55. Hiromi Mizuno, *Science for the Empire: Scientific Nationalism in Modern Japan* (Stanford, CA: Stanford University Press, 2009), 47–49.

56. Bambang Hidayat, "Under a Tropical Sky: A History of Astronomy in Indonesia," *Journal of Astronomical History and Heritage* 3, no. 1 (2000): 45–58.

57. Ethan Mark, "Appealing to Asia: Nation, Culture, and the Problem of Imperial Modernity in Japanese-Occupied Java, 1942–1945" (PhD diss., Columbia University, 2003).

58. My description of Buitenzorg in 1942 is drawn from M. J. van Steenis-Kruseman, *Verwerkt Indisch verleden* (Oegstgeest: N.p., 1988), 11–20.

59. Baas Becking was in Holland on a short trip meant to recruit new staff for the gardens when the Germans invaded in May of 1940, and Van Slooten acted in his place.

60. Van Steenis to Verdoorn, November 28, 1945, Van Steenis file, Institute for History and Foundations of Science, Utrecht University.

61. C. G. G. J. van Steenis, "Memorandum," October 18, 1945, file 8, Van Steenis Papers, Nationaal Archief, The Hague.

62. Steenis-Kruseman, *Verwerkt Indisch verleden*, 23–26.

63. H. C. D. de Wit, "Miscellaneous Information," *Flora Malesiana Bulletin* 5 (1949): 131. At the end of the notice, De Wit concluded, "We met him as enemies. The course of events excluded any intimate or even friendly relation. It would have been good to have known him as a friend."

64. C. G. G. C. van Steenis, "Blown by the Wind, Learnings through the Years: Autobiographical Notes," March 1986, file 30, Van Steenis Papers, Nationaal Archief, The Hague.

65. M. J. van Steenis-Kruseman, "Herbarium and Library: Research Work and Investigation under Nippon Government," September 18, 1945, file 92, Van Steenis Papers, Nationaal Archief, The Hague.

66. C. G. G. J. van Steenis, "Odoardo Beccari," 1944, file 54, Van Steenis Papers, Nationaal Archief, The Hague.

67. Steenis-Kruseman, *Verwerkt Indisch verleden*, 29.

68. Van Steenis to Verdoorn, September 12, 1945, Van Steenis file, Institute for History and Foundations of Science, Utrecht University.

69. Benedict Anderson, *Java in a Time of Revolution: Occupation and Resistance, 1944-1946* (Ithaca, NY: Cornell University Press, 1972), 168.

70. Van Steenis to Baas Becking, December 17, 1945, file 872, Archief Algemeene Secretarie, 1944-1950, ANRI, Jakarta; C. G. G. J. van Steenis, "Report on the State of the Scientific Institutes Buitenzorg," January 25, 1946, file 243, ICWO Collection, Nationaal Archief, The Hague; F. S. V. Donnison, *British Military Administration in the Far East, 1943-1946* (London: Her Majesty's Stationary Office, 1956), 427-29; Anderson, *Java in a Time of Revolution*, 297.

71. Van Steenis-Kruseman, *Verwerkt Indisch Verleden*, 36.

72. Koolhaas and Reisma, "Bezoek aan Buitenzorg op 28 Mei van Lt. Kolonel Eggleton, Lt. Kolonel Reitsma, Lt. Zweerts en Dr. D. R. Koolhaas," May 29, 1946, file 872, Archief Algemeene Secretarie, 1944-1950, ANRI, Jakarta.

73. No one with the title of governor general was appointed after 1945. For more about the Dutch motivations after 1945, see the studies in P. J. Drooglever and M. J. B. Schouten, eds., *De Leeuw en de Banteng* (The Hague: Instituut voor Nederlandse Geschiedenis, 1997).

74. Among the nationalists the social revolutionaries were outmaneuvered and defeated by more pragmatic nationalists, who concentrated on winning independence from the Dutch. Rudolf Mrázek, "Sjahrir and the Left Wing at the Time of the Indonesian Revolution," in *De Leeuw en de Banteng*, ed. P. J. Drooglever and M. J. B. Schouten (The Hague: Instituut voor Nederlandse Geschiedenis, 1997), 109-30.

75. James Siegel, *Fetish, Recognition, Revolution* (Princeton, NJ: Princeton University Press, 1997), 183-86.

76. C. G. G. J. van Steenis, "Herbarium en Museum voor Systematische Botanie," 1946, file 197, Van Steenis Papers, Nationaal Archief, The Hague.

77. John Smail, *Bandung in the Early Revolution, 1945-1946: A Study in the Social History of the Indonesian Revolution* (Ithaca, NY: Cornell University Southeast Asia Program, 1964), 147-53.

78. Van Steenis-Kruseman, *Verwerkt Indisch verleden*, 3-37.

79. L. G. M. Baas Becking, "Coordinatie van Natuurwetenschappen," March 10, 1946, file 871, Archief Algemeene Secretarie, 1944-1950, ANRI, Jakarta.

80. H. M. J. Hart, "De Beteekenis van het Wetenschappelijk Onderzoek voor Nederlandsch-Indië," January 11, 1946, file 205, ICWO Collection, Nationaal Archief, The Hague.

81. V. J. Koningsberger, "Een Natuurwetenschappelijk Rijksdienst voor Nederlandsch-Indie," 1946, file 197, Van Steenis Papers, Nationaal Archief, The Hague.

Van Steenis's plan for a massive flora of the archipelago, known as the Flora Malesiana, was tangential to Koningsberger and Honig's plan, but because Van Steenis was one of the very few biologists who had been working in Buitenzorg in the last few years it remained an active proposal. Van Steenis to P. Honig, May 10, 1946, and P. Honig to Van Steenis, June 22, 1946, file 197, Van Steenis Papers, Nationaal Archief, The Hague.

82. Van Mook, October 11, 1945, file 871, and Baas Becking to Van Mook, October 30, 1945, file 872, Archief Algemeene Secretarie, 1944–1950, ANRI, Jakarta. Baas Becking's powerful job was formalized in March of 1946. Bt. Lt. Gouverneur-General, March 21, 1946/1, file 871, Archief Algemeene Secretarie, 1944–1950, ANRI, Jakarta.

83. "Bespreking op 5 December 1945 ten kantore van Prof. Baas Becking om 2.5 u. n.m.," file 249, ICWO Collection, The Hague.

84. J. J. P. de Jong, "H. J. van Mook als vraagteken: Het Nederlands beleid en de overgangsperiode," in *De Leeuw en de Banteng*, ed. P. J. Drooglever and M. J. B. Schouten (The Hague: Instituut voor Nederlandse Geschiedenis, 1997), 190–200.

85. Doorn, *De laatste eeuw van Indië*, 247–59. See also H. W. van den Doel, *Afscheid van Indië: De val van het Nederlandse imperium in Azië* (Amsterdam: Prometheus, 2001), 156–59; and Yong Mun Cheong, *H. J. van Mook and Indonesian Independence: A Study of His Role in Dutch-Indonesian Relations, 1945–1948* (The Hague: Martinus Nijhoff, 1982), 86–90.

86. "Bespreking over de Coordinatie-Commissie op Zaterdag, 9 Maart 1946 's morgens om 10 uur ten Paleize Koningsplein," file 206, ICWO Collection, Nationaal Archief, The Hague.

87. H. J. van Mook, *Indonesië, Nederland, en de Wereld* (Amsterdam: De Bezige Bij, 1949), 142–47.

88. Baas Becking and Hart to Van Mook, August 12, 1946/3661/XX, "Nota betreffende: Een Nederlansch-Indischen Natuurwetenschappelijken Dienst," "Ontwerp-Besluit tot het instellen van een Nederlansch-Indischen Natuurwetenschappelijken Dienst," and "Memorie van Toelichting betreffende de ontwerp-ordonnantie tot het instellen van een Nederlandsch-Indischen Natuurwetenschappelijken Dienst," file 871, Archief Algemeene Secretarie, 1944–1950, ANRI, Jakarta.

89. "Nota betreffende: Een Nederlandsch-Indischen Natuurwetenschappelijke Dienst."

90. Honig to Van Mook, November 1946, file 871, Archief Algemeene Secretarie, 1944–1950, ANRI, Jakarta.

91. "Ordonnantie op de Instelling, Inrichting en Bevoegdheid van de Organisatie voor Natuurwetenschappelijk Onderzoek," 1947; "Memorie van Toelichting," 1947; "Kort verslag van de vergardering van 3 April ten Paleize Koningsplein betreffende de ontwerp ordonnantie Organisatie Wetenschappelijk Onderzoek"; H. J. van Mook, "Nota van den Luitenant Gouverneur-General," April 5, 1948; Bt. May 1, 1948, no. 1, file 871, Archief Algemeene Secretarie, 1944–1950, ANRI, Jakarta.

92. Baas Becking, "Kort overzicht van de werkzaamheden door L. G. M. Baas Becking verricht voor de Nederlands Indische Regering en voor de federale interim regering," December 31, 1948, file 872, Archief Algemeene Secretarie, 1944–1950, ANRI, Jakarta.

93. It moved to a prominent address on Koningsplein Zuid, near the site where the American embassy stands today. Honig to Van Mook, August 16, 1948/1031, file 872, Archief Algemeene Secretarie, 1944–1950, ANRI, Jakarta.

94. For the Linggajati Agreement and its breakdown, see Anthony Reid, *The Indonesian National Revolution, 1945–1950* (Hawthorne, Australia: Longman, 1974), 109–12.

95. V. J. Koningsberger et al. to Minister van Overzeesche Gebiedsdelen, June 10, 1947, file 871, Archief Algemeene Secretarie, 1944–1950, ANRI, Jakarta.

96. Doel, *Afscheid van Indië*, 241.

97. Baas Becking, "Kort overzicht."

98. *Buitenzorg Scientific Centre* (Buitenzorg: Archipel, 1948), 15.

99. Van Slooten to Van Steenis, June 8, 1948, file 8, Van Steenis Papers, Nationaal Archief, The Hague.

100. Baas Becking, "Kort Overzicht."

101. Zainal Mahmud et al., eds., *Sejarah Penelitian Pertanian Indonesia* (Jakarta: Badan Penelitian dan Pengembangan Pertanian, 1996), 33. Wisaksono in the 1950s became a senior administrator at the University of Indonesia.

102. D. F. van Slooten, "Wetenschap: Het Reserve-Kapitaal der Maatschappij," *Chronica Naturae* 105, nos. 7–8 (1949): 186–90; D. F. van Slooten, "Kebun Raya Indonesia: Doelstellingen en werkzaamheden," March 29, 1950, file 256, Van Steenis Papers, Nationaal Archief, The Hague.

103. Pramoedya Ananta Toer, *The Mute's Soliloquy: A Memoir* (New York: Hyperion East, 1999), 195.

104. Adam Messner, "Effects of the Indonesian National Revolution and Transfer of Power on the Scientific Establishment," *Indonesia* 58 (1994): 41–68.

105. C. G. G. J. van Steenis, "Instructions for Cooperators," *Flora Malesiana Bulletin* 1 (1947): 2–29.

106. C. G. G. J. van Steenis, "Gespecifeerde uitgaven in Nederland verbonden aan het werk Flora Malesiana," 1948, file 869, Archief Algemeene Secretarie, 1944–1950, ANRI, Jakarta.

107. Donk to Van Steenis, August 27, 1948, and Van Slooten to Van Steenis, June 8, 1948, file 6, Van Steenis Papers, Nationaal Archief, The Hague.

108. Bloembergen to Van Steenis, February 9, 1949, file 5, Van Steenis Papers, Nationaal Archief, The Hague.

109. C. G. G. J. van Steenis, "Introduction," in *Flora Malesiana*, vol. 4, series 1 (Batavia: Noordhoff-Kolff, 1948), ix. According to notes found in his papers, this section of the introduction was one of the last pieces he wrote before it went to press in December of 1948. C. G. G. J. van Steenis, "The Importance of the 'Flora Malesiana' to Tropical Botany, Forestry, and Agriculture," November 1948, file 41, Van Steenis Papers, Nationaal Archief, The Hague.

110. Van Steenis to De Wit, June 25, 1949, file 9, Van Steenis Papers, Nationaal Archief, The Hague.

111. Van Steenis to Schuurman, August 6, 1949, file 8, Van Steenis Papers, Nationaal Archief, The Hague.

112. Van Steenis to Schuurman, September 21, 1949, file 8, Van Steenis Papers, Nationaal Archief, The Hague.

113. Van Steenis to De Wit, March 16, 1950, file 9, Van Steenis Papers, Nationaal Archief, The Hague.

114. Van Steenis to De Wit, August 27, 1950, file 9, Van Steenis Papers, Nationaal Archief, The Hague.

115. Koesnoto, "Kebun Raya Indonesia Bogor, Rentjana Pekerdjaan 1950," file 178, Van Steenis Papers, Nationaal Archief, The Hague.

116. Herbert Feith, *The Decline of Constitutional Democracy in Indonesia, 1950–1957* (Ithaca, NY: Cornell University Press, 1962), 32–37.

117. Van Steenis to De Wit and Lam, July 4, 1950, file 9, Van Steenis Papers, Nationaal Archief, The Hague.

118. "Explanatory Memorandum to the Deed Establishing the Foundation 'Flora Malesiana,'" October 21, 1950, file 160, Van Steenis Papers, Nationaal Archief, The Hague.

119. L. A. van der Woerd, "Some Questions Which Arise to the Pharmacologist in Indonesia," *OSR News* 3, no. 4 (1951): 54–55.

120. "Annual Report of the General Secretary to the President of the Board of the Organization for Scientific Research in Indonesia for 1949," *OSR News* 2, no. 5 (1950): 50.

121. H. Vlugter, "The Academic Year, 1949/1950," *OSR News* 2, no. 12 (1950): 171–72.

122. Koesnoto to Donk and others, October 12, 1951/236, file 178, Van Steenis Papers, Nationaal Archief, The Hague.

123. "Bogor Scientific Centre," *Organization for Scientific Research Bulletin* 12 (July 1952): 1–35.

## Chapter 7. Desk Science

1. Rudolf Mrázek, "Bridges of Hope: Senior Citizens' Memories," *Indonesia* 70 (October 2000): 49.

2. Ibid., 50.

3. The classic statement is in Benedict Anderson, "Old State, New Society: Indonesia's New Order in Comparative Historical Perspective," *Journal of Asian Studies* 42, no. 3 (1983): 477–96.

4. See Daniel S. Lev, *The Transition to Guided Democracy: Indonesian Politics, 1957–1959* (Ithaca, NY: Cornell University Southeast Asia Program, 1966); and Ulf Sundhaussen, *Road to Power: Indonesian Military Politics, 1945–1967* (Kuala Lumpur: Oxford University Press, 1982). The transition to Guided Democracy was the first in a series of political changes that ten years later led to an authoritarian Indonesian state. Scholars have explained this change by charting the rise of the Partai Komunis Indonesia (Indonesian Communist Party, PKI) and army under Sukarno's Guided Democracy, the coup and countercoup of September 30 and October 1, 1965, the violent suppression of the communists, and the imposition of Suharto's authoritarian New Order. Harold Crouch, *The Army and Politics in Indonesia*, rev. ed. (Ithaca, NY: Cornell University Press, 1988); R. E. Elson, *Suharto: A Political Biography* (Cambridge: Cambridge University Press, 2001); John Roosa, *Pretext for Mass Murder: The September 30th Movement and Suharto's Coup d'État in Indonesia* (Madison: University of Wisconsin Press, 2006).

5. Herbert Feith, *The Decline of Constitutional Democracy in Indonesia, 1950–1957* (Ithaca, NY: Cornell University Press, 1962).

6. Donald Hindley, *The Communist Party of Indonesia, 1951–1963* (Berkeley: University of California Press, 1964).

7. David Reeve, *Golkar of Indonesia: An Alternative to the Party System* (Singapore: Oxford University Press, 1985), 130–40.

8. Hindley, *The Communist Party*, 280–81.

9. D. Dwidjoseputro, *Review of the State of Biology in Indonesia as a Scientific Discipline* (Malang: Institut Keguruan dan Ilmu Pendidikan, 1970), 2.

10. Adam Messer, "Effects of the Indonesian National Revolution and Transfer of Power on the Scientific Establishment," *Indonesia* 58 (1994): 41.

11. Zainal Mahmud et al., eds., *Sejarah penelitian pertanian Indonesia* (Jakarta: Badan Penelitian dan Pengembangan Pertanian, 1996), 32–33.

12. Koesnoto, "De toekomst van de Regeringsinstellingen voor natuurwetenschappelijk onderzoek te Bogor," July 13, 1951, file 246, Van Steenis Papers, Nationaal Archief, The Hague.

13. Koesnoto to Van Steenis, January 30, 1951, file 178, Van Steenis Papers, Nationaal Archief, The Hague.

14. M. Jacobs, "Weergave van de vergadering van stafleden van KRI op 28 Mei 1955," file 7, Van Steenis Papers, Nationaal Archief, The Hague.

15. Ruinen to Van Steenis, February 1, 1955, file 51, Van Steenis Papers, Nationaal Archief, The Hague. For a later statement of the low morale of the European biologists, see J. van Borssum Waalkes to Lam, August 8, 1955, Confidential Correspondence Box, 1950–1990, NHN, Leiden.

16. Sampurno Kadarsan, interview with the author, May 19, 2001, Bogor.

17. Dwidjoseputro, *Review of the State of Biology in Indonesia*, 11. One exception was Soetomo Soerohaldoko. Just after graduating from the agricultural school at Gadjah Mada University in Yogyakarta in 1955, he joined the staff at the botanical gardens as a research scientist. Soetomo Soerohaldoko, interview with the author, January 18, 2001, Bogor.

18. Koesnoto to Van Steenis, February 17, 1955, file 51, Van Steenis Papers, Nationaal Archief, The Hague.

19. *Laporan tahun 1953 dari Djawatan Penjelidikan Alam (Kebun Raya Indonesia)* (Bogor: Kebun Raya Indonesia, 1957).

20. M. Jacobs, "Weergave van de vergadering van Stafleden van KRI op 28 Mei 1955." See also Koesnoto Setyodiwiryo, "Pendirian Akademi Biologi oleh Kementerian Pertanian Bogor," May 18, 1955/416, file 232, Van Steenis Papers, Nationaal Archief, The Hague.

21. J. Ruinen to Koesnoto, May 28, 1955, file 232, Van Steenis Papers, Nationaal Archief, The Hague.

22. Jacobs to Van Steenis, June 3, 1955, file 7, Van Steenis Papers, Nationaal Archief, The Hague. On June 9, 1955, the Department of Agriculture established a preparatory committee for the academy. See Surat Keputusan Menteri Pertanian, no. 88/Um/55, July 6, 1955, file 1051, Arsip Kabinet Presiden, 1950–1959, ANRI, Jakarta.

23. Soetomo Soerohaldoko, "Pendirian Akademi Biologi Sebagai Cikal Bakal Akademi Kementerian Pertanian," pamphlet, 1998, 6.

24. The other members of the planning committee, besides Koesnoto, were from the Department of Agriculture. Warsito, "Sejarah Akademi Departemen Pertanian Ciawi-Bogor," photocopy, 2000.

25. Jacobs to Van Steenis, June 18, 1955, file 7, Van Steenis Papers, Nationaal Archief, The Hague.

26. Surat Keputusan Menteri Pertanian, 166/Um/55, December 7, 1955, file 1051, Arsip Kabinet Presiden, 1950–1959, ANRI, Jakarta.

27. Soerohaldoko, "Pendirian Akademi Biologi," 7–8.

28. Mien Rifai, interview with the author, June 25, 2001, Jakarta.

29. Soenartono Adisoemarto, "Di Tangan Koesnoto Lilin itu Menyala," in *Bunga-Bunga Pun Bermekaran*, ed. Setijati Sastrapradja (Bogor: Badan Penelitian dan Pengembangan Pertanian, 1999), 3–10.

30. Warsito, "Sejarah Akademi," 13–14.

31. Koesnoto Setyodiwirjo, "Limas Hajati, Bagaimana memelihara dan mempertahankannja," October 10, 1955, reprinted in Warsito, "Sejarah Akademi."

32. Jacobs to Van Steenis, October 25, 1955, file 7, Van Steenis Papers, Nationaal Archief, The Hague.

33. Setijati Sastrapradja, interview with the author, May 10, 2001, Cibinong.

34. Achmad M. Fagi, "Kata Pengantar," in *Bunga-Bunga Pun Bermekaran*, ed. Setijati Sastrapradja (Bogor: Badan Penelitian dan Pengembangan Pertanian, 1999), xi–xii.

35. Adisoemarto, "Di Tangan Koesnoto Lilin itu Menyala." For a history of the Cibodas gardens, see Soetomo Soerohaldoko et al., *Kebun Raya Cibodas, 11 April 1852–11 April 2000* (Bogor: LIPI, 2000).

36. Jacobs to Van Steenis, October 25, 1955.

37. Soenartono Adisoemarto, "Setinggi-tinggi Terbang Bangau," in *Bunga-Bunga Pun Bermekaran*, ed. Setijati Sastrapradja (Bogor: Badan Penelitian dan Pengembangan Pertanian, 1999), 17–36.

38. Sri Koesdijati Soepomo, "Liku-Liku Sebuah Perjalanan," in *Bunga-Bunga Pun Bermekaran*, ed. Setijati Sastrapradja (Bogor: Badan Penelitian dan Pengembangan Pertanian, 1999), 216.

39. Setijati Sastrapradja, interview with the author, May 10, 2001, Cibinong.

40. Setijati Sastrapradja, "Ketika Langit Bertabur Bintang," in *Bunga-Bunga Pun Bermekaran*, ed. Setijati Sastrapradja (Bogor: Badan Penelitian dan Pengembangan Pertanian, 1999), 134.

41. Mien Rifai, interview with the author, June 25, 2001, Jakarta. These assignments were not always appreciated. See, for example, Soeharnis's ambivalence about his assignment as the director of an agricultural high school in Ambon in 1971: Soeharnis, "Hari-hari Penuh Tantangan," in *Bunga-Bunga Pun Bermekaran*, ed. Setijati Sastrapradja (Bogor: Badan Penelitian dan Pengembangan Pertanian, 1999), 207.

42. Warsito, "Sejarah Akademi," 32. No new students were accepted in 1956 due to lack of funds. *Annual Report of 1956, Lembaga Pusat Penjelidikan Alam* (Bogor: LPPA, n.d.), 100.

43. Soerohaldoko, "Pendirian Akademi Biologi," 12.

44. Ibid., 18.

45. I have only read an occasional personal or professional letter from the first generation of Indonesian scientists in either Indonesia or the Netherlands. While in Indonesia in 2001, I was told there were no known extant archives of scientific institutions dating from the 1950s and 1960s. As a consequence the material in this chapter is based on published sources and interviews, as well as the correspondence of Dutch and Dutch-educated scientists now in archives in the Netherlands.

46. Ruth McVey, "The Case of the Disappearing Decade," in *Democracy in Indonesia, 1950s and 1990s*, ed. David Bourchier and John Legge (Clayton, Australia: Monash University Centre of Southeast Asian Studies, 1994), 3–15.

47. Adnan Buyung Nasution, *The Aspiration for Constitutional Government in Indonesia: A Socio-legal Study of the Indonesian Konstituante* (Jakarta: Pustaka Sinar Harapan, 1992).

48. Kostermans received his doctorate in botany from Utrecht University in the 1930s under the direction of A. A. Pulle. He arrived in the Indies during the late 1930s but had trouble getting his biology career going. Soon after arriving he spent more than a year in jail after being convicted of pederasty. During the Japanese occupation he worked on the Burma Railroad. After the war he found work at the Forestry Institute, where he stayed on after independence and got to know Dilmy. After the expulsion of Dutch residents in 1959, he became an Indonesian citizen. The best biographical treatment is Mien A. Rifai, "Kostermans: The Man, His Work, His Legacy," in *Plant Diversity in Malesia III: Proceedings of the Third International Flora Malesiana Symposium, 1995*, ed. J. Dransfield, M. J. E. Coode, and D. A. Simpson (Kew: Royal Botanic Gardens, 1997), 225–30. For further biographical material, see M. Jacobs, "Kostermans Seventy-Five," *Reinwardtia* 10, no. 1 (1982): 9–20; and Van Steenis's vituperative response to it, "Biograpische Notities over A. J. G. H. Kostermans," March 23, 1986, Confidential Correspondence Box, 1946–1976, NHN, Leiden.

49. Jacobs to Van Steenis, January 6, 1957, file 92, Van Steenis Papers, Nationaal Archief, The Hague.

50. Dilmy to Van Steenis, March 1, 1955/L/82, file 92, Van Steenis Papers, Nationaal Archief, The Hague.

51. *Annual Report for 1957, Lembaga Pusat Penjelidikan Alam* (Bogor: LPPA, n.d.), 10–11. See also Dilmy to Van Steenis, April 12, 1957, file 92, Van Steenis Papers, Nationaal Archief, The Hague.

52. A. J. G. H. Kostermans, "In Memorium Anwari Dilmy," *Reinwardtia* 10, no. 1 (1982): 5–7.

53. Dilmy to Van Steenis, Feburary 2, 1961/L/35, file 92, Van Steenis Papers, Nationaal Archief, The Hague.

54. Kostermans to Van Steenis, July 25, 1960, file 76, Van Steenis Papers, Nationaal Archief, The Hague.

55. For example, see A. Dilmy and A. J. G. H. Kostermans, "Research on the Vegetation of Indonesia," in *Study of Tropical Vegetation: Proceedings of the Candy Symposium*, ed. C. H. Holmes (Paris: UNESCO, 1958), 28–30.

56. Sastrapradja, "Ketika Langit Bertabur Bintang," 138; Setijati Sastrapradja, interview with the author, May 10, 2001, Cibinong.

57. Brotonegoro, "Senyum Kelegaan Dalam Pengabdian," 49; Kostermans to Van Steenis, July 25, 1960, file 76, Van Steenis Papers, Nationaal Archief, The Hague.

58. Aprilani Sugiarto, interview with the author, May 9, 2001, Jakarta.

59. While many of the students from the first Ciawi cohort eventually went overseas for graduate-level education, they were not the first to do so even among biologists. In the mid-1950s, S. Somadikarta went to Germany and Otto Soemarwoto and Sampurno Kadarsan went to the United States. All three had worked at the botanical gardens during the 1950s and would work there again after returning from abroad.

60. Sastrapradja, "Ketika Langit Bertabur Bintang," 139.

61. Koesnoto to Donk and others, October 12, 1951/236, file 178, Van Steenis Papers, Nationaal Archief, The Hague.

62. Messer, "Effects of the Indonesian National Revolution."

63. Sarwono Prawirohardjo, "Laporan Singkat Panitia Persiapan Pembentukan Madjelis Ilmu Pengetahuan Nasional," *Berita M.I.P.I.* (1956): 14–20; "Undang-Undang No. 6 Tahun 1956 Tentang Pembentukan Madjelis Ilmu Pengetahuan Indonesia," *Berita M.I.P.I.* (1956): 3–13; Sarwono Prawirohardjo, "The Scientific Development of Indonesia," *Berita M.I.P.I.* 5, no. 5 (1961): 5–14; "Memori Pendjelasan Mengenai Usul Undang-Undang Tentang Pembentukan Madjelis Ilmu Pengetahuan Indonesia," file 1e, Arsip Kabinet Presiden, 1950–1959, ANRI, Jakarta.

64. Sarwono, "The Scientific Development of Indonesia," 7.

65. Suzanne Moon, "Takeoff of Self-Sufficiency? Ideologies of Development in Indonesia, 1957–1961," *Technology and Culture* 39, no. 2 (1998): 187–212.

66. Sarwono Prawirohardjo, "Madjelis Ilmu Pengetahuan Indonesia dan Universitas-Universitas di Indonesia," *Berita M.I.P.I.* 2, no. 1 (1958): 32–38.

67. Sarwono Prawirohardjo, "Tugas dan Rentjana Kerdja Madjelis Ilmu Pengetahuan Indonesia Dalam Masa Depan," *Berita M.I.P.I.* 1, no. 2 (1957): 16–18; Koesnoto Setyodiwiryo, "Prasaran Perihal Ikut Sertanja Indonesia Dalam Kongres-Kongres Ilmu Pengetahuan Internasional dan Perihal Mengelenggarakan Kongres-Kongres Ilmu Pengetahuan Nasional," *Berita M.I.P.I.* 1, no. 2 (1957): 18–23.

68. "Kesimpulan-kesimpulan Pertemuan Jang Pertama Antara Madjelis Ilmu Pengetahuan Indonesia dengan Lembaga-Lembaga Ilmu Pengetahuan," *Berita M.I.P.I.* 1, no. 2 (1957): 29–30.

69. Koesnoto Setyodiwiryo, "Kebun Raya Indonesia (Lembaga Pusat Penjelidikan Alam)," *Berita M.I.P.I.* 1, no. 1 (1957): 5–29. This was apparently a translation of an English brochure about the gardens published the previous year, *The Botanic Gardens of Indonesia* (Jakarta: Ministry of Agriculture, 1956).

70. J. A. Kaligis, "Surat dari Redaksi," *Berita M.I.P.I.* 1, no. 1 (1957): 4.

71. Kostermans to Van Steenis, March 10, 1961, file 76, Van Steenis Papers, Nationaal Archief, The Hague.

72. On power sharing in the 1950s, see Feith, *The Decline of Constitutional Democracy*.

73. Manipol (Manifesto Politik) indicates the political manifesto of the 1945 constitution, and USDEK is an acronomy for Undang-undang 1945; Sosialisme à la Indonesia; Demokrasi Terpimpin; Ekonomi Terpimpin; and Kepribadian Indonesia (1945 Constitution; Indonesian socialism; Guided Democracy; Guided Economy; and Indonesian Identity).

74. Still the best overview of Guided Democracy is Herbert Feith, "Dynamics of Guided Democracy," in *Indonesia*, ed. Ruth T. McVey, rev. ed. (New Haven, CT: Human Relations Area Files Press, 1967), 309–409. For the official handbook of Manipol-USDEK, see H. Roeslan Abdulgani, *Pendjelasan Manipol dan USDEK* (Jakarta: Departemen Penergangan RI, 1960). For excerpts from Sukarno's important speeches explaining Manipol-USDEK and Guided Democracy, see Herbert Feith and Lance Castles, eds., *Indonesian Political Thinking, 1945–1965* (Ithaca, NY: Cornell University Press, 1970), 98–116.

75. Feith, "Dynamics of Guided Democracy," 370–71.

76. Ibid., 376–77. Sukarno borrowed the English word *retooling*. He defined it as "changing the means, changing the tools and instruments that no longer fit with the thinking of Guided Democracy, with new ways, with new tools and instruments, which are appropriate with the new outlook. Retooling also means economizing all the means

and tools still in use, so that the tools can perhaps be improved and even returned to their original sharpness." Abdulgani, *Manipol dan USDEK*, 50.

77. For economic decolonization, see J. Thomas Lindblad, *Bridges to Business: The Economic Decolonization of Indonesia* (Leiden: KITLV Press, 2008). See also Leslie Palmer, *Indonesia and the Dutch* (Oxford: Oxford University Press, 1962).

78. C. G. G. J. Van Steenis, "Fourteenth Annual Report of Flora Malesiana Foundation July 1957–December 1958," file 61, Van Steenis Papers, Nationaal Archief, The Hague.

79. Dilmy to Van Steenis, October 20, 1959/L/358, file 92, Van Steenis Papers, Nationaal Archief, The Hague.

80. Kostermans to Jacobs, January 28, 1960, file 92, Van Steenis Papers, Nationaal Archief, The Hague.

81. For a summary of Sadikin's career, see Ibrahim Manwan, "Penabur Benih Yang Tak Kenal Letih," in *Bunga-Bunga Pun Bermekaran*, ed. Setijati Sastrapradja (Bogor: Badan Penelitian dan Pengembangan Pertanian, 1999), 253–61.

82. Feith, "Dynamics of Guided Democracy," 398.

83. Sukarno, "Membangun Dunia Kembali," in *Haluan Politik dan Pembangunan Negara* (Jakarta: Departemen Penerangan RI, 1960), 180.

84. Guy J. Pauker, "Indonesia's Eight-Year Development Plan," *Pacific Affairs* 34, no. 2 (1961): 126–27.

85. Sarwono, "The Scientific Development of Indonesia," 12; Kostermans to Van Steenis, March 16, 1962, file 76, Van Steenis Papers, Nationaal Archief, The Hague. The Department of Agriculture was not just active in research at the botanical gardens. For a history of scientific research during the 1950s and 1960s at the Department of Agriculture, see Auzay Hamid, "Penelitian Pertanian Zaman Pra PELITA," in *Sejarah Penelitian Pertanian di Indonesia* (Jakarta: Badan Penelitian dan Pengembangan Pertanian, 1995), 141–225.

86. M. Makagiansar, "The Organization of Scientific Research in Indonesia and Significant Developments and Trends since 1960," *Berita M.I.P.I.* 6, no. 3 (1962): 131.

87. Kostermans to Van Steenis, April 26, 1964, file 76, Van Steenis Papers, Nationaal Archief, The Hague.

88. Kostermans to Van Steenis, March 16, 1962, file 76, Van Steenis Papers, Nationaal Archief, The Hague.

89. Kostermans to Van Steenis, June 1963, Kostermans correspondence archive, 1956–1966, NHN, Leiden.

90. Ibid.; Kostermans to Van Steenis, May 24, 1960 file 76, Van Steenis Papers, Nationaal Archief, The Hague; Kostermans to Van Steenis, September 5, 1960, file 76, Van Steenis Papers, Nationaal Archief, The Hague.

91. Kostermans, "Anwari Dilmy," 6; Kostermans to Van Steenis, May 24, 1968, file 76, Van Steenis Papers, Nationaal Archief, The Hague.

92. Kostermans to Van Steenis, June 1963, Kostermans correspondence archive, 1956–1966, NHN, Leiden.

93. Kostermans to Van Steenis, January 17, 1963, file 232, Van Steenis Papers, Nationaal Archief, The Hague.

94. Soedjono D. Poesponegoro, "Pidato J. M. Menteri Research Nasional Pada Peresmian Pusat Dokumentasi Ilmiah-Nasional," *Berita M.I.P.I.* 9, nos. 3–4 (1965): 79.

95. Roosa, *Pretext for Mass Murder*.

96. Bakri Abbas, "Kebebasan Mimbar di Indonesia," *Berita M.I.P.I.* 10, nos. 3–4 (1966): 85.

97. Bakri Abbas, "Konsolidasi Organisasi Ilmu Pengetahuan dan Research," *Berita M.I.P.I.* 11, nos. 1–2 (1967): 1.

98. Hamzah Machmudy, interview with the author, July 31, 2001, Bogor.

99. *Indonesian Institute of Sciences* (Jakarta: Indonesian National Scientific Documentation Center, 1970).

100. Sarwono Prawirohardjo, "Sambutan Ketua Madjelis Ilmu Pengetahuan Indonesia pada Pembukaan Pameran Biologi Lembaga Biologi Nasional," *Berita M.I.P.I.* 11, nos. 3–4 (1967): 6–7.

101. Otto Soemarwoto and D. A. Tisna Amidjaja, *Perkembangan Biologi, 1945–1965* (Jakarta: Balai Pustaka, 1965), 8.

102. Sampurno Kadarsan, "Pidato Ketua Panitia Penjelenggara Peringatan 150 Tahun Kebun Raya Lembaga Biologi Nasional," *Berita M.I.P.I.* 11, nos. 3–4 (1967): 1–2.

103. Otto Soemarwoto, "Development of the National Biological Institute," in *Indonesia: Resources and Their Technological Development*, ed. Howard Beers (Lexington: University Press of Kentucky, 1970), 53–69.

104. Otto Soemarwoto, "Lembaga Biologi Nasional dalam Perspektip," *Berita M.I.P.I.* 11, nos. 3–4 (1967): 9–17.

105. Soemarwoto, "Development of the National Biological Institute," 64–65.

106. Maxwell S. Doty and Aprilany Soegiarto, "The Development of Marine Resources in Indonesia," in *Indonesia: Resources and Their Technological Development*, ed. Howard Beers (Lexington: University Press of Kentucky, 1970), 70–89; Maxwell Doty, "The American Botanist and the Biological Institute of Indonesia," *Bioscience* 14, no. 12 (1964): 36–41.

107. Kostermans to Van Steenis, May 24, 1968, file 76, Van Steenis Papers, Nationaal Archief, The Hague.

108. At this time he resigned as assistant director of LBN, although he remained director of the gardens. Sastrapradja, "Sekali Langkah Terayunkan," 123.

109. He had to spend ten years in Canada because his Canadian adviser knew a Ciawi degree was not equivalent to a bachelor of science. Adisoemarto, "Setinggi-tinggi Terbang Bangau," 17. Didin Sastrapradja spent only three and a half years at the University of Hawai'i and went straight into the PhD program without having obtained a bachelor's or master's degree. Didin Sastrapradja, interview with the author, August 6, 2001, Bogor.

110. Aprilani Soegiarto, "Menyibak Ombak Menyelam Dalam," in *Bunga-Bunga Pun Bermekaran*, ed. Setijati Sastrapradja (Bogor: Badan Penelitian dan Pengembangan Pertanian, 1999), 190–91.

111. Dilmy to Soemarwoto, June 29, 1968/68, file 93, Van Steenis Papers, Nationaal Archief, The Hague.

112. Dilmy to Van Steenis, April 26, 1970, file 92, Van Steenis Papers, Nationaal Archief, The Hague.

113. Mien Rifai, interview with the author, June 25, 2001, Jakarta.

114. Dilmy and others, "Laporan rapat anggota staff Herbarium tentang keadaan koleksi Herbarium Bogoriense," June 25, 1968, file 93, Van Steenis Papers, Nationaal Archief, The Hague.

115.  Hartono had been in this post since May of 1967. He came from the Army Corps of Engineers. "Current Data on the Indonesian Army Elite," *Indonesia* 7 (1969): 195–201.

116.  M. Jacobs, "Kostermans Seventy-Five," 17.

117.  Otto Soemarwoto, interview with the author, June 14, 2001, Bandung.

118.  For a good introduction to the political culture of the early New Order, see Donald K. Emmerson, "Understanding the New Order: Bureaucratic Pluralism in Indonesia," *Asian Survey* 23, no. 11 (1983): 1220–41.

119.  M. H. van Raalte, "Verslag van een reis naar Bogor," December 1965, file 116, Van Steenis Papers, Nationaal Archief, The Hague.

120.  Soepadmo to Van Steenis, October 31, 1966, Soepadmo correspondence archive, 1956–1966, NHN, Leiden (emphasis added).

121.  M. Jacobs to Rohani Ramly, Feburary 14, 1974/286, and Van Steenis to Soepadmo, July 21, 1967, Soepadmo correspondence archive, 1967–1976, NHN, Leiden.

122.  Van Steenis to Soemarwoto, December 1, 1966, no. 2406, Soemarwoto correspondence archive, 1956–1966, NHN, Leiden.

123.  Dilmy, Soegeng, and Kostermans to Van Steenis, December 10, 1966, Soemarwoto correspondence archive, 1956–1966, NHN, Leiden.

124.  Soepadmo to Van Steenis, July 6, 1972, Soepadmo correspondence archive, 1967–1976, NHN, Leiden; Soepadmo to Van Steenis, September 10, 1971, Van Steenis Papers, Nationaal Archief, The Hague.

125.  Kostermans to Van Steenis, November 22, 1964, file 76, Van Steenis Papers, Nationaal Archief, The Hague.

126.  K. Heyne, *De nuttige planten van Nederlandsch-Indië*, 2nd ed., 3 vols. (Batavia: Ruygrok, 1927).

127.  Soemarwoto to Kostermans, May 11, 1967, file 116, Van Steenis Papers, Nationaal Archief, The Hague.

128.  Otto Soemarwoto, "BIOTROP: Its Objectives and Program," in *Laporan Seminar Biologi Kedua, 18–20 Pebruari 1970 Ciawi-Bogor*, vol. 1 (Bogor: LBN, 1970), 34–36.
34–36; Dwidjoseputro, *Review of the State of Biology in Indonesia*, 19–21.

129.  On the location in the gardens, see "Gedung Baru BIOTROP Diresmikan Pemakainnja," *Berita L.I.P.I.* 16, no. 3 (1972): 46.

130.  Otto Soemarwoto, interview with the author, June 14, 2001, Bandung.

131.  Aprilani Sugiarto, interview with the author, May 9, 2001, Jakarta.

132.  *Review of the Work of the Lembaga Biologi Nasional—LIPI (The National Biological Institute), 1 April 1969–31 March 1974* (Bogor: LBN, 1975), 1–2.

133.  Ibid., 5–13.

134.  S. Sastrapradja, S. Adisumarto, and M. A. Rifai, "Elements of Life and the Classrooms," *Berita L.I.P.I.* 15, no. 1 (1971): 25–34; S. Sastrapradja, M. A. Rifai, and S. Adisumarto, "Ilmu Pengetahuan di Indonesia: Tantangan buat Para Pendidik," *Berita L.I.P.I.* 15, no. 2 (1971): 7–14; D. S. Sastrapradja, "Multi Fungsi Kebun Raya di Indonesia," *Berita L.I.P.I.* 15, no. 3 (1971): 13–19; S. Sastrapradja, "Peranan Lembaga Biologi Nasional dalam Pengembangan Pendidikan Biologi di Indonesia," *Berita L.I.P.I.* 18, no. 3 (1974): 23–32.

135.  The other goal of REPELITA I was to curb the unwieldy population increase. During REPELITA I, technocrats under the patronage of top military commanders

directed economic policy. Martin Rudner, "The Indonesian Military and Economic Policy: The Goals and Performance of the First Five-Year Development Plan, 1969–1974," *Modern Asian Studies* 10, no. 2 (1976): 249–84.

136. In 1986 there were 202 researchers with a college degree compared to 40 in 1973. S. Sastrapradja, "Catatan Masa Jabatan Direktur Lembaga Biologi Nasional, 1 Maret 1973–25 Juni 1986," pamphlet, LBN, 1986, 2.

137. Ibid., 3.

138. S. Sastrapradja to Van Steenis, April 25, 1980/291, file 220, Van Steenis Papers, Nationaal Archief, The Hague.

139. See, for example, D. S. Sastrapradja, "Multi Fungsi Kebun Raya di Indonesia"; M. A. Rifai, "Demam Penelitian," *Berita L.I.P.I.* 17, no. 1 (1973): 3–9; S. Sastrapradja, "Hak Hidup dan Kewajiban Peneliti di Indonesia," *Berita L.I.P.I.* 17, no. 3 (1973): 3–8; M. A. Rifai, S. Sastrapradja, and S. Adisoemarto, "Morfologi Lembaga Penelitian," *Berita L.I.P.I.* 19, no. 1 (1975): 1–10; and A. Soegiarto, "Lembaga Penelitian yang Ideal," *Berita L.I.P.I.* 19, no. 1 (1975): 11–19.

140. Celia Lowe, *Wild Profusions: Biodiversity Conservation in an Indonesian Archipelago* (Princeton, NJ: Princeton University Press, 2006), 164.

## Conclusion

1. Charles Keyes, "'The Peoples of Asia': Science and Politics in the Classification of Ethnic Groups in Thailand, China, and Vietnam," *Journal of Asian Studies* 61, no. 4 (2002): 1163–1203.

2. James C. Scott, *Seeing Like a State: How Certain Schemes to Improve the Human Condition Have Failed* (New Haven, CT: Yale University Press, 1998).

3. See for example, Lyn Schumaker, *Africanizing Anthropology: Fieldwork, Networks, and the Making of Cultural Knowledge in Central Africa* (Durham, NC: Duke University Press, 2001).

4. Roy MacLeod, "Introduction," in "Nature and Empire: Science and the Colonial Enterprise," ed. Roy MacLeod, *Osiris* 15 (2000): 1–13.

5. Lewis Pyenson in his history of Indonesian science fails to investigate what Dutch colonial officials thought or did about science, choosing instead to guess that they considered it "elaborate and sacred theatre." Lewis Pyenson, *Empire of Reason: Exact Sciences in Indonesia, 1840–1940* (Leiden: Brill, 1989), 178.

6. For the professionalization of early-nineteenth-century British Indian naturalists, see Richard H. Grove, *Green Imperialism: Colonial Expansion, Tropical Island Edens, and the Origins of Environmentalism, 1600–1860* (Cambridge: Cambridge University Press, 1995), 309–79. For the career of an Australian naturalist in the 1840s and 1850s, see Robert A. Stafford, "The Long Arm of London: Sir Roderick Murchison and Imperial Science in Australia," in *Australian Science in the Making*, ed. R. W. Home (Cambridge: Cambridge University Press, 1988), 69–101.

7. R. W. Home, "The Beginnings of an Australian Physics Community," in *Scientific Colonialism: A Cross-Cultural Comparison*, ed. Nathan Reingold and Marc Rothenberg (Washington, DC: Smithsonian Institution Press, 1987), 3–34.

8. Ian Inkster and Jan Todd, "Support for the Scientific Enterprise, 1850–1900," in *Australian Science in the Making*, ed. R. W. Home (Cambridge: Cambridge University Press, 1988), 102–32.

9. Roy MacLeod, "Imperial Reflections in the Southern Seas: The Funafuti Expeditions, 1896–1904," in *Nature in Its Greatest Extent: Western Science in the Pacific*, ed. Roy MacLeod and Philip F. Rehbock (Honolulu: University of Hawai'i Press, 1988), 159–91.

10. Ron Johnston and Jean Buckley, "The Shaping of Contemporary Scientific Institutions," in *Australian Science in the Making*, ed. R. W. Home (Cambridge: Cambridge University Press, 1988), 374–98.

11. S. C. B. Gascoigne, "Australian Astronomy since the Second World War," in *Australian Science in the Making*, ed. R. W. Home (Cambridge: Cambridge University Press, 1988), 345–73; Woodruff T. Sullivan, *Cosmic Noise: A History of Early Radio Astronomy* (Cambridge: Cambridge University Press, 2009), 118–54.

12. Deepak Kumar, *Science and the Raj, 1857–1905* (Delhi: Oxford University Press, 1995).

13. Kavita Philip, *Civilizing Natures: Race, Resources, and Modernity in Colonial South India* (New Brunswick, NJ: Rutgers University Press, 2004).

14. J. S. Furnivall, *Netherlands India: A Study of Plural Economy* (Cambridge: Cambridge University Press, 1944), 260. See also his expanded essay on a comparison of the two colonies: J. S. Furnivall, *Colonial Policy and Practice: A Comparative Study of Burma and Netherlands India* (New York: New York University Press, 1956).

15. C. Fasseur and D. H. A. Kolff, "Some Remarks on the Development of Colonial Bureaucracies in India and Indonesia," *Itinerario* 10, no. 1 (1986): 31–55.

16. Gyan Prakash, *Another Reason: Science and the Imagination of Modern India* (Princeton, NJ: Princeton University Press, 1999); David Arnold, *Science, Technology, and Medicine in Colonial India*, vol. 3, pt. 5, of *The New Cambridge History of India* (Cambridge: Cambridge University Press, 2000).

17. V. V. Krishna, "The 'Colonial' Model and the Emergence of National Science in India: 1876–1920," in *Science and Empires: Historical Studies about Scientific Development and European Expansion*, ed. Patrick Petitjean, Catherine Jami, and Anne Marie Moulin (Dordrecht: Kluwer, 1992), 57–72.

18. Deepak Kumar, "Reconstructing India: Disunity in the Science and Technology for Development Discourse, 1900–1947," in "Nature and Empire: Science and the Colonial Enterprise," ed. Roy MacLeod, *Osiris* 15 (2000): 241–57.

19. Scott, *Seeing Like a State*, 309–41.

20. Otto Soemarwoto, *Environmental Education and Research in Indonesian Universities* (Singapore: Maruzen Asia, 1981).

21. Otto Soemarwoto, *Ekologi Lingkungan Hidup dan Pembangunan* (Jakarta: Jambatan, 1983).

22. Otto Soemarwoto, interview with the author, June 14, 2001, Bandung.

23. Ibid. For more about Soemarwoto and BIOTROP, see chapter 7.

24. Peter Boomgaard, "Oriental Nature, Its Friends and Its Enemies: Conservation of Nature in Late-Colonial Indonesia, 1889–1949," *Environment and History* 5 (1999): 257–92; Robert Cribb, *The Politics of Environmental Protection* (Clayton, Australia: Monash University Centre of Southeast Asian Studies, 1988), 3–6.

25. Otto Soemarwoto, *Atur-Diri-Sendiri: Paradigma Baru Pengeloloaan Lingkungan Hidup* (Yogyakarta: Gadjah Mada University Press, 2001), 96.

26. Ibid., 170.

27. Ibid., 126–30.

28. During the New Order, public intellectuals who ran a variety of NGOs, news outlets, and semigovernmental organizations in between Indonesian citizens and the state continued to exist in a constrained capacity, and many political scientists studying New Order Indonesia in the 1980s and 1990s were cautiously optimistic about the potential of civil society institutions to promote democratization. See the studies in Arief Budiman, ed., *State and Civil Society in Indonesia* (Clayton, Australia: Monash University Centre of Southeast Asian Studies, 1990); Philip Eldridge, *Non-government Organizations and Democratic Participation in Indonesia* (Kuala Lumpur: Oxford University Press, 1995); Douglas Ramage, *Politics in Indonesia: Democracy, Islam, and the Ideology of Tolerance* (London: Routledge, 1995); R. William Liddle, *Leadership and Culture in Indonesian Politics* (Sydney: Allen and Unwin, 1997); and Robert Hefner, *Civil Islam: Muslims and Democratization in Indonesia* (Princeton, NJ: Princeton University Press, 2000).

29. The historian Robert Elson has pointed out that since 1998 democracy has been "appropriated by those already in power." R. E. Elson, *The Idea of Indonesia: A History* (Cambridge: Cambridge University Press, 2008), 303. For more about politics after the fall of Suharto, see Chris Manning and Peter van Diemen, *Indonesia in Transition: Social Dimensions of the Reformasi and the Economic Crisis* (London: Zed Books, 2000); and Edward Aspinall, *Opposing Suharto: Compromise, Resistance, and Regime Change in Indonesia* (Stanford, CA: Stanford University Press, 2005).

30. Soemarwoto, *Atur-Diri-Sendiri*, 94.

31. Boomgaard, "Oriental Nature."

32. Cribb, *The Politics of Environmental Protection.*

33. Robert Cribb, "Environmental Policy and Politics in Indonesia," in *Ecological Policy and Politics in Developing Countries: Economic Growth, Democracy, and Environment,* ed. Uday Desai (Albany: State University of New York Press, 1998), 65–85.

34. Budy Resosudarmo, ed., *The Politics and Economics of Indonesia's Natural Resources* (Singapore: Institute of Southeast Asian Studies, 2005).

35. A recent report about strategies to combat deforestation in tropical countries, written as a collaboration between the World Bank and the Bogor-based Center for International Forestry Research, among others, argues that future environmental policies should be shaped by governments, which would manage input from local stakeholders. Kenneth M. Chomitz et al., *At Loggerheads? Agricultural Expansion, Poverty Reduction, and Environment in the Tropical Forests* (Washington, DC: World Bank, 2007).

# Bibliography

## Archives

Nationaal Archief (National Archives), The Hague, the Netherlands
  Archief Ministerie van Koloniën, 1815–1849, Openbaar Verbaal
  Archief Ministerie van Koloniën, 1850–1900, Openbaar Verbaal
  Archief Ministerie van Koloniën, 1901–1953, Openbaar Verbaal
  Archief Ministerie van Koloniën, 1901–1940, Geheim Verbaal
  Archief Indische Comité voor Wetenschappelijke Onderzoekingen
  C. G. G. J. van Steenis papers
  W. H. de Vriese papers
Arsip Nasional Republik Indonesia (National Archives of the Republic of Indonesia),
  Jakarta, Indonesia
  Archief Algemeene Secretarie, 1817–1890
  Archief Algemeene Secretarie, 1891–1942
  Archief Algemeene Secretarie, 1944–1950
  Arsip Kabinet Presiden, 1950–1959
National Herbarium of the Netherlands, Leiden University branch
  Confidential correspondence boxes
  Correspondence archives: Buitenzorg Botanical Gardens, A. J. G. H. Koster-
    mans, Otto Soemarwoto, Engkik Soepadmo
  H. C. D. de Wit and C. G. G. J. van Steenis, eds., "Correspondentie hoofdza-
    kelijk van Teysmann en Binnendrijk voorts Wigman, Lovink, Treub, enz.
    met de Hortulanus H. Witte te Leiden, 1847–1885," December 1948
  H. C. D. de Wit and C. G. G. J. van Steenis, eds., "Correspondentie hoofza-
    kelijk van Teysmann en Binnendijk met Prof. Dr. W.H. de Vriese, 1847–
    1954," December 1948
Boerhaave Museum, Leiden, the Netherlands
  Correspondence archives: W. M. Docters van Leeuwen, M. Treub
Koninklijk Instituut voor de Tropen (Royal Institute for the Tropics), Amsterdam, the
  Netherlands
  J. C. Koningsberger, "Herinneringen aan, 1907–1923," November 1929
  J. C. Koningsberger, "Herinneringen aan Melchior Treub," 1945
Naturalis Museum, Leiden, the Netherlands
  M. A. Lieftinck correspondence archive
Institute for History and Foundations of Science, Utrecht University, the Netherlands
  C. G. G. J. van Steenis archive

Utrecht Universiteit Bibliotheek (Utrecht University Library), Utrecht, the Netherlands
    Miquel Papers, HS 1873
Staatsbibliothek zu Berlin, Handschriftenabteilung
    Melchior Treub papers, Lb 1875

## Official Publications

*Handelingen der Staten-Generaal.* The Hague. 1903–4.

*Handelingen van de Volksraad.* Batavia. 1939–40.

*Jaarboek van het Departement van Landbouw in Nederlandsch-Indië.* Buitenzorg. 1905–10.

*Jaarboek van het Departement van Landbouw, Nijverheid, en Handel in Nederlandsch-Indië.* Buitenzorg. 1911–29.

*Regeerings-Almanac.* Batavia. 1815–1942.

*Staatsblad van Nederlandsch-Indië.* Batavia. 1816–1942.

*Statistisch Jaaroverzicht van Nederlandsch-Indië.* Batavia. 1922–39.

*Verslag omtrent den staat van 's Lands Plantentuin te Buitenzorg.* Buitenzorg. 1877–1904.

## Contemporary Periodicals and Newspapers

*Annales du Jardin Botanique de Buitenzorg* (1876–1908)

*Bangoen* (1937)

*Berita M.I.P.I.* (1956–67)

*Berita L.I.P.I.* (1968–75)

*Djawa-Moeda* (1916)

*Het Tijdschrift* (1912)

*Hindia Poetra* (1916)

*Indische Gids* (1894–1910)

*Java-Bode* (1853)

*Jong Indië* (1909)

*Kemadjoean-Hindia* (1923)

*Koloniale Studiën* (1917–38)

*Kopiïst* (1842–43)

*Madjallah Boelanan Oentoek Kemadjoean Ra'jat* (1937–38)

*Nationale Commentaren* (1938–40)

*Natuurkundig Tijdschrift voor Nederlandsch-Indië* (1850–73)

*O.S.R. News/Bulletin* (1949–52)

*Pandji Poestaka* (1927)

*Penindjauan* (1934)

*Soeloeh Indonesia* (1927)

*Soerat Chabar Boelanan oentoek Kemadjoean dan Keamanan Kaoem dan Negeri* (1926)

*Tijdschrift voor het Binnenlandsch Bestuur* (1904)

*Tijdschrift voor Nederlandsch-Indië* (1838–62)

*Verhandelingen van het Bataviaasch Genootschap* (1843–53)

## Printed Works, Theses, and Other Sources

Abbas, Bakri. "Kebebasan Mimbar di Indonesia." *Berita M.I.P.I.* 10, nos. 3–4 (1966): 85
———. "Konsolidasi Organisasi Ilmu Pengetahuan dan Research." *Berita M.I.P.I.* 11, nos. 1–2 (1967): 1
Abdulgani, H. Roeslan. *Pendjelasan Manipol dan USDEK.* Jakarta: Departemen Penergangan RI, 1960.
Abdullah, Taufik. "The Beginning of the Padri Movement." In *Papers of the Dutch-Indonesian Historical Conference,* ed. A. B. Lapian and Anthony Day, 143–53. Leiden: Bureau of Indonesian Studies, 1978.
Abeyasekere, Susan. *One Hand Clapping: Indonesian Nationalists and the Dutch, 1939–1942.* Clayton, Australia: Monash University Centre of Southeast Asian Studies, 1976.
———. "The Soetardjo Petition." *Indonesia* 15 (1973): 81–107.
Adisoemarto, Soenartono. "Di Tangan Koesnoto Lilin itu Menyala." In *Bunga-Bunga Pun Bermekaran,* ed. Setijati Sastrapradja, 3–10. Bogor: Badan Penelitian dan Pengembangan Pertanian, 1999.
———. "Setinggi-tinggi Terbang Bangau." In *Bunga-Bunga Pun Bermekaran,* ed. Setijati Sastrapradja, 17–36. Bogor: Badan Penelitian dan Pengembangan Pertanian, 1999.
Allen, David. *The Naturalist in Britain: A Social History.* London: A. Lane, 1976.
Anderson, Benedict. *Imagined Communities: Reflections on the Origin and Spread of Nationalism.* Rev. ed. London: Verso, 1991.
———. *Java in a Time of Revolution: Occupation and Resistance, 1944–1946.* Ithaca, NY: Cornell University Press, 1972.
———. *Language and Power: Exploring Political Cultures in Indonesia.* Ithaca, NY: Cornell University Press, 1990.
———. "Old State, New Society: Indonesia's New Order in Comparative Historical Perspective." *Journal of Asian Studies* 42, no. 3 (1983): 477–96.
———. "A Time of Darkness and a Time of Light: Transpositions in Early Indonesian Nationalist Thought." In *Language and Power: Exploring Political Cultures in Indonesia,* 241–70. Ithaca, NY: Cornell University Press, 1990.
"Annual Report of the General Secretary to the President of the Board of the Organization for Scientific Research in Indonesia for 1949." *OSR News* 2, no. 5 (1950): 49–57.
*Annual Report of 1956, Lembaga Pusat Penjelidikan Alam.* Bogor: LPPA, n.d.
*Annual Report for 1957, Lembaga Pusat Penjelidikan Alam.* Bogor: LPPA, n.d.
Arimbo. "Volkshoofden en z.g. intellectueele leiders." *Soeloeh Indonesia* 2, no. 8 (August 1927): 7–10.
Arnold, David. *Science, Technology, and Medicine in Colonial India.* Vol. 3, pt. 5, of *The New Cambridge History of India.* Cambridge: Cambridge University Press, 2000.
Aspinall, Edward. *Opposing Suharto: Compromise, Resistance, and Regime Change in Indonesia.* Stanford, CA: Stanford University Press, 2005.
"Bagaimana Kita Dapat Memateri Sendiri?" *Madjallah Boelanan Oentoek Kemadjoean Ra'jat* 2, no. 9 (1938): 164–66.
Becking, J. H. *Henri Jacob Victor Sody (1892–1959), His Life and Work: A Biographical and Bibliographical Study.* Leiden: Brill, 1989.
"The beteekenis van de woorden Boedi Oetama." *Indische Gids* 32, no. 1 (1910): 514–15.

Benda, Harry. "Christiaan Snouck Hurgronje and the Foundations of Dutch Islamic Policy in Indonesia." *Journal of Modern History* 30, no. 4 (1958): 338-47.

———. "The Pattern of Administrative Reforms in the Closing Years of Dutch Rule in Indonesia." *Journal of Asian Studies* 25, no. 4 (1966): 589-605.

Bennett, Tony. *The Birth of the Museum: History, Theory, Politics.* London: Routledge, 1995.

Bernelot Moens, J. C. *De Kinacultuur in Azië, 1854 t/m 1882.* Batavia: Ernst, 1882.

———. "Onderzoek van eenige kina-basten van Java." *Natuurkundig Tijdschrift voor Nederlandsch Indië* 31 (1870): 165-76.

Bleeker, P. "Algemeen verslag der werkzaamheden van de Natuurkundige Vereeniging in Nederlandsch Indië." *Natuurkundig Tijdschrift voor Nederlandsch Indië* 2 (1851): 1-24.

———. "Algemeen verslag der werkzaamheden van de Natuurkundige Vereeniging in Nederlandsch Indië over het jaar 1851." *Natuurkundig Tijdschrift voor Nederlandsch Indië* 3 (1852): 1-16.

———. "Levensbericht van P. Bleeker." *Jaarboek van de Koninklijke Akademie van Wetenschappen* (1877): 11-158.

———. "Overzigt der geschiedenis van het Bataviaasch Genootschap van Kunsten en Wetenschappen van 1778-1853." *Verhandelingen van het Bataviaasch Genootschap* 25 (1853): 1-24.

Blume, C. L. *Antwoord op den Heer W. H. de Vriese.* Leiden: Couvée, 1850.

———. *Opheldering van de inlichtingen van den Heer Fr. Junghuhn.* Leiden: Couvée, 1850.

———. *Rumphia sive commentationes botanicae de plantis Indiae orientalis* 3 (1849): 219-20.

Boerlage, J. G. *Handleiding tot de kennis der Flora van Nederlandsch Indië.* 3 vols. Leiden: Brill, 1890-1900.

"Bogor Scientific Centre." *Organization for Scientific Research Bulletin* 12 (July 1952): 1-35.

Bonneuil, Christophe. "The Manufacture of Species: Kew Gardens, the Empire and the Standardisation of Taxonomic Practices in Late Nineteenth-Century Botany." In *Instruments, Travel and Science: Itineraries of Precision from the Seventeenth to the Twentieth Century*, ed. Marie-Noëlle Bourguet, Christian Liccope, and H. Otto Sibum, 189-215. London: Routledge, 2002.

Boomgaard, Peter. *Children of the Colonial State: Population Growth and Economic Development in Java, 1795-1880.* Amsterdam: Free University Press, 1989.

———. "Colonial Forest Policy in Java in Transition, 1865-1916." In *The Late Colonial State in Indonesia: Political and Economic Foundations of the Netherlands Indies, 1880-1942*, ed. Robert Cribb, 117-37. Leiden: KITLV Press, 1994.

———. "Oriental Nature, Its Friends and Its Enemies: Conservation of Nature in Late-Colonial Indonesia, 1889-1949." *Environment and History* 5 (1999): 257-92.

———. "The Welfare Services in Indonesia, 1900-1942." *Itinerario* 10, no. 1 (1986): 57-81.

Boorsma, W. G. "Toespraak, door den Directeur der school gehouden bij de opening van het nieuwe gebouw op 3 Juni 1916." In *Middelbare Landbouwschool te Buitenzorg, programma voor het leerjaar, 1916-1917*, 36-42. Buitenzorg: LNH, 1916.

Booth, Anne. "The Evolution of Fiscal Policy and the Role of Government in the Colonial Economy." In *Indonesian Economic History in the Dutch Colonial Era*, ed. Anne Booth, W. J. O'Malley, and Anna Weidemann, 210-43. New Haven, CT: Yale University Southeast Asian Studies Program, 1990.

————. *The Indonesian Economy in the Nineteenth and Twentieth Centuries: A History of Missed Opportunities.* London: Macmillan, 1998.

Bossenbroek, Martin. "Joannes Benedictus van Heutsz en de leer van het functionele geweld." In *In de vaart der volken: Nederlanders rond 1900*, ed. Herman Beliën, Martin Bossenbroek, and Gert Jan van Setten, 87-96. Amsterdam: Bert Bakker, 1998.

*The Botanic Gardens of Indonesia.* Jakarta: Ministry of Agriculture, 1956.

Bowler, Peter J. *Evolution: The History of an Idea.* Rev. ed. Berkeley: University of California Press, 1989.

Braam, J. S. van. *Een landbouwdepartement in Indië.* Batavia: Kolff, 1903.

Brascamp, E. H. B. "Dr. S. H. Koorders." *Tectona* 13, no. 5 (1920): 377-504.

Brockway, Lucille. *Science and Colonial Expansion: The Role of the British Royal Botanic Gardens.* New York: Academic Press, 1979.

Brotonegoro, Sutaryo. "Senyum Kelegaan Dalam Pengabdian." In *Bunga-Bunga Pun Bermekaran*, ed. Setijati Sastrapradja, 49-55. Bogor: Badan Penelitian dan Pengembangan Pertanian, 1999.

Brugmans, I. J. *Geschiedenis van het onderwijs in Nederlandsch-Indië.* Groningen: Wolters, 1938.

Budiman, Arief, ed. *State and Civil Society in Indonesia.* Clayton, Australia: Monash Univeristy Centre of Southeast Asian Studies, 1990.

*Buitenzorg Scientific Centre.* Buitenzorg: Archipel, 1948.

Burck, W. "Het Herbarium en Museum van 's Lands Plantentuin." In *'s Lands Plantentuin te Buitenzorg, 18 Mei 1817-18 Mei 1892*, 218-33. Batavia: Landsdrukkerij, 1892.

————. "Sur les diptérocarpées des Indes Néerlandaises." *Annales du Jardin Botanique de Buitenzorg* 6 (1887): 145-48.

————. "Sur les Sapotacées des Indes Néerlandaises et les origins botaniques de la gutta-percha." *Annales du Jardin Botanique de Buitenzorg* 5 (1886): 1-86.

————. "Sur l'organisation florale chez quelques Rubiacées." *Annales du Jardin Botanique de Buitenzorg* 3 (1883): 105-19.

Cannon, Susan Faye. "Humboldtian Science." In *Science in Culture: The Early Victorian Period*, 73-110. New York: Dawson, 1978.

Chomitz, Kenneth M., Piet Buys, Giacomo De Luca, Timothy S. Thomas, and Sheila Wertz-Kanounnikoff. *At Loggerheads? Agricultural Expansion, Poverty Reduction, and Environment in the Tropical Forests.* Washington, DC: World Bank, 2007.

Cittadino, Eugene. *Nature as the Laboratory: Darwinian Plant Ecology in the German Empire, 1880-1900.* Cambridge: Cambridge University Press, 1990.

Clark, William, Jan Golinski, and Simon Schaffer, eds. *The Sciences in Enlightened Europe.* Chicago: University of Chicago Press, 1999.

Colijn, H. *Koloniale vraagstukken van heden en morgen.* Amsterdam: De Standaard, 1928.

Couperus, Louis. *The Hidden Force.* Amherst: University of Massachusetts Press, [1900] 1985.

Creutzberg, P., ed. *Het Ekonomisch Beleid in Nederlandsch-Indië: Capita Selecta—Een Bronnenpublicatie.* Vol. 1. Groningen: Wolters-Noordhoff, 1972.

Cribb, Robert. "Development Policy in the Early 20th Century." In *Development and Social Welfare: Indonesia's Experiences under the New Order*, ed. Frans Hüsken, Mario Rutten, and Jan-Paul Dirkse, 225-45. Leiden: KITLV Press, 1993.

————. "Environmental Policy and Politics in Indonesia." In *Ecological Policy and Politics*

*in Developing Countries: Economic Growth, Democracy, and Environment*, ed. Uday
Desai, 65-85. Albany: State University of New York Press, 1998.

——, ed. *The Late Colonial State in Indonesia: Political and Economic Foundations of the
Netherlands Indies, 1880-1942*. Leiden: KITLV Press, 1994.

——. *The Politics of Environmental Protection*. Clayton, Australia: Monas University
Centre of Southeast Asian Studies, 1988.

Crouch, Harold. *The Army and Politics in Indonesia*. Rev. ed. Ithaca, NY: Cornell University Press, 1988.

"Current Data on the Indonesian Army Elite." *Indonesia* 7 (1969): 195-201.

Curtin, P. D. "'The White Man's Grave': Image and Reality, 1780-1850." *Journal of
British Studies* 1 (1961): 94-110.

Dahm, Bernard. *Sukarno and the Struggle for Indonesian Independence*. Ithaca, NY:
Cornell University Press, 1969.

Dammerman, K. W. "'Het werk van Melchior Treub na 25 jaren,' een antwoord aan Dr.
V. J. Koningsberger." *Koloniale Studiën* 18, no. 2 (1934): 402-12.

——. "History of the Visitors' Laboratory ('Treub Laboratorium') of the Botanic
Gardens, Buitenzorg, 1884-1934." In *Science and Scientists in the Netherlands Indies*,
ed. Pieter Honig and Frans Verdoorn, 59-75. New York: Board for the Netherlands
Indies, Surinam and Curaçao, 1945.

——. "'s Lands Plantentuin te Buitenzorg." *Koloniale Studiën* 22, no. 1 (1938): 33-57.

Daum, Andreas. "Science, Politics, and Religion: Humboldtian Thinking and the
Transformation of Civil Society in Germany, 1830-1870." In "Science and Civil
Society," ed. Lynn K. Nyhart and Thomas H. Broman. *Osiris* 17 (2002): 107-40.

Day, Tony. *Fluid Iron: State Formation in Southeast Asia*. Honolulu: University of
Hawai'i Press, 2002.

Delden, E. E. van. *Klein Repertorium: Index op tijdschriftartikelen met betrekking tot voormalig Nederlands-Indië*. Vol. 5, *Tijdschrift voor Nederlandsch Indië, 1838-1866*. Amsterdam: KIT, 1990.

——. *Klein Repertorium: Index op tijdschriftartikelen met betrekking tot voormalig
Nederlands-Indië*. Vol. 7, *Acht algemene tijdschriften, 1834-1864*. Amsterdam: KIT, 1993.

"Departement van Economische Zaken." In *Regeerings Almanak voor Nederlandsch Indië
1935* (Batavia: Landsdrukkerij, 1935), 1:243-46.

Dettelbach, Michael. "Global Physics and Aesthetic Empire: Humboldt's Physical
Portrait of the Tropics." In *Visions of Empire: Voyages, Botany, and Representations of
Nature*, ed. David Philip Miller and Peter Hanns Reill, 258-92. Cambridge:
Cambridge University Press, 1996.

——. "Humboldtian Science." In *Cultures of Natural History*, ed. N. Jardine, J. A.
Secord, and E. C. Spary, 287-304. Cambridge: Cambridge University Press, 1996.

Dick, Howard, ed. *The Emergence of a National Economy: An Economic History of Indonesia, 1800-2000*. Honolulu: University of Hawai'i Press, 2002.

Dijk, Kees van. *The Netherlands Indies and the Great War, 1914-1918*. Leiden: KITLV
Press, 2007.

Dilmy, A., and A. J. G. H. Kostermans. "Research on the Vegetation of Indonesia." In
*Study of Tropical Vegetation: Proceedings of the Candy Symposium*, ed. C. H. Holmes,
28-30. Paris: UNESCO, 1958.

Doel, H. W. van den. *Afscheid van Indië: De val van het Nederlandse imperium in Azië*.
Amsterdam: Prometheus, 2001.

————. *De stille macht: Het Europese binnenlands bestuur op Java en Madoera, 1808–1942*. Amsterdam: Bert Bakker, 1994.

————. "Practical Agricultural Education in the Netherlands East Indies: The Transfer of Agricultural Knowledge to the Indigenous Population of Java, 1875–1920." *Journal of the Japan-Netherlands Institute* 6 (1996): 78–94.

Donnison, F. S. V. *British Military Administration in the Far East, 1943–1946*. London: Her Majesty's Stationary Office, 1956.

Doorn, J. A. A. van. *De laatste eeuw van Indië: Ontwikkeling en ondergang van een koloniaal project*. Amsterdam: Ooievaar, 1996.

Doty, Maxwell. "The American Botanist and the Biological Institute of Indonesia." *Bioscience* 14, no. 12 (1964): 36–41.

Doty, Maxwell, and Aprilany Soegiarto. "The Development of Marine Resources in Indonesia." In *Indonesia: Resources and Their Technological Development*, ed. Howard Beers, 70–89. Lexington: University Press of Kentucky, 1970.

Douwes Dekker, E. F. E. "Onderwijs." *Het Tijdschrift* 1, no. 9 (1912): 312–16.

Drayton, Richard. *Nature's Government: Science, Imperial Britain, and the "Improvement" of the World*. New Haven, CT: Yale University Press, 2000.

Drees, W., and N. Ant. Jansen. *Eerste koloniaal onderwijscongres: Stenografisch verslag*. The Hague: N.p., 1916.

Drooglever, P. J. *De Vaderlandse Club: Totoks en de Indische politiek*. Franeker: T. Wever, 1980.

Drooglever, P. J., and M. J. B. Schouten, eds. *De Leeuw en de Banteng*. The Hague: Instituut voor Nederlandse Geschiedenis, 1997.

Duara, Prasenjit. *Sovereignty and Authenticity: Manchukuo and the East Asian Modern*. Lanham, MD: Rowman and Littlefield, 2003.

Dwidjoseputro, D. *Review of the State of Biology in Indonesia as a Scientific Discipline*. Malang: Institut Keguruan dan Ilmu Pendidikan, 1970.

Eldridge, Philip. *Non-government Organizations and Democratic Participation in Indonesia*. Kuala Lumpur: Oxford University Press, 1995.

Elson, R. E. *The Idea of Indonesia: A History*. Cambridge: Cambridge University Press, 2008.

————. *Suharto: A Political Biography*. Cambridge: Cambridge University Press, 2001.

Emmerson, Donald K. "Understanding the New Order: Bureaucratic Pluralism in Indonesia." *Asian Survey* 23, no. 11 (1983): 1220–41.

"Faedahnja Pendidikan dan Pengetahoean Jang Loeas." *Madjallah Boelanan Oentoek Kemadjoean Ra'jat* 2, no. 9 (1938): 163.

Fagi, Achmad M. "Kata Pengantar." In *Bunga-Bunga Pun Bermekaran*, ed. Setijati Sastrapradja, xi–xii. Bogor: Badan Penelitian dan Pengembangan Pertanian, 1999.

Fairchild, David. "An American Plant Hunter in the Netherlands Indies: Buitenzorg and Doctor Treub." In *Science and Scientists in the Netherlands Indies*, ed. Pieter Honig and Frans Verdoorn, 79–99. New York: Board for the Netherlands Indies, Surinam and Curaçao, 1945.

Fairchild, David, and Thomas Barbour. "The Crisis at Buitenzorg." *Science* 80 (1934): 33–34.

Fan, Fa-ti. *British Naturalists in Qing China: Science, Empire, and Cultural Encounter*. Cambridge, MA: Harvard University Press, 2004.

Fasseur, C. *De Indologen: Ambtenaren voor het Oost, 1825–1950*. Amsterdam: Bert Bakker, 1994.

———. *De weg naar het paradijs en andere Indische geschiedenissen*. Amsterdam: Bert Bakker, 1995.

———. "Een koloniale paradox: De Nederlandse expansie in de Indische archipel (1830-1870)." In *De Weg naar het paradijs en andere Indische geschiedenissen*, 47-73. Amsterdam: Bert Bakker, 1995.

———. "Indische persperikelen, 1847-1860." *Bijdragen en Mededelingen betrefende de Geschiedenis der Nederlanden* 91, no. 1 (1976): 56-75.

———. *The Politics of Colonial Exploitation: Java, the Dutch, and the Cultivation System*. Ithaca, NY: Cornell University Southeast Asia Program, 1992.

———. "Purse or Principle: Dutch Colonial Policy in the 1860s and the Decline of the Cultivation System." *Modern Asian Studies* 25, no. 1 (1991): 33-52.

Fasseur, C., and D. H. A. Kolff. "Some Remarks on the Development of Colonial Bureaucracies in India and Indonesia." *Itinerario* 10, no. 1 (1986): 31-55.

Feith, Herbert. *The Decline of Constitutional Democracy in Indonesia, 1950-1957*. Ithaca, NY: Cornell University Press, 1962.

———. "Dynamics of Guided Democracy." In *Indonesia*, ed. Ruth T. McVey, 309-409. Rev. ed. New Haven, CT: Human Relations Area Files Press, 1967.

Feith, Herbert, and Lance Castles, eds. *Indonesian Political Thinking, 1945-1965*. Ithaca, NY: Cornell University Press, 1970.

Findling, John E., and Kimberly Pelle, eds. *Historical Dictionary of World's Fairs and Expositions, 1851-1988*. New York: Greenwood Press, 1990.

Furnivall, J. S. *Colonial Policy and Practice: A Comparative Study of Burma and Netherlands India*. New York: New York University Press, 1956.

———. *Netherlands India: A Study of Plural Economy*. Cambridge: Cambridge University Press, 1944.

Gascoigne, S. C. B. "Australian Astronomy since the Second World War." In *Australian Science in the Making*, ed. R. W. Home, 345-73. Cambridge: Cambridge University Press, 1988.

*Gedenkboek Franz Junghuhn, 1809-1909*. The Hague: Nijhoff, 1910.

"Gedung Baru BIOTROP Diresmikan Pemakainnja." *Berita L.I.P.I.* 16, no. 3 (1972): 46.

Gelderen, J. van. *The Recent Development of Economic Foreign Policy in the Netherlands East Indies*. London: Longmans, 1939.

Gorkom, K. W. van. "Cinchona Cultivation after Junghuhn's Death." In *Science and Scientists in the Netherlands East Indies*, ed. Pieter Honig and Frans Verdoorn, 196-203. New York: Board for the Netherlands Indies, Surinam and Curaçao, 1945.

———. "Cinchona in Java from 1872 to 1907." *Agricultural Ledger* 17, no. 4 (1912): 35-106.

———. *A Handbook of Cinchona Culture*. Amsterdam: J. H. Bussy, 1883.

———. "Levensbericht van Rudolph Herman Christiaan Carel Scheffer." *Jaarboek van de Koninklijke Akademie van Wetenschappen* (1880): 1-21.

———. "Verslag nopens de Kina-Kultuur op Java over het jaar 1868." *Natuurkundig Tijdschrift voor Nederlandsch-Indië* 31 (1870): 147-62.

———. "Verslag nopens de Kina-Kultuur op Java over het jaar 1869." *Natuurkundig Tijdschrift voor Nederlandsch-Indië* 32 (1873): 1-22.

Goss, Andrew. "Decent Colonialism? Pure Science and Colonial Ideology in the Netherlands East Indies, 1910-1929." *Journal of Southeast Asian Studies* 40, no. 1 (2009): 187-214.

Gr., M. "Schets van Hasskarl's leven en werken: 1811-1894." *Indische Gids* 16, no. 1 (1894): 290-99.

Gramiccia, Gabriele. *The Life of Charles Ledger (1819-1905): Alpacas and Quinine.* London: Macmillan, 1988.

Greenhalgh, Paul. *Ephemeral Vistas: The* Expositions Universelles, *Great Exhibitions, and World's Fairs, 1851-1939.* Manchester: Manchester University Press, 1988.

Groot, Hans. *Van Batavia naar Weltevreden: Het Bataviaasch Genootschap van Kunsten en Wetenschappen, 1778-1867.* Leiden: KITLV Press, 2009.

Groothoff, Arnold. *De Kinacultuur.* 2nd ed. Haarlem: Tjeenk Willink, 1915.

———. *Rationeele Exploitatie van Kina-Plantsoenen.* Haarlem: Tjeenk Willink, 1919.

Grove, Richard H. *Green Imperialism: Colonial Expansion, Tropical Island Edens, and the Origins of Environmentalism, 1600-1860.* Cambridge: Cambridge University Press, 1995.

Guenst, F. "Levensschets van dr. Franz Wilhelm Junghuhn." In F. W. Junghuhn, *Licht- en schaduwbeelden uit de binnenlanden van Java.* 6th ed. Amsterdam: Guenst, 1867.

Haberlandt, Gottlieb. *Eine Botanische Tropenreise: Indo-Malaysiche Vegetationsbilder und Reiseskizzen.* 2nd ed. Leipzig: Verlag von Wilhelm Engelmann, 1910.

Haeckel, Ernst. *A Visit to Ceylon.* New York: Peter Eckler, n.d.

Hamid, Auzay. "Penelitian Pertanian Zaman Pra PELITA." In *Sejarah Penelitian Pertanian di Indonesia,* 141-225. Jakarta: Badan Penelitian dan Pengembangan Pertanian, 1995.

Hasskarl, J. K. *Tweede Catalogus der in 's Lands Plantentuin te Buitenzorg gekweekte gewassen.* Batavia: Landsdrukkerij, 1844.

Headrick, Daniel. *The Tentacles of Progress: Technology Transfer in the Age of Imperialism, 1850-1940.* New York: Oxford University Press, 1988.

———. *The Tools of Empire: Technology and European Imperialism in the Nineteenth Century.* New York: Oxford University Press, 1981.

Hefner, Robert W. *Civil Islam: Muslims and Democratization in Indonesia.* Princeton, NJ: Princeton University Press, 2000.

Henley, David. *Nationalism and Regionalism in a Colonial Context: Minahasa in the Dutch East Indies.* Leiden: KITLV Press, 1996.

"Het tweede congres van Boedi Oetomo." *Indische Gids* 32, no. 1 (1910): 94-96.

Heyne, K. *De nuttige planten van Nederlandsch-Indië.* 2nd ed. 3 vols. Batavia: Ruygrok, 1927.

Hidayat, Bambang. "Under a Tropical Sky: A History of Astronomy in Indonesia." *Journal of Astronomical History and Heritage* 3, no. 1 (2000): 45-58.

Hindley, Donald. *The Communist Party of Indonesia, 1951-1963.* Berkeley: University of California Press, 1964.

Hoëvell, W. R. van. "De Kina-Kultuur op Java beoordeeld in den vreemde." *Tijdschrift voor Nederlandsch-Indië* 24, no. 2 (1862): 360-74.

———. "Geschiedkunding Overzigt van de beofening van Kunsten en Wetenschappen in Neêrland's Indië." *Tijdschrift voor Neêrlands Indië* 2, no. 1 (1839): 1-115.

Hogstad, Anton. "The Three Hundredth Anniversary of the First Recognized Use of Cinchona." In *Proceedings of the Celebration of the Three Hundredth Anniversary of the First Recognized Use of Cinchona,* 1-3. Saint Louis: Missouri Botanical Gardens, 1931.

Holt, Claire. *Art in Indonesia: Continuities and Change.* Ithaca, NY: Cornell University Press, 1967.

Home, R. W., ed. *Australian Science in the Making.* Cambridge: Cambridge University Press, 1988.

———. "The Beginnings of an Australian Physics Community." In *Scientific Colonialism: A Cross-Cultural Comparison*, ed. Nathan Reingold and Marc Rothenberg, 3-34. Washington, DC: Smithsonian Institution Press, 1987.

Honig, Pieter, and Frans Verdoorn, eds. *Science and Scientists in the Netherlands Indies.* New York: Board for the Netherlands Indies, Surinam and Curaçao, 1945.

Hooper-Greenhill, E. *Museums and the Shaping of Knowledge.* London: Routledge, 1992.

Idenburg, A. W. F. "Oprichting van een Landbouw departement in Ned.-Indië." *Tijdschrift voor het Binnenlandsch Bestuur* 26 (1904): 394-417.

"Ilmoe Chewan." *Madjallah Boelanan Oentoek Kemadjoean Ra'jat* 1, no. 11 (1937): 172.

"Ilmoe Djiwa." *Penindjauan* 1, no. 4 (1934): 83.

*Indonesian Economics: The Concept of Dualism in Theory and Policy.* The Hague: Van Hoeve, 1961.

*Indonesian Institute of Sciences.* Jakarta: Indonesian National Scientific Documentation Center, 1970.

Ingleson, John. *Road to Exile: The Indonesian Nationalist Movement, 1927-1934.* Singapore: Heinemann, 1979.

———. "Sutomo, the Indonesian Study Club, and Organised Labour in Late Colonial Surabaya." *Journal of Southeast Asian Studies* 39, no. 1 (2008): 31-57.

"Introduction: Indonesia, to Our Readers Abroad." *Bangoen* 1, no. 1 (1937): 4.

Jacobs, M. "Kostermans Seventy-Five." *Reinwardtia* 10, no. 1 (1982): 9-20.

Johnston, Ron, and Jean Buckley. "The Shaping of Contemporary Scientific Institutions." In *Australian Science in the Making*, ed. R. W. Home, 374-98. Cambridge: Cambridge University Press, 1988.

Jong, J. J. P. de. "H. J. van Mook als vraagteken: Het Nederlands beleid en de overgangsperiode." In *De Leeuw en de Banteng*, ed. P. J. Drooglever and M. J. B. Schouten, 190-200. The Hague: Instituut voor Nederlandse Geschiedenis, 1997.

Junghuhn, F. *Atlas van platen, bevattende elf pittoreske gezigten, behoordende tot het werk* Java. The Hague: Mieling, 1854.

———. "Een woord over den boom *Sambinoer* op Sumatra, betrekkelijk deszelfs botanische bepaling volgens C. L. Blume." *Nederlandsch Kruidkundig Archief* 2 (1850): 261-75.

———. "Inlichtingen, aangeboden aan het publiek over zeker geschrift van den Heer C. L. Blume en antwoord aan dien Heer." *Algemeene Konst- en Letterbode*, October 11, 1850, 232-40.

———. *Java, deszelfs gedaante, bekleeding, en inwendige structuur.* 4 vols. Amsterdam: Van Kampen, 1850-53.

———. *Licht- en schaduwbeelden uit de binnenlanden van Java.* 6th ed. Amsterdam: Guenst, [1854] 1867.

———. "Oproeping en beleefd verzoek aan Nederlansch-Indië's ingezetenen." *De Kopiïst* 2 (1843): 358-62.

———. *Terugreis van Java naar Europa.* Zalt-Bommel: John Noman en Zoon, 1851

———. "Toestand der aangekweekte kinaboomen op het eiland Java tijdens het bezoek van zijne Excellentie den Gouverneur Generaal Chs. F. Pahud in het laatst der maand Julij 1857." *Natuurkundig Tijdschrift voor Nederlandsch-Indië* 15 (1858): 23-138.

Junghuhn, F., and J. E. de Vrij. "De Kinakultuur op Java op het einde van het jaar 1859." *Natuurkundig Tijdschrift voor Nederlandsch-Indië* 21 (1860): 179-275.

Kadarsan, Sampurno. "Pidato Ketua Panitia Penjelenggara Peringatan 150 Tahun Kebun Raya Lembaga Biologi Nasional." *Berita M.I.P.I.* 11, nos. 3-4 (1967): 77-82.

Kadarsan, Sampurno, Machfudz Djajasasmita, Pranowo Martodihardjo, and Soekarja Somadikarta. *Satu Abad Museum Zoologi Bogor, 1894-1994.* Bogor: Pusat Penelitian dan Pengembangan Biologi, LIPI, 1994.

Kaligis, J. A. "Surat dari Redaksi." *Berita M.I.P.I.* 1, no. 1 (1957): 4.

"Kamerdebatten over de oprichting van een Departement van Landbouw in Nederlandsch-Indië." *Tijdschrift voor het Binnenlandsch Bestuur* 27 (1904): 151-73, 259-84, 367-431.

Kartini, Raden Adjeng. *Letters of a Javanese Princess.* New York: Norton, 1964.

Kartodirdjo, Sartono. *The Peasant Revolt of Banten in 1888.* The Hague: Nijhoff, 1966.

*Katalogus der tentoonstelling van produkten der natuur en der industrie van den Indischen archipel, te houden te Batavia in Oktober en November 1853.* Batavia: Lange, 1853.

Kerbosch, M. "Cinchona Culture in Java: Its History and Development." In *Proceedings of the Celebration of the Three Hundredth Anniversary of the First Recognized Use of Cinchona,* 181-209. Saint Louis: Missouri Botanical Gardens, 1931.

"Kesimpulan-kesimpulan Pertemuan Jang Pertama Antara Madjelis Ilmu Pengetahuan Indonesia dengan Lembaga-Lembaga Ilmu Pengetahuan." *Berita M.I.P.I.* 1, no. 2 (1957): 29-30.

Keyes, Charles. "'The Peoples of Asia': Science and Politics in the Classification of Ethnic Groups in Thailand, China, and Vietnam." *Journal of Asian Studies* 61, no. 4 (2002): 1163-1203.

Kluiver, D. J. W. J. *Het Landbouw en Medisch Onderwijs in Nederlandsch-Indië.* Batavia: Volkslectuur, 1935.

Koerner, Lisbet. *Linnaeus: Nature and Nation.* Cambridge, MA: Harvard University Press, 1999.

Koesnoto Setyodiwiryo. "Kebun Raya Indonesia (Lembaga Pusat Penjelidikan Alam)." *Berita M.I.P.I.* 1, no. 1 (1957): 5-29.

———."Prasaran Perihal Ikut Sertanja Indonesia Dalam Kongres-Kongres Ilmu Pengetahuan Internasional dan Perihal Mengelenggarakan Kongres-Kongres Ilmu Pengetahuan Nasional." *Berita M.I.P.I.* 1, no. 2 (1957): 18-23.

Kohler, Robert E. *Landscapes and Labscapes: Exploring the Lab-Field Border in Biology.* Chicago: University of Chicago Press, 2002.

Kol, H. van. *Uit onze Koloniën.* Leiden: Sijthoff, 1903.

Koningsberger, J. C. "Landbouwschool en Cultuurtuin." In *Jaarboek van het Departement van Landbouw in Nederlandsch-Indië 1907,* 62-69. Batavia: Landsdrukkerij, 1908.

Koningsberger, V. J. "Het werk van Melchior Treub na 25 jaren." *Koloniale Studiën* 18, no. 1 (1934): 249-58.

Koorders, S. H., and Th. Valeton. *Bijdragen tot de kennis der boomsoorten van Java.* 13 vols. Buitenzorg: 's Lands Plantentuin/Departement van Landbouw, 1894-1910.

Kossmann, E. H. *De Lage Landen, 1780-1940: Anderhalve eeuw Nederland en België.* Amsterdam: Elsevier, 1976.

Kostermans, A. J. G. H. "In Memorium Anwari Dilmy." *Reinwardtia* 10, no. 1 (1982): 5-7.

Krishna, V. V. "The 'Colonial' Model and the Emergence of National Science in India: 1876-1920." In *Science and Empires: Historical Studies about Scientific Development and European Expansion,* ed. Patrick Petitjean, Catherine Jami, and Anne Marie Moulin, 57-72. Dordrecht: Kluwer, 1992.

Kumar, Deepak. "Reconstructing India: Disunity in the Science and Technology for Development Discourse, 1900-1947." In "Nature and Empire: Science and the Colonial Enterprise," ed. Roy MacLeod. *Osiris* 15 (2000): 241-57.

——. *Science and the Raj, 1857-1905.* Delhi: Oxford University Press, 1995.

Kworo, Noto. "Oude en Nieuwe Methoden in de Verloskundige Hulp op Java." In "Soembangsih: Gedenkboek Boedi-Oetomo 1908-20 Mei 1918," ed. Noto Soeroto. Special issue, *Nederlandsch Indië Oud en Nieuw* (1918): 65-75.

Lam, H. J. "Conspectus of Institutions of Pure and Applied Science in or Concerning the Netherlands East Indies." In *Science in the Netherlands East Indies*, ed. L. M. R. Rutten, 383-432. Amsterdam: Koninklijke Akademie van Wetenschappen, 1929.

*Laporan Tahun 1953 dari Djawatan Penjelidikan Alam (Kebun Raya Indonesia).* Bogor: Kebun Raya Indonesia, 1957.

Leersum, P. van. "Junghuhn and Cinchona Cultivation." In *Science and Scientists in the Netherlands East Indies*, ed. Pieter Honig and Frans Verdoorn, 190-96. New York: Board for the Netherlands Indies, Surinam and Curaçao, 1945.

——. "Kina." In *Dr. K. W. van Gorkom's Oost-Indische Cultures*, ed. H. C. Prinsen Geerligs, 3:165-275. Amsterdam: De Bussy, 1919.

Legge, John. *Sukarno: A Political Biography.* Sydney: Allen and Unwin, 1972.

Lev, Daniel S. *The Transition to Guided Democracy: Indonesian Politics, 1957-1959.* Ithaca, NY: Cornell University Southeast Asia Program, 1966.

Liddle, R. William. *Leadership and Culture in Indonesian Politics.* Sydney: Allen and Unwin, 1997.

Lieftinck, M. A., and A. C. V. van Bemmel. "The Development of the Zoological Museum at Buitenzorg." In *Science and Scientists in the Netherlands Indies*, ed. Pieter Honig and Frans Verdoorn, 226-31. New York: Board for the Netherlands Indies, Surinam and Curaçao, 1945.

Lindblad, J. Thomas. *Bridges to Business: The Economic Decolonization of Indonesia.* Leiden: KITLV Press, 2008.

——. "The Late Colonial State and Economic Expansion, 1900-1930s." In *The Emergence of a National Economy: An Economic History of Indonesia, 1800-2000*, ed. Howard Dick, 111-52. Honolulu: University of Hawai'i Press, 2002.

Locher-Scholten, Elsbeth. *Ethiek in fragmenten: Vijf studies over koloniaal denken en doen van Nederlanders in de Indonesische Archipel, 1877-1942.* Utrecht: Hes Publishers, 1981.

Lotsy, J. P. "Levensbericht van Melchior Treub." *Levensberichten der afgestorven medeleden van de Maatschappij der Nederlandsche Letterkunde te Leiden* (1911-12): 1-26.

Lowe, Celia. *Wild Profusions: Biodiversity Conservation in an Indonesian Archipelago.* Princeton, NJ: Princeton University Press, 2006.

L. P., Dr. "Een Britsch Indische Natuurkundige benoemd tot Nederlandsche Hoogleraar." *Nationale Commentaren* 1, no. 13 (1938): 212.

Maat, Harro. *Science Cultivating Practice: A History of Agricultural Science in the Netherlands and Its Colonies, 1863-1986.* Dordrecht: Kluwer, 2001.

MacLeod, Roy. "Imperial Reflections in the Southern Seas: The Funafuti Expeditions, 1896-1904." In *Nature in Its Greatest Extent: Western Science in the Pacific*, ed. Roy MacLeod and Philip F. Rehbock, 159-91. Honolulu: University of Hawai'i Press, 1988.

——. "Introduction." In "Nature and Empire: Science and the Colonial Enterprise," ed. Roy MacLeod. *Osiris* 15 (2000): 1-13.

"Madame Curie." *Penindjauan* 1, no. 32 (1934): 755.

"Madjallah oentoek Kemadjoean Ra'jat." *Madjallah Boelanan Oentoek Kemadjoean Ra'jat* 1, no. 12 (1937): 178.

Mahmud, Zainal, Djiman Sitepu, Sampurno Kadarsan, Elna Karmawati, and Hobir, eds. *Sejarah Penelitian Pertanian Indonesia*. Jakarta: Badan Penlitian dan Pengembangan Pertanian, 1996.

Makagiansar, M. "The Organization of Scientific Research in Indonesia and Significant Developments and Trends since 1960." *Berita M.I.P.I.* 6, no. 3 (1962): 130–38.

Manning, Chris, and Peter van Diemen. *Indonesia in Transition: Social Dimensions of the Reformasi and the Economic Crisis*. London: Zed Books, 2000.

Manwan, Ibrahim. "Penabur Benih Yang Tak Kenal Letih." In *Bunga-Bunga Pun Bermekaran*, ed. Setijati Sastrapradja, 253–61. Bogor: Badan Penelitian dan Pengembangan Pertanian, 1999.

Mansvelt, W. M. F. "De omzetting van 's Lands Plantentuin tot Departement van Landbouw." *Koloniale Studiën* 21, no. 1 (1937): 115–33.

Mark, Ethan. "Appealing to Asia: Nation, Culture, and the Problem of Imperial Modernity in Japanese-Occupied Java, 1942–1945." Ph.D. diss., Columbia University, 2003.

Markham, Clements. *Travels in Peru and India While Superintending the Collections of Chinchona Plants and Seeds in South America and Their Introduction into India*. London: John Murray, 1862.

McNeely, Ian F. *The Emancipation of Writing: German Civil Society in the Making, 1790s–1820s*. Berkeley: University of California Press, 2003.

McVey, Ruth. "The Case of the Disappearing Decade." In *Democracy in Indonesia, 1950s and 1990s*, ed. David Bourchier and John Legge, 3–15. Clayton, Australia: Monash University Centre of Southeast Asian Studies, 1994.

———. *The Rise of Indonesian Communism*. Ithaca, NY: Cornell University Press, 1965.

———. "Taman Siswa and the Indonesian National Awakening." *Indonesia* 4 (1967): 128–49.

Merrill, Elmer D. *Report on Investigations Made in Java in the Year 1902*. Manila: Bureau of Public Printing, 1903.

Messer, Adam. "Effects of the Indonesian National Revolution and Transfer of Power on the Scientific Establishment." *Indonesia* 58 (1994): 41–68.

*Middelbare Landbouwschool te Buitenzorg, programma voor het leerjaar, 1915–1916*. Buitenzorg: Departement van Landbouw, Nijverheid, en Handel, 1915.

*Middelbare Landbouwschool te Buitenzorg, programma voor het leerjaar, 1920–1921*. Buitenzorg: Departement van Landbouw, Nijverheid, en Handel, 1920.

Miert, Hans van. "Benepen stemmen in het zwembassin: Boedi Oetoemo in de eerste Volksraad 1918-1921." *Jambatan* 11, nos. 1–2 (1993): 3–35.

———. *Bevlogenheid en onvermogen: Mr. J. H. Abendanon en de ethische richting in het Nederlandse kolonialisme*. Leiden: KITLV Press, 1991.

Miller, David Philip, and Peter Hanns Reill, eds. *Visions of Empire: Voyages, Botany, and Representations of Nature*. Cambridge: Cambridge University Press, 1996.

Miquel, F. A. W. *De Cinchonae speciebus quibusdam, adiectis iis quae in Java coluntur*. Amsterdam: Van der Post, 1869.

Mizuno, Hiromi. *Science for the Empire: Scientific Nationalism in Modern Japan*. Stanford, CA: Stanford University Press, 2009.

Mook, H. J. van. *De organisatie van de Indische Regeering*. Batavia: Albrecht, 1932.

———. *Indonesië, Nederland, en de Wereld*. Amsterdam: De Bezige Bij, 1949.

Moon, Suzanne. "Constructing 'Native Development': Technological Change and the Politics of Colonization in the Netherlands East Indies, 1905-1930." Ph.D. diss., Cornell University, 2000.

———. "Takeoff of Self-Sufficiency? Ideologies of Development in Indonesia, 1957-1961." *Technology and Culture* 39, no. 2 (1998): 187-212.

———. *Technology and Ethical Idealism: A History of Development in the Netherlands East Indies*. Leiden: CNWS, 2007.

Morwood, M. J., et al. "Archaeology and Age of a New Hominin from Flores in Eastern Indonesia." *Nature* 431 (2004): 1087-91.

Mrázek, Rudolf. "Bridges of Hope: Senior Citizens' Memories." *Indonesia* 70 (October 2000): 37-51.

———. *Engineers of Happy Land: Technology and Nationalism in a Colony*. Princeton, NJ: Princeton University Press, 2002.

———. "Sjahrir and the Left Wing at the Time of the Indonesian Revolution." In *De Leeuw en de Banteng*, ed. P. J. Drooglever and M. J. B. Schouten, 109-30. The Hague: Instituut voor Nederlandse Geschiedenis, 1997.

Müller, Carl. "Die Verpflauzung des Chinabaumes und seine Cultur." *Unsere Zeit: Deutsche Revue der Gegenwart* 9, no. 2 (1873): 62-74, 186-215, 258-73.

Multatuli. *Max Havelaar*. London: Penguin, [1860] 1987.

Nagazumi, Akira. *The Dawn of Indonesian Nationalism: The Early Years of Budi Utomo 1908-1918*. Tokyo: Institute of Developing Economies, 1972.

Nasution, Adnan Buyung. *The Aspiration for Constitutional Government in Indonesia: A Socio-legal Study of the Indonesian Konstituante, 1956-1959*. Jakarta: Pustaka Sinar Harapan, 1992.

Nicolson, Malcolm. "Historical Introduction." In Alexander von Humboldt, *Personal Narrative of a Journey to the Equinoctial Regions of the New Continent*, ix-xxxiv. London: Penguin, 1995.

Niel, Robert van. *The Emergence of the Modern Indonesian Elite*. The Hague: Van Hoeve, 1960.

———. "Government Policy and the Civil Administration in Java during the Early Years of the Cultivation System." In *Java under the Cultivation System: Collected Writings*, 87-106. Leiden: KITLV Press, 1992.

Nieuwenhuys, Rob, and Frits Jaquet. *Java's Onuitputtelijke Natuur: Reisverhalen, tekeningen en fotografieen van Franz Wilhelm Junghuhn*. Alphen aan de Rijn: Sijthoff, 1980.

"No. 3000." *Penindjauan* 1, no. 32 (1934): 750.

Noer, Deliar. *The Modernist Muslim Movement in Indonesia, 1900-1942*. Singapore: Oxford University Press, 1973.

Nyhart, Lynn K. *Biology Takes Form: Animal Morphology and the German Universities, 1800-1900*. Chicago: University of Chicago Press, 1995.

———. *Modern Nature: The Rise of the Biological Perspective in Germany*. Chicago: University of Chicago Press, 2009.

Onghokham. *Runtuhnya Hindia Belanda*. Jakarta: Gramedia, 1987.

"Orang Prempoean dan Kemadjoean Djaman." *Madjallah Boelanan Oentoek Kemadjoean Ra'jat* 1, no. 1 (1937): 14.

Osborne, Michael A. "Acclimatizing the World: A History of the Paradigmatic Colonial Science." In "Nature and Empire: Science and the Colonial Enterprise," ed. Roy MacLeod. *Osiris* 15 (2001): 135-51.

———. *Nature, the Exotic, and the Science of French Colonialism.* Bloomington: Indiana University Press, 1994.

Palmer, Leslie. *Indonesia and the Dutch.* London: Oxford University Press, 1962.

Pauker, Guy J. "Indonesia's Eight-Year Development Plan." *Pacific Affairs* 34, no. 2 (1961): 115-30.

Peluso, Nancy Lee. *Rich Forests, Poor People: Resource Control and Resistance in Java.* Berkeley: University of California Press, 1992.

"Pemandangan Doenia." *Madjallah Boelanan Oentoek Kemadjoean Ra'jat* 1, no. 3 (1937): 36.

Pemberton, John. *On the Subject of "Java."* Ithaca, NY: Cornell University Press, 1994.

"Pembahagian dari Isinja." *Madjallah Boelanan Oentoek Kemadjoean Ra'jat* 2, no. 24 (1938): 474-76.

"Pendidikan Adjunct-Landbouw Consulent." *Pandji Poestaka* 5, no. 38 (1927): 610-22.

"Pengetahoean tentang Hoekoem Negeri." *Madjallah Boelanan Oentoek Kemadjoean Ra'jat* 2, no. 2 (1938): 31.

"Pengetahoean tentang Mendidik anak-anak." *Madjallah Boelanan Oentoek Kemadjoean Ra'jat* 1, no. 1 (1937): 8.

"Perkawinan dari Prinses Juliana dan Prins Bernhard." *Madjallah Boelanan Oentoek Kemadjoean Ra'jat* 1, no. 1 (1938): 4.

Philip, Kavita. *Civilizing Natures: Race, Resources, and Modernity in Colonial South India.* New Brunswick, NJ: Rutgers University Press, 2004.

Poesponegoro, Soedjono D. "Pidato J. M. Menteri Research Nasional Pada Peresmian Pusat Dokumentasi Ilmiah-Nasional." *Berita M.I.P.I.* 9, nos. 3-4 (1965): 77-82.

Pols, Hans. "European Physicians and Botanists, Indigenous Herbal Medicine in the Dutch East Indies, and Colonial Networks of Mediation." *East Asia Science, Technology, and Society: An International Journal* 3 (2009): 173-208.

Prakash, Gyan. *Another Reason: Science and the Imagination in Modern India.* Princeton, NJ: Princeton University Press, 1999.

Pramoedya Ananta Toer. *The Mute's Soliloquy: A Memoir.* New York: Hyperion East, 1999.

Pranata, Djaja. "Orgaan Doentoekkan Kemadjoean Hindia dan Anak Boemi." *Djawa-Moeda* 1, no. 1 (1916): 1.

Pratt, Mary Louise. *Imperial Eyes: Travel Writing and Transculturation.* London: Routledge, 1992.

Prawirotaroemo, F. X. In "Soembangsih: Gedenkboek Boedi-Oetomo, 1908-20 Mei 1918," ed. Noto Soeroto. Special issue, *Nederlandsch Indië Oud en Nieuw* (1918): 133-34.

Prins, J. A. "Clay, Jacob." In *Dictionary of Scientific Biography*, ed. Charles Gillispie, 3: 312-3. New York: Scribner's, 1971.

"Programma voor de tentoonstelling." *Natuurkundig Tijdschrift voor Nederlandsch Indië* 3 (1852): 648-52.

"Prospectus van den natuurkundig tijdschrift voor Nederlandsch Indie." *Natuurkundig Tijdschrift voor Nederlandsch Indië* 1 (1850): 1-3.

Pyenson, Lewis. "Assimilation and Innovation in Indonesian Science." In "Beyond Joseph Needham: Science, Technology, and Medicine in East and Southeast Asia," ed. Morris F. Low. *Osiris* 13 (1998): 34-47.

———. *Empire of Reason: Exact Sciences in Indonesia, 1840–1940.* Leiden: Brill, 1989.

Raby, Peter. *Alfred Russel Wallace: A Life.* Princeton, NJ: Princeton University Press, 2001.

Ramage, Douglas. *Politics in Indonesia: Democracy, Islam, and the Ideology of Tolerance.* London: Routledge, 1995.

Ratoe Langie, S. S. J. G. "De Indonesische Universiteit." Pt. 1. *Nationale Commentaren* 1, no. 19 (1938): 323–25.

———. De Indonesische Universiteit." Pt. 2. *Nationale Commentaren* 1, no. 20 (1938): 340–42.

———. "Waarom geen strijd?" *Hindia Poetra* 1, no. 4 (1916): 89–91.

Ravensteijn, Wim, and Jan Kop, eds. *For Profit and Prosperity: The Contributions Made by Dutch Engineers to Public Works in Indonesia, 1800–2000.* Leiden: KITLV Press, 2008.

Reeve, David. *Golkar of Indonesia: An Alternative to the Party System.* Singapore: Oxford University Press, 1985.

Reid, Anthony. *The Indonesian National Revolution, 1945–1950.* Hawthorne, Australia: Longman, 1974.

Resosudarmo, Budy, ed. *The Politics and Economics of Indonesia's Natural Resources.* Singapore: Institute for Southeast Asian Studies, 2005.

*Review of the Work of the Lembaga Biologi Nasional—LIPI (The National Biological Institute), 1 April 1969–31 March 1974.* Bogor: LBN, 1975.

Ridsdale, C. E. "Hasskarl's Cinchona Barks, Historical Review." *Reinwardtia* 10, no. 2 (1985): 245–64.

Rifai, M. A. "Demam Penelitian." *Berita L.I.P.I.* 17, no. 1 (1973): 3–9.

———. "Kostermans: The Man, His Work, His Legacy." In *Plant Diversity in Malesia III: Proceedings of the Third International Flora Malesiana Symposium, 1995,* ed. J. Dransfield, M. J. E. Coode, and D. A. Simpson, 225–30. Kew: Royal Botanic Gardens, 1997.

Rifai, M. A., S. Sastrapradja, and S. Adisoemarto. "Morfologi Lembaga Penelitian." *Berita L.I.P.I.* 19, no. 1 (1975): 1–10.

Rijkens, H. H. "Nog eens: Over het slibbezwaar van eenige rivieren in het Serajoedal en daarmede in verband staande onderzoekingen door Dr. E. C. Jul. Mohr, 1908, Batavia, G. Kolff en Co., Mededeelingen, uitgaande van het Dep. v.d. Landbouw, no. 5, 1908." *Landbouwkundig Tijdschrift* 22 (1910): 262–71.

Rijnberg, Theo F. *'s Lands Plantentuin, Buitenzorg, 1817–1992.* Enschede: Johanna Oskamp, 1992.

"Riwajat Orang Pandai, Tersohor: Benjamin Franklin (7)." *Madjallah Boelanan Oentoek Kemadjoean Ra'jat* 2, no. 1 (1938): 8.

Robbins, William J. "Elmer Drew Merrill (1876–1956)." *Biographical Memoirs* 32 (1958): 273–333.

Roosa, John. *Pretext for Mass Murder: The September 30th Movement and Suharto's Coup d'État in Indonesia.* Madison: University of Wisconsin Press, 2006.

Rudner, Martin. "The Indonesian Military and Economic Policy: The Goals and Performance of the First Five-Year Development Plan, 1969–1974." *Modern Asian Studies* 10, no. 2 (1976): 249–84.

Ruinen, J. "L. G. M. Baas Becking in Memoriam." *Vakblad voor biologen* 43, no. 3 (1963): 41–47.

Rumphius, G. E. *The Ambonese Curiosity Cabinet.* Trans. E. M. Beekman. New Haven, CT: Yale University Press, 1999.

Rutten, L. M. R., ed. *Science in the Netherlands East Indies.* Amsterdam: Koninklijke Akademie van Wetenschappen, 1929.

Sarwono Prawirohardjo. "Laporan Singkat Panitia Persiapan Pembentukan Madjelis Ilmu Pengetahuan Nasional." *Berita M.I.P.I.* (1956): 14-20.

———. "Madjelis Ilmu Pengetahuan Indonesia dan Universitas-Universitas di Indonesia." *Berita M.I.P.I.* 2, no. 1 (1958): 32-38.

———. "Peranan Lembaga Biologi Nasional dalam Pengembangan Pendidikan Biologi di Indonesia." *Berita L.I.P.I.* 18, no. 3 (1974): 23-32.

———. "Sambutan Ketua Madjelis Ilmu Pengetahuan Indonesia pada Pembukaan Pameran Biologi Lembaga Biologi Nasional." *Berita M.I.P.I.* 11, nos. 3-4 (1967): 6-7.

———. "The Scientific Development of Indonesia." *Berita M.I.P.I.* 5, no. 5 (1961): 5-14.

———. "Tugas dan Rentjana Kerdja Madjelis Ilmu Pengetahuan Indonesia Dalam Masa Depan." *Berita M.I.P.I.* 1, no. 2 (1957): 16-18.

Sastrapradja, D. S. "Multi Fungsi Kebun Raya di Indonesia." *Berita L.I.P.I.* 15, no. 3 (1971): 13-19.

Sastrapradja, Setijati. "Catatan Masa Jabatan Direktur Lembaga Biologi Nasional, 1 Maret 1973-25 Juni 1986." Pamphlet. LBN, 1986.

———. "Hak Hidup dan Kewajiban Peneliti di Indonesia." *Berita L.I.P.I.* 17, no. 3 (1973): 3-8.

———. "Ketika Langit Bertabur Bintang." In *Bunga-Bunga Pun Bermekaran,* ed. Setijati Sastrapradja, 129-48. Bogor: Badan Penelitian dan Pengembangan Pertanian, 1999.

Sastrapradja, Setijati, ed. *Bunga-Bunga Pun Bermekaran.* Bogor: Badan Penelitian dan Pengembangan Pertanian, 1999.

Sastrapradja, S., S. Adisumarto, and M. A. Rifai. "Elements of Life and the Classrooms." *Berita L.I.P.I.* 15, no. 1 (1971): 25-34.

Sastrapradja, S., M. A. Rifai, and S. Adisumarto. "Ilmu Pengetahuan di Indonesia: Tantangan buat Para Pendidik." *Berita L.I.P.I.* 15, no. 2 (1971): 7-14.

Sato, Shigeru. *War, Nationalism, and Peasants: Java under the Japanese Occupation, 1942-1945.* Armonk, NY: M. E. Sharpe, 1994.

Scheffer, R. H. C. C. "Sur deux espèces du genre *Genocaryum* Miq." *Annales du Jardin Botanique de Buitenzorg* 1 (1876): 96-102.

———. "Sur quelques Palmiers du groupe des Arécinées." *Natuurkundig Tijdschrift van Nederlandsch-Indië* 32 (1873): 149-93.

Schiebinger, Londa. *The Mind Has No Sex? Women in the Origins of Modern Science.* Cambridge, MA: Harvard University Press, 1989.

Schiff, S. D. "Cirkulaire van de kommissie to het beheer der Tentoonstelling, te houden te Batavia in de maand September van het jaar 1853." *Natuurkundig Tijdschrift voor Nederlandsch-Indië* 3 (1852): 128-31.

Schmidt, Max C. P. *Franz Junghuhn: Biographische Beiträge zur hundersten Wiederkehr Seines Geburtstages.* Leipzig: Verlag der Duerr'schen Buchhandlung, 1909.

Schoor, Wim van der. "Biologie en Landbouw: F. A. F. C. Went en de Indische Proefstations." *Gewina* 17 (1994): 145-61.

———. "Pure Science and Colonial Agriculture: The Case of the Private Java Sugar Experiment Stations (1885-1940)." *Journal of the Japan-Netherlands Institute* 6 (1996): 68-77.

Schrieke, B. J. O., ed. *Report of the Scientific Work Done in the Netherlands on Behalf of the Dutch Overseas Territories during the Period between Approximately 1918 and 1943*. Amsterdam: North-Holland, 1948.

Schumaker, Lyn. *Africanizing Anthropology: Fieldwork, Networks, and the Making of Cultural Knowledge in Central Africa*. Durham, NC: Duke University Press, 2001.

Scott, James C. *Seeing Like a State: How Certain Schemes to Improve the Human Condition Have Failed*. New Haven, CT: Yale University Press, 1998.

Sears, Laurie J. *Shadows of Empire: Colonial Discourse and Javanese Tales*. Durham, NC: Duke University Press, 1996.

Secord, Anne. "Science in the Pub: Artisan Botanists in Early Nineteenth-century Lancashire." *History of Science* 32 (1994): 269–315.

*Sejarah Penelitian Pertanian di Indonesia*. Jakarta: Badan Penelitian dan Pengembangan Pertanian, 1995.

Sep, Peter. "De receptie van *Licht- en schaduwbeelden uit de binnenlanden van Java* van F. W. Junghuhn." *Indische Letteren* 2 (1987): 53–64.

Sewojo, Dwidjo. "Boedi Oetomo: Rede gehouden voor de Defltsche Studenten-vereeniging 'Onze Koloniën' op 2 Mei 1917." *Nederlandsch-Indië Oud en Nieuw* 2 (1917-18): 67–72.

Shiraishi, Takashi. *An Age in Motion: Popular Radicalism in Java, 1912–1926*. Ithaca, NY: Cornell University Press, 1990.

———. "The Disputes between Tjipto Mangoenkoesoemo and Soetatmo Soeriokoesoemo: Satria vs. Pandita." *Indonesia* 32 (1981): 93–108.

Shteir, Ann B. *Cultivating Women, Cultivating Science: Flora's Daughters and Botany in England, 1760 to 1860*. Baltimore: Johns Hopkins University Press, 1996.

Siegel, James. *Fetish, Recognition, Revolution*. Princeton, NJ: Princeton University Press, 1997.

Simoen D., B. "Kewadjiban Nederland dan jang haroes diterima Hindia." *Kemadjoean-Hindia*, April 18, 1923.

Sirks, M. J. *Indisch natuuronderzoek*. Amsterdam: Ellermans/Harms, 1915.

Scidmore, Eliza. *Java: The Garden of the East*. Singapore: Oxford University Press, [1897] 1984.

*'s Lands Plantentuin te Buitenzorg, 18 Mei 1817–18 Mei 1892*. Batavia: Landsdrukkerij, 1892.

Sliggers, B. C., and M. H. Besselink, eds. *Het verdwenen museum: Natuurhistorische verzamelingen, 1750–1850*. Blaricum: V+K Publishing, 2000.

Slooten, D. F. van. "Biologisch en landbouwkundig werk in Nederlandsch Indië: Het Herbarium en Museum voor Systematsiche Botanie van 's Lands Plantentuin." *Vakblad voor Biologen* 14, no. 9 (1933): 161–74.

———. "De Nederlandsch Indische Natuur-Historische Vereeniging en *de Tropische Natuur*." In "Jubileum Uitgave." Special issue, *De Tropische Natuur* (1936): 3–8.

———. "Wetenschap: Het Reserve-Kapitaal der Maatschappij." *Chronica Naturae* 105, nos. 7-8 (1949): 186–90.

Smail, John. *Bandung in the Early Revolution, 1945–1946: A Study in the Social History of the Indonesian Revolution*. Ithaca, NY: Cornell University Southeast Asia Program, 1964.

Smith, J. J. *Geillustreerde Gids voor "'s Lands Plantentuin" te Buitenzorg*. Buitenzorg: Departement van LNH, 1917.

Snelders, H. A. M. "Gerrit Jan Mulders bemoeienissen met het natuurwetenschappelijk onderzoek in Nederlands Indië." *Tijdschrift voor de Geschiedenis der Geneeskunde, Natuurwetenschappen, Wiskunde, en Techniek* 13, no. 4 (1990): 253–64.

Soegiarto, A. "Lembaga Penelitian yang Ideal." *Berita L.I.P.I.* 19, no. 1 (1975): 11–19.

———. "Menyibak Ombak Menyelam Dalam." In *Bunga-Bunga Pun Bermekaran*, ed. Setijati Sastrapradja, 187–97. Bogor: Badan Penelitian dan Pengembangan Pertanian, 1999.

Soeharnis. "Hari-hari Penuh Tantangan." In *Bunga-Bunga Pun Bermekaran*, ed. Setijati Sastrapradja, 201–11. Bogor: Badan Penelitian dan Pengembangan Pertanian, 1999.

Soeharto, Pitot, and A. Zainoel Ihsan, eds. *Cahaya di Kegelapan.* Jakarta: Jayasakti, 1981.

Soemarwoto, Otto. *Atur-Diri-Sendiri: Paradigma Baru Pengeloloaan Lingkungan Hidup.* Yogyakarta: Gadjah Mada University Press, 2001.

———. "BIOTROP: Its Objectives and Program." *Laporan Seminar Biologi Kedua, 18–20 Pebruari 1970 Ciawi-Bogor*, 1:34–36. Bogor: LBN, 1970.

———. "Development of the National Biological Institute." In *Indonesia: Resources and Their Technological Development*, ed. Howard Beers, 53–69. Lexington: University Press of Kentucky, 1970.

———. *Ekologi Lingkungan Hidup dan Pembangunan.* Jakarta: Jambatan, 1983.

———. *Environmental Education and Research in Indonesian Universities.* Singapore: Maruzen Asia, 1981.

———. "Lembaga Biologi Nasional dalam Perspektip." *Berita M.I.P.I.* 11, nos. 3–4 (1967): 9–17.

Soemarwoto, Otto, and D. A. Tisna Amidjaja. *Perkembangan Biologi, 1945–1965.* Jakarta: Balai Pustaka, 1965.

Soepomo, Sri Koesdijati. "Liku-Liku Sebuah Perjalanan." In *Bunga-Bunga Pun Bermekaran*, ed. Setijati Sastrapradja, 213–32. Bogor: Badan Penelitian dan Pengembangan Pertanian, 1999.

Soerohaldoko, Soetomo. "Pendirian Akademi Biologi Sebagai Cikal Bakal Akademi Kementerian Pertanian." Pamphlet. 1998.

Soerohaldoko, Soetomo, et al. *Kebun Raya Cibodas, 11 April 1852–11 April 2000.* Bogor: LIPI, 2000.

Soeroto, Noto, ed. "Soembangsih: Gedenkboek Boedi-Oetomo, 1908–20 Mei–1918." Special issue, *Nederlandsch Indië Oud en Nieuw* (1918).

Soetomo. *Toward a Glorious Indonesia: Reminiscences and Observations of Dr. Soetomo.* Ed. Paul W. van der Veur. Athens: Ohio University Center for International Studies, 1987.

Spary, E. C. *Utopia's Garden: French Natural History from Old Regime to Revolution.* Chicago: University of Chicago Press, 2000.

Stafford, Robert A. "The Long Arm of London: Sir Roderick Murchison and Imperial Science in Australia." In *Australian Science in the Making*, ed. R. W. Home, 69–101. Cambridge: Cambridge University Press, 1988.

Stafleu, F. A. *F. A. W. Miquel, Netherlands Botanist.* Utrecht: Botanisch Museum en Herbarium, 1966.

Stapelkamp, Herman. "De rol van Van Hoëvell in de Bataviase Mei-Beweging van 1848." *Jambatan* 4, no. 3 (1986): 11–20.

Steenis, C. G. G. J. van, ed. *Flora Malesiana.* Vol. 4, series 1. Batavia: Noordhoff-Kolff, 1948.

——. "Instructions for Cooperators." *Flora Malesiana Bulletin* 1 (1947): 2-29.

——. "Introduction." In *Flora Malesiana*, 4:i-xl. Series 1. Batavia: Noordhoff-Kolff, 1948.

——. "Treub, Melchior." In *Dictionary of Scientific Biography*, ed. Charles Gillispie, 13:458-60. New York: Scribner's, 1976.

Steenis-Kruseman, M. J. van. *Verwerkt Indisch Verleden*. Oegstgeest: N.p., 1988.

Stokvis, J. E. "Geen universitair, wel hooger vakonderwijs." *Het Tijdschrift* 1, no. 9 (1912): 276-78.

Stoler, Ann Laura. *Along the Archival Grain: Epistemic Anxieties and Colonial Common Sense*. Princeton, NJ: Princeton University Press, 2008.

——. *Carnal Knowledge and Imperial Power: Race and the Intimate in Colonial Rule*. Berkeley: University of California Press, 2002.

Stoler, Ann Laura, and Frederick Cooper. "Between Metropole and Colony: Rethinking a Research Agenda." In *Tensions of Empire: Colonial Cultures in a Bourgeois World*, ed. Frederick Cooper and Ann Laura Stoler, 1-56. Berkeley: University of California Press, 1997.

Stuurman, Siep. *Wacht op onze daden: Het liberalisme en de vernieuwing van de Nederlandse staat*. Amsterdam: Bert Bakker, 1992.

Sukarno. "Membangun Dunia Kembali." In *Haluan Politik dan Pembangunan Negara*, 163-95. Jakarta: Departemen Penerangan RI, 1960.

Sullivan, Woodruff T. *Cosmic Noise: A History of Early Radio Astronomy*. Cambridge: Cambridge University Press, 2009.

Sundhaussen, Ulf. *Road to Power: Indonesian Military Politics, 1945-1967*. Kuala Lumpur: Oxford University Press, 1982.

Sutherland, Heather. *The Making of a Bureaucratic Elite: The Colonial Transformation of the Javanese Priyayi*. Singapore: Heinemann, 1979.

Tagliacozzo, Eric. *Secret Trades, Porous Borders: Smuggling and States along a Southeast Asian Frontier, 1865-1915*. New Haven, CT: Yale University Press, 2005.

Taylor, Jean Gelman. *The Social World of Batavia: European and Eurasian in Dutch Asia*. Madison: University of Wisconsin Press, 1983.

Taylor, Norman. "Chapters in the History of Cinchona: Modern Developments." In *Science and Scientists in the Netherlands East Indies*, ed. Pieter Honig and Frans Verdoorn, 203-7. New York: Board for the Netherlands Indies, Surinam and Curaçao, 1945.

——. *Cinchona in Java: The Story of Quinine*. New York: Greenberg, 1945.

Tempelaars, A. M. *Inventaris van het archief van Prof. Dr. W. H. de Vriese (1806-1862) betreffende zijn onderzoek naar de kultures in Nederlands-Indië, 1857-1862, met retroacta vanaf 1817*. Schaarsbergen: Algemeen Rijksarchief, 1977.

"Tentoonstelling." *Natuurkundig Tijdschrift voor Nederlandsch Indië* 4 (1853): 429-32.

"Tentoonstelling te Batavia, te houden in 1853." *Natuurkundig Tijdschrift voor Nederlandsch Indië* 3 (1852): 647-48.

Termorshuizen, Gerard, ed. *In de Binnenlanden van Java: Vier negentiende-eeuwse reisverhalen*. Leiden: KITLV Press, 1993.

——. *Journalisten en heethoofden: Een geschiedenis van de Indisch-Nederlandse dagbladpers, 1744-1905*. Amsterdam: Nijgh en Van Ditmar, 2001.

Teysmann, J. E. "Bijdrage tot de geschiedenis der Kina-Kultuur op Java." *Natuurkundig Tijdschrift voor Nederlandsch-Indië* 25 (1863): 47-80.

———. "Botanische reis van den heer Teijsmann." *Natuurkundig Tijdschrift voor Nederlandsch-Indië* 4 (1853): 206.

———. "Kritische opmerkingen op de bijdragen van Doctor Junghuhn." *Tijdschrift voor Neerlandsch-Indië* 5, no. 1 (1843): 486-508.

———. "'s Lands Plantentuin te Buitenzorg in 1850." In *Verslag van het beheer en den staat der Nederlandsche Bezittingen en Kolonien, 1850*, 93-97. Utrecht: Kemink, 1858.

Theunissen, Bert. *Eugène Dubois en de Aapmens van Java: Een bijdrage tot de geschiedenis van de paleoantropologie.* Amsterdam: Rodopi, 1985.

———. *"Nut en nog eens nut": Wetenschapsbeelden van Nederlandse natuuronderzoekers, 1800-1900.* Hilversum: Verloren, 2000.

Thomas, Th. "Een hoogeschool voor Nederlandsch-Indië." Pts. 1 and 2. *Jong Indië* 1, no. 49 (1909): 537-38; 1, no. 50 (1909): 549-51.

Tjipto Mangoenoesoemo. "De vrees voor Demos." *Het Tijdschrift* 5 (1911): 154-58.

Tollenaar, D. "Ontwikkeling en toekomst van het natuurwetenschappelijk onderzoek voor Nederlandsch-Indië." *Koloniale Studiën* 21, no. 3 (1937): 237-53.

Treub, M. *Der botanische garten: "'s Lands Plantentuin" zu Buitenzorg auf Java—Festschrift zur Feier seines 75, jährigen bestehens (1817-1892).* Leipzig: Engelmann, 1893.

———. "Études sur les Lycopodiacées." *Annales du Jardin Botanique de Buitenzorg* 4 (1884): 107-37.

———. "Études sur les Lycopodiacées." *Annales du Jardin Botanique de Buitenzorg* 5 (1886): 87-139.

———. "Études sur les Lycopodiacées." *Annales du Jardin Botanique de Buitenzorg* 7 (1888): 141-50.

———. *Geschiedenis van 's Lands Plantentuin: Eerste gedeelte.* Batavia: Landsdrukkerij, 1889.

———. "Inleiding." In *Jaarboek van het Departement van Landbouw in Nederlandsch-Indië 1907*, v. Batavia: Landsdrukkerij, 1908.

———. "Korte Geschiedenis van 's Lands Plantentuin." In *'s Lands Plantentuin te Buitenzorg, 18 Mei 1817-18 Mei 1892*, 1-60. Batavia: Landsdrukkerij, 1892.

———. *Landbouw, Januari 1905-October 1909.* Amsterdam: Scheltema and Holkema, 1910.

———. *Over de Taak en den Werkkring van 's Lands Plantentuin te Buitenzorg.* Buitenzorg: 's Lands Plantentuin, 1899.

———. "Some Words on the Life-History of Lycopods." *Annals of Botany* 1, no. 2 (1887): 119-23.

———. "A Tropical Botanic Garden." In *Annual Report of the Board of Regents of the Smithsonian Institution to July, 1890*, 389-406. Washington, DC: Government Printing Office, 1891.

———. "Un jardin botanique tropical." *Revue des Deux Mondes* 97 (1890): 162-83.

———. "Voorwoord." In J. G. Boerlage, *Handleiding tot de kennis der Flora van Nederlandsch Indië*, 1:i-iv. Leiden: Brill, 1890-1900.

Tsuchiya, Kenji. *Democracy and Leadership: The Rise of the Taman Siswa Movement in Indonesia.* Honolulu: University of Hawai'i Press, 1987.

———. "Kartini's Image of Java's Landscape." *East Asian Cultural Studies* 25, nos. 1-4 (1986): 59-86.

"Undang-Undang No. 6 Tahun 1956 Tentang Pembentukan Madjelis Ilmu Pengetahuan Indonesia." *Berita M.I.P.I.* (1956): 3-13.

Veer, Paul van 't. "Een revolutiejaar, Indische stijl: Wolter Robert baron van Hoëvell,

1812–1879." In *Geen blad voor de mond: Vijf radicalen uit de negentiende eeuw*, 101–44. Amsterdam: Arbeiderspers, 1963.

———. *Geen blad voor de mond: Vijf radicalen uit de negentiende Eeuw*. Amsterdam: Arbeiderspers, 1963.

———. "Het einde van een eeuw: Ir. Henri Hubertus van Kol, 1852–1925." In *Geen blad voor de mond: Vijf radicalen uit de negentiende eeuw*, 183–217. Amsterdam: Arbeiderspers, 1963.

———. *Het leven van Multatuli*. Amsterdam: Arbeiderspers, 1979.

———. "In de schaduw van de kinaboom: Franz Wilhelm Junghuhn, 1809–1864." In *Geen blad voor de mond: Vijf radicalen uit de negentiende eeuw*, 54–100. Amsterdam: Arbeiderspers, 1958.

Veth, H. J. *Overzicht van hetgeen, in het bijzonder door Nederlands, gedaan is voor de kennis der fauna van Nederlandsch-Indië*. Leiden: Doesburgh, 1879.

Veur, Paul W. van der, ed. *Education and Social Change in Indonesia*. Athens: Ohio University Center for International Studies, 1969.

———. "Introduction." In Soetomo, *Towards a Glorious Indonesia: Reminiscences and Observation by Dr. Soetomo*, ed. Paul W. van der Veur, xv–lxii. Athens: Ohio University Center for International Studies, 1987.

———. "Progress and Procrastination in Education Prior to World War II." In *Education and Social Change in Indonesia*, ed. Paul W. van der Veur, 1–17. Athens: Ohio University Center for International Studies, 1969.

———. "The Social and Geographical Origins of Dutch-Educated Indonesians." In *Education and Social Change in Indonesia*, ed. Paul W. van der Veur, 18–49. Athens: Ohio University Center for International Studies, 1969.

Vickers, Adrian. *A History of Modern Indonesia*. Cambridge: Cambridge University Press, 2005.

Vlugter, H. "The Academic Year, 1949/1950." *OSR News* 2, no. 12 (1950): 171–72.

Vriese, W. H. de. *De Kina-Boom uit Zuid-Amerika overgebragt naar Java onder de Regering van Koning Willem III*. The Hague: Mieling, 1855.

———, ed. *Reis naar het Oostelijk Gedeelte van den Indischen Archipel in het jaar 1821, door C. G. C. Reinwardt*. Amsterdam: Frederick Muller, 1858.

———. *Wetenschap en beschaving, de grondslagen van welvaart der landen en volken van den Indischen archipel*. Leiden: Hazenberg, 1861.

Wal, S. L. van der. *De opkomst van de nationalistische beweging in Nederlands-Indië*. Groningen: Wolters, 1967.

Wallace, Alfred. *The Malay Archipelago*. New York: Harper, 1869.

Walters, Dirk R., and David J. Keil. *Vascular Plant Taxonomy*. 4th ed. Dubuque, IA: Kendall/Hunt, 1996.

Warsito. "Sejarah Akademi Departemen Pertanian Ciawi-Bogor." Photocopy, 2000.

Weber-van Bosse, A. *Een jaar aan boord H. M. Siboga*. Amsterdam: Atlas, [1904] 2000.

Weitzel, A. W. P. *Batavia in 1858: Schetsen en beelden uit de hoofdstad van Nederlandsch Oost Indië*. Gorinchem: Noorduijn, 1860.

Wellenstein, E. P. *Nederlandsch-Indië in de Wereldcrisis*. The Hague: Martinus Nijhoff, 1931.

———. "Remarks on 'Dualistic Economics.'" In *Indonesian Economics: The Concept of Dualism in Theory and Policy*, 195–213. The Hague: Van Hoeve, 1961.

Werner, Johannes, ed. *The Love Letters of Ernst Haeckel*. New York: Harper, 1930.

Wertheim, W. F. *Indonesian Society in Transition: A Study of Social Change.* 2nd ed. The Hague: Van Hoeve, 1964.

Winter, C. F. "Romo: Een Javaansch gedicht, naar de bewerking van Joso Dhipoero." *Verhandelingen van het Bataviaasch Genootschap* 21, no. 2 (1846-47): 1-596.

Wit, H. C. D. de. "De K. N. V. en botanie in Indië." In *Een eeuw natuurwetenschap in Indië, 1850-1950: Gedenkboek Koninklijke Natuurkundige Vereeniging,* ed. P. J. Willikes MacDonald, 153-77. Bandung: KNV, 1950.

———. "Miscellaneous Information." *Flora Malesiana Bulletin* 5 (1949): 131.

Woerd, L. A. van der. "Some Questions Which Arise to the Pharmacologist in Indonesia." *OSR News* 3, no. 4 (1951): 54-55.

Yong Mun Cheong. *H. J. van Mook and Indonesian Independence: A Study of His Role in Dutch-Indonesian Relations, 1945-1948.* The Hague: Martinus Nijhoff, 1982.

Zeijlstra, H. H. *Melchior Treub: Pioneer of a New Era in the History of the Malay Archipelago.* Amsterdam: KIT, 1959.

Zuidervaart, Huib J., and Rob H. Van Gent. "'A Bare Outpost of Learned European Culture on the Edge of the Jungles of Java': Johan Maurits Mohr (1716-1775) and the Emergence of Instrumental and Institutional Science in Dutch Colonial Indonesia." *Isis* 95, no. 1 (2004): 1-33.

# Index

Note: Page numbers in italics indicate an illustration.

Humboldt of Java, *18*, 18–21, 30, 58, 117; as Pahud's client, 33–34, 37–39, 41; political pressure on, 42–46, 79; quarrel with Blume, 184n33; quarrel with Teysmann, 41, 187n85

Kadarsan, Sampurno, 147, 215n16, 217n59
Kanehira, R., 128–29
Kartini, 113
Kartowisastro, Hermen, 138
Kebun Raya Indonesia (The Great Indonesian Gardens). *See* Botanical Gardens (Buitenzorg/Bogor)
*Kemadjoean Ra'jat*, 112–14, 117
*kemajuan*, 99, 100, 103, 109, 110, 111, 112, 113
Kemp, P. H. van der, 82
Keyes, Charles, 170
Ki Hadjar Dewantoro. *See* Soewardi Soerjaningrat
Koesnoto Setyodiwiryo, 136–37, 138, 141, 142, 145–54, 157
Kol, Henri van, 87–88, 90
Kolff, D. H. A., 173
Koningsberger, J. C., 75, 86–87, 89, 92, 107, 122, 135
Koningsberger, V. J., 122, 132, 135, 211–12n81
Koorders, S. H., 85, 175
Koster, Josephina, *84*
Kostermans, A. J. G. H.: biographical details, 217n48; at the Bogor herbarium, 151–52, 155, *155*, 162–63, 165, 166; as source about 1960s, 158, 159
Kusumasumantri, Iwas, 156
Kuswata Kartiwinata, 155, *155*

Lam, H. J., 120, *121*, 135
LBN. *See* Lembaga Biologi Nasional
Ledger, Charles, 51, 192n79
Leiden University, 30, 75, 118
Lembaga Biologi Nasional (LBN): founding of, 158; leading Indonesian biology, 161, 166–67, 169; as part of New Order bureaucracy, 161–62, 164–65, 166–68, 175. *See also* Botanical Gardens (Buitenzorg/Bogor)
Lembaga Ilmu Pengetahuan Indonesia (LIPI), 161–63, 165–68, 169, 175
liberalism: economic agenda of, 34, 43–46, 52; as political movement, 33–35, 47, 78

Liberal Party (Dutch), 9, 21, 34–35, 38, 42–46, 48
Linggajati agreement, 131, 134–35, 213n94
Linnaeus, Carl, 3, 17, 179n1
LIPI. *See* Lembaga Ilmu Pengetahuan Indonesia
Loon, J. W. van, 55, 192n91
Lovink, H. J., 79, 92–94, 106, 201n75
Lowe, Celia, 169

Maat, Harro, 193n4
Machmudy, Hamzah, 220n98
Madjelis Ilmu Pengetahuan Indonesia (MIPI): bureaucratization of, 158, 159; coordinating Indonesian science, 153–55, 157, 163–64; founding of, 140, 153; transformation into LIPI, 160, 161
malaria, 35
Malawar range, 39
Malaya, 79, 159, 165
Manipol-USDEK, 156, 218n74. *See also* Guided Democracy
Markham, Clements, 36, 42, 45, 189–90n35
Massart, Jean, 71
medical colleges: Batavia, 26, 59, 88, 96, 98, 106, 181n22; Surabaya, 103, 139, 181n22
Mellsop, John, 130
*merdeka*, 109, 113
Merkus, P., 183n16
Merrill, Elmer, 71–73
midwifery, 101, 102, 203n22
Mijer, P., 38, 44
MIPI. *See* Madjelis Ilmu Pengetahuan Indonesia
Miquel, F. A. W.: as cinchona expert, 42–43, 44, 48–49, 191n67, 192n78; and Ledger's seeds, 52, 53; as Utrecht university professor, 57, 62
Mohr, J. M., 8–9
Monchy, P. de, 82
Mook, H. J. van: as colonial reformer, 10, 119, 124–27: as lieutenant governor general, 130–31, 132–35, 140, 211n73, 212nn84–85
Moon, Suzanne, 118, 193n4
Mrázek, Rudolf, 76, 113, 142
Muhammadiyah, 203n27
Mulder, G. J., 34, 38, 42–43, 44, 48, 188n5
Multatuli. *See* Douwes Dekker, E.

# NEW PERSPECTIVES IN
# SOUTHEAST ASIAN STUDIES

*From Rebellion to Riots: Collective Violence on Indonesian Borneo*
Jamie S. Davidson

*The Floracrats: State-Sponsored Science and the Failure of the Enlightenment
in Indonesia*
Andrew Goss

*Amazons of the Huk Rebellion: Gender, Sex, and Revolution in the Philippines*
Vina A. Lanzona

*Policing America's Empire: The United States, the Philippines, and the Rise
of the Surveillance State*
Alfred W. McCoy

*An Anarchy of Families: State and Family in the Philippines*
Edited by Alfred W. McCoy

*The Hispanization of the Philippines: Spanish Aims and Filipino Responses,
1565–1700*
John Leddy Phelan

*Pretext for Mass Murder: The September 30th Movement and Suharto's
Coup d'État in Indonesia*
John Roosa

*The Social World of Batavia: Europeans and Eurasians in Colonial Indonesia,
second edition*
Jean Gelman Taylor

*Việt Nam: Borderless Histories*
Edited by Nhung Tuyet Tran and Anthony Reid

*Modern Noise, Fluid Genres: Popular Music in Indonesia, 1997–2001*
Jeremy Wallach